The Ecology and Evolution of Ant-Plant Interactions

Interspecific Interactions

A Series Edited by John N. Thompson

ALSO IN THIS SERIES

Induced Responses to Herbivory by Richard Karban and Ian T. Baldwin

Parasitism by Claude Combes

Ecological Niches by Jonathan M. Chase and Mathew A. Leibold

Antipredator Defenses in Birds and Mammals by Tim Caro

The Geographic Mosaic of Coevolution by John N. Thompson

The Ecology and Evolution of Ant-Plant Interactions

VICTOR RICO-GRAY AND
PAULO S. OLIVEIRA

THE UNIVERSITY OF CHICAGO PRESS CHICAGO AND LONDON

VICTOR RICO-GRAY is a senior research scientist at the Instituto de Ecología, A.C., in Mexico. PAULO S. OLIVEIRA is professor of ecology at the Universidade Estadual de Campinas in Brazil.

The University of Chicago Press, Chicago 60637
The University of Chicago Press, Ltd., London

Printed in the United States of America

16 15 14 13 12 11 10 09 08 07   1 2 3 4 5

ISBN-13: 978-0-226-71347-2 (cloth)
ISBN-13: 978-0-226-71348-9 (paper)
ISBN-10: 0-226-71347-4 (cloth)
ISBN-10: 0-226-71348-2 (paper)

Library of Congress Cataloging-in-Publication Data

Rico-Gray, Victor.
The ecology and evolution of ant-plant interactions / Victor Rico-Gray and Paulo S. Oliveira.
p. cm.
Includes bibliographical references and index.
ISBN-13: 978-0-226-71347-2 (cloth : alk. paper)
ISBN-10: 0-226-71347-4 (cloth : alk. paper)
ISBN-13: 978-0-226-71348-9 (pbk. : alk. paper)
ISBN-10: 0-226-71348-2 (pbk. : alk. paper)
1. Ants—Ecology. 2. Insect-plant relationships. I. Oliveria, Paulo S. II. Title.
QL568 .F7R53 2007
595.79′6—dc22

2006036173

♾ The paper used in this publication meets the minimum requirements of the American National Standard for Information Sciences—Permanence of Paper for Printed Library Materials, ANSI Z39.48-1992.

For
Alex Rico-Palacios
and
Maristela Oliveira

Contents

Preface

Ants and flowering plants are ubiquitous and represent dominant features in most terrestrial landscapes, and their evolutionary histories have been crossing paths for at least 100 million years. The study of the interactions between ants and plants received a significant and determinant boost in the mid-1960s with the publication of experimental research by Daniel H. Janzen (1966, 1967b) on *Acacia* plants and *Pseudomyrmex* ants. The development of this field over the past 40 years owes much to his effort. We are confident that many researchers in addition to us were inspired by Janzen's work. The first volume dealing broadly with the subject of ant-plant interactions is Andrew J. Beattie's *The Evolutionary Ecology of Ant-Plant Mutualisms,* produced a little over twenty years ago (1985). The other most influential publication is *The Ants,* a book by Bert Hölldobler and Edward O. Wilson (1990), where they present an extensive treatise on what ants are, how their colonies are organized, what they do, whom they interact with, and their ecological importance. The book serves as a permanent source of information.

The research field of ant-plant interactions has progressed enormously over a wide range of topics. One can find hundreds of journal articles and book chapters dealing, for instance, with plant defense by ants, the association of ants with exudate-producing Hemiptera and lepidopteran larvae, pollination by ants, ant-fed plants, seed dispersal by ants, granivory, leaf-cutter ants, ant gardens, and all sorts of combinations among these topics. There have also been several attempts at reviewing this literature, ranging from reviews in journals and book chapters to books dealing with different topics on ant-plant interactions such as leaf-cutter ants, seed dispersal by ants, and ant-plant mutualisms in general. The main question we faced when we decided to write a book on ant-plant interactions was how to organize this wealth of information into a coherent readable volume. The ideas presented in John N. Thompson's book *Interaction and Coevolution* (1982) were seminal for our development in the field of interspecific interactions, and we agreed that its organization would provide us with a basic framework to present the information on how and why ants and plants interact.

Since species interact in an astonishing variety of ways, we needed to synthesize general patterns in the ecology and evolution of ant-plant interactions. Our aim was to present patterns that go beyond taxonomic boundaries, that is, patterns that derive more from the mode of interaction than from the biology of a particular group. Thus, we developed our approach based on modes of interaction: In the first part of the book (chapters 1 through 8) we assemble in an ecological way the great variety found in interactions between ants and plants. We then treat three general topics that round out and explore that information in different ways, namely canopy-dwelling ants and plant and insect exudates, variation in ant-plant interactions, and ant-plant interactions in agroecosystems (chapters 9 through 11). In chapter 12 we provide a general overview and present a series of perspectives that range from studies that should be attempted in future research, to topics that are emerging, to the association between conservation studies and ant-plant interaction studies. The amount of information available on ant-plant interactions is already vast and is increasing at a very fast rate. For each of the topics covered we did our best to refer to all the relevant literature—more than a thousand references are cited in the volume.

This book would not have been completed without the guidance, patience, and encouragement of the science editors of the University of Chicago Press, Christie Henry and the late Susan Abrams, and the scientific editor for the Interspecific Interactions series, John N. Thompson. Lois R. Crum, Pete Beatty, and Leslie Keros provided invaluable help at final stages of the editing process. We heartily thank them all. The final version of the book manuscript was substantially improved by valuable suggestions of five external reviewers. We appreciate the time they took to give critical reviews; their technical advice helped us considerably to shape the scope of the volume, adjust chapter contents, and clarify our writing.

We are indebted to the many colleagues who generously shared thoughts, discussions, reprints, figures, comments, and encouragement during different stages in the writing of this book and thereby helped to make it as up-to-date as possible: Anurag A. Agrawal, Adolfo

Alvarez, Nico Blüthgen, Judie L. Bronstein, Emilio Bruna, Mariana Cuautle, Alain Dejean, Kleber Del-Claro, Margaret S. Devall, Cecilia Díaz-Castelazo, Alejandro Farji-Brener, Patricia Folgarait, Carlos R. Fonseca, Alina Freire, André V. L. Freitas, Eduardo Galante, José Gómez, Arturo Gómez-Pompa, Aaron D. Gove, Sergio Guevara, Paulo R. Guimarães Jr., Martin Heil, Carlos M. Herrera, Carol C. Horvitz, Pedro Jordano, Michael Kaspari, Suzanne Koptur, Christian Kost, Conrad C. Labandeira, Jonathan D. Majer, Ma. Angeles Marcos, Doyle McKey, Olga Martha Montiel, Helena C. Morais, Jorge Navarro-Alberto, Victor Parra-Tabla, Rod Peakall, Ivette Perfecto, Marco A. Pizo, Peter H. Raven, Leticia Rios-Casanova, Gustavo Romero, Juan Carlos Serio-Silva, Leonard B. Thien, Jill Thompson, John N. Thompson, Heraldo L. Vasconcelos, David A. White, and Alex Wild.

We thank Arturo Piña for redoing many of the graphs, and Katsumi and Miguel Flores for the illustrations. We thank Alejandra Valencia, Araceli Toga, and Edith Rebolledo for providing logistic support through the years. Special thanks go to Lorraine and Leonard Thien for sharing their house with V. Rico-Gray during his sabbatical year in New Orleans.

We are most grateful to the organizations that provided the funds for our research and collaborations with others during the past 15 years. In Mexico, we thank the Consejo Nacional de Ciencia y Tecnología (CONACYT) and the Departmento de Ecologia Funcional and Departamento de Ecología Aplicada of the Instituto de Ecología, A.C. In Brazil, financial support was provided by the Conselho Nacional de Desenvolvimento Científico e Tecnológico (CNPq), the Fundação de Amparo à Pesquisa do Estado de São Paulo (FAPESP), and the Fundo de Apoio ao Ensino, à Pesquisa e à Extensão of the Universidade Estadual de Campinas. We would also like to thank the different institutions that shared space for the writing and bibliographic research during this project: the Missouri Botanical Garden, the Department of Cell and Molecular Biology of Tulane University, the Departamento de Ecologia of the Universidade de Brasilia, CIBIO of the Universidad de Alicante, CITRO of the Universidad Veracruzana, the Organization for Tropical Studies, and Washington State University.

Finally, we are deeply grateful to our wives, Mónica Palacios-Rios and Maristela Oliveira, for their support and encouragement throughout the duration of this project. Background music provided by the Rolling Stones and Antonio Carlos Jobim.

CHAPTER ONE

Ant-Plant Interactions

Ants (Hymenoptera: Formicidae) are probably the most dominant insect group on earth, both ecologically and numerically. They are so abundant that approximately 8 million individuals live underground in one hectare of Amazonian rain forest (Hölldobler and Wilson 1990), and ants are estimated to represent 10% to 15% of the entire animal biomass in many terrestrial ecosystems (Beattie and Hughes 2002). On this basis alone, the study of the ecology and evolution of the ants would be important for understanding the ecology of terrestrial biological communities. The outstanding hallmark of ants is that all species are eusocial; this trait is probably what caused their tremendous success (Wilson 1987b). Moreover, the wingless workers can easily penetrate minute cavities, ants maintain sustained population densities, and their array of glands and secretions enables sophisticated chemical communication, such that they can rapidly recruit workers for sequestering food, they can defend their colony, especially with soldiers and stinging workers, and they are even able to subdue large prey (Hölldobler and Wilson 1990).

Eusociality has actually evolved 12 times in the Hymenoptera: once in the ants, and all the other times in the Vespoidea (social/paper wasps) and the Apoidea (bees). Moreover, the evolution of mutualisms is positively associated with the richness of social behavior in species such as ants (Beattie 1985), so that (1) mutualisms should be more common among the more social species within a taxon, and (2) the richness of social behaviors within a species may be partly an evolutionary result of mutualisms that allow species to spend less time foraging for food (Thompson 1982). As discussed in the forthcoming chapters, mutualistic associations between ants and plants are far more common than antagonistic associations between them. However, we should always consider that antagonistic and most mutualistic interactions are strongly related (Thompson 1982; Holland et al. 2005; see below).

Angiosperms, or flowering plants, the dominant plant group on land for more than 100 million years, comprise about 250,000–300,000 species, by far the largest number of species of any plant group (Raven, Evert, and Eichhorn 1986; Schneider et al. 2004 and references therein). The most distinctive characteristic of angiosperms is the flower. However, the evolutionary success of angiosperms cannot be ascribed solely to benefits conferred by possessing flowers; in addition there is an array of interspecific interactions (e.g., pollination, seed dispersal, and herbivory) that have also helped to shape their great diversity (Niklas 1997). Angiosperms include the Monocotyledones (ca. 65,000 species) and the Dicotyledones (ca. 170,000 species); ants interact with members of both groups as well as with some species of ferns (Pterydophyta; see chapter 6). Angiosperms make up much of the visible world of modern plants, and the study of their evolution and ecology, like that of ants, is important to understand the ecology of terrestrial biological communities.

The suggestion of mutually beneficial interactions between ants and plants is present in classic studies (Belt 1874; Delpino 1875; Trelease 1881), but support for this view was based mostly on anatomical and behavioral observations rather than experimental evidence, stimulating a great amount of debate among early naturalists (Wheeler 1910). Although the biology and the geographic distribution of myrmecophilous plants and their ants have long been documented in considerable detail (e.g., Bequaert 1922; Wheeler and Bequaert 1929; Wheeler 1942), it was not until the pioneering experimental field studies of Daniel H. Janzen (1966, 1967b) that a burst of research on ant-plant associations began in virtually all types of habitats around the globe.

During the mid-1960s, Janzen studied the interaction between the plant *Acacia cornigera* (Fabaceae) and the ant *Pseudomyrmex ferrugineus* (Pseudomyrmecinae) in eastern Mexico (Janzen 1966, 1967a, 1967b, 1969b, 1973b). He demonstrated beyond doubt what happened to plants when their defenses against herbivores were removed, since certain acacias associated with *Pseudomyrmex* ants do not possess a significant chemical defense arsenal; they depend heavily on ant activity for their defense. The *Acacia-Pseudomyrmex* association is probably the best-known and most widely used example of ant-plant mutualism in which a plant offers all categories of resources (i.e., extrafloral nectar, Beltian food bodies, and domatia in hollow thorns) in exchange for defense from herbivores and encroaching vines. The *Acacia-Pseudomyrmex* association has been described in many textbooks and reviews on ant-plant interactions (e.g., Howe and Westley 1988; Keeler 1989; Begon, Harper, and Townsend 1990) and has stimulated a number of studies on acacias and their associated ants (e.g., Keeler 1981d; Knox et al. 1986; Ward 1991, 1993; Willmer and Stone 1997; Young, Stubblefield, and Isbell 1997; Suarez, de Moraes, and Ippolito 1998), as well as work on mutualism, coevolution, and an array of topics of general interest on animal-plant interactions in many contexts (see also chapter 6).

The study of ant-plant associations has developed enormously since Janzen's work, and numerous reviews on different aspects of the natural history and evolutionary ecology of these systems have been published (e.g., Buckley 1982a; Beattie 1985; Benson 1985; Jolivet 1986; Hölldobler and Wilson 1990; Huxley and Cutler 1991; Davidson and McKey 1993; Bronstein and Barbosa 2002; Beattie and Hughes 2002; Gorb and Gorb 2003; Heil and McKey 2003; Wirth et al. 2003; Rico-Gray et al. 2004). In these 40 years more than 900 journal articles and book chapters and 9 books on topics of ant-plant interaction have appeared (fig. 1.1), and ant-plant protection mutualisms have continued to be systems for testing aspects of plant-defense theory.

Ant-plant interactions (i.e., direct ant-plant interactions, or indirect interactions mediated by Hemiptera or Lepidoptera) are geographically widespread and common in many plant communities and have been shown to be important in plant defense against herbivores (Schupp and Feener 1991; Koptur 1992a; Rico-Gray, García-Franco, et al. 1998; Oliveira and Freitas 2004). In the case of the study of ant-plant mutualism, for example, by developing approaches to measure benefits, costs and net outcomes, and the explicit consideration of their variability, the study of ant-plant interactions has played a major role in shaping our broad understanding of mutualism (Bronstein 1998). The study of ant-plant interactions also offers an excellent opportunity to analyze the effects of both historical and ecological factors on the evolution of mutualisms (McKey and Davidson 1993). Since ants and flowering plants are two very important groups in biological communities, it seems clear that

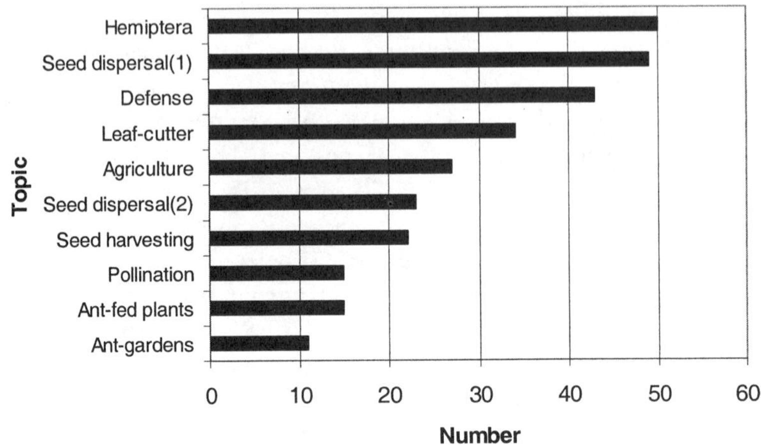

FIGURE 1.1. Number of journal articles and book chapters on different topics of ant-plant interactions, 1966–2005. The publications counted for this chart are those that address only one topic; there are many others, of course, that consider more than one topic. Hemiptera = ant-plant-Hemiptera interactions; Seed dispersal (1) = ants as primary seed-dispersal agents; Defense = plant defense by ants; Leaf-cutter = plant-leaf-cutter-ant systems; Agriculture = ants and agricultural systems; Seed dispersal (2) = ants as secondary seed-dispersal agents; Seed harvesting = seed-harvesting ants; Pollination = pollination by ants.

understanding the ecology of ant-plant interactions can allow a much better understanding of the unparalleled success of these two remarkable groups and of interspecific interactions in general.

In this chapter we explore the origin and early evolution of ant-plant associations, describe possible phylogenetic associations between the groups, present some of the current evidence on the latter, and discuss some general ideas on coevolution and interspecific interactions, specifically related to ant-plant interactions.

The Origin and Early Evolution of Ant-Plant Associations

Insects and angiosperms have different fossil histories. Insects originated during the Devonian, 390 million years ago, while angiosperms first appeared as fossils during the lower Cretaceous and have dominated the earth's vegetation since the mid-Cretaceous (90 million years ago) (Crane, Friis, and Pedersen 1995; Qiu et al. 1999; Schneider et al. 2004). The first stage for angiosperm evolution is *Amborella,* a shrub of the monotypic New Caledonian family Amborellaceae, which is the sister group to all other angiosperms (Qiu et al. 1999; Soltis, Soltis, and Chase 1999; Soltis et al. 2000; Graham and Olmstead 2000; Davies et al. 2004). The angiosperm lineage diverged from closely related gymnosperms early in the Mesozoic, but it is unlikely that angiosperms were diverse during the Triassic and the Jurassic (Friis, Chaloner, and Crane 1987). Based on the description of fossil flowers, the initial major diversification of angiosperms took place during the early Cretaceous (Crane 1998).

Several studies have suggested that the extraordinary diversity of living insects is fueled by the diversity of angiosperms. In particular, tissues such as leaves and flowers of these advanced land plants have provided an expanded spectrum of ecological resources that could be exploited by various guilds of herbivorous and pollinating insects (Bernhardt and Thien 1987; Pellmyr et al. 1990; Bergstrom et al. 1991; Labandeira and Sepkoski 1993; Sanderson and Donoghue 1994; Crepet 1996; Grimaldi 1999; and references therein). However, virtually all major insect feeding types were in place considerably before angiosperms became serious contenders in terrestrial ecosystems, and evidence from the fossil record of vascular plant–insect interaction seems to support this inference (Labandeira and Sepkoski 1993).

The angiosperms experienced a tremendous radiation in all geographic regions during the mid-Cretaceous. Their diversity exceeded that of other groups of land plants, and their rapid evolution continued through the late Cretaceous; by that time, 44% of extant angiosperm orders already had Cretaceous fossil records, including most living lineages (Labandeira and Sepkoski 1993; Wilf et al. 2000; and references therein). There is no signature of this event in the family-level fossil record of insects, whose great radiation began 245 million years ago (Labandeira and Sepkoski 1993). Such an analysis based on diversity patterns at the family level, however, excludes the possibility that radiation within a family of insects might have been significant in angiosperm diversification, for example, the Apidae within the Hymenoptera related to antophilous flowers (Sanderson and Donoghue 1994; Crepet 1996; Grimaldi 1999). For example, Cretaceous radiations of leaf beetles (Coleoptera: Chrysomelidae) occurred during an extended interval of evolutionary innovation for angiosperms, suggesting the possibility of plant-beetle coevolution or of adaptive beetle radiations that closely followed the ongoing rapid evolution and diversification of their angiosperm hosts (Farrell 1998; Wilf et al. 2000).

Radiations of the major anthophilic groups of insects took place in the late part of the lower Cretaceous to the upper Cretaceous, including bees (Apoidea/Apidae), pollen wasps (Vespidae: Masarinae), various families of brachyceran flies (Acroceridae, Apioceridae, Bombyliidae, Empididae, Nemestrinidae, Stratiomyidae, and Syrphidae), and the Lepidoptera (Grimaldi 1999). The pattern of diversification of these insects, centered in the mid-Cretaceous, is consistent with the chronology of appearance of entomophilous syndromes in Cretaceous flowers (Crepet 1996; Grimaldi 1999), and not with the model of late Jurassic or earliest Cretaceous diversification of pollinating insects (Labandeira and Sepkoski 1993).

Based on a large-scale molecular phylogeny of ants, Moreau et al. (2006) suggest that modern "crown group" ants last shared a common ancestor during the Early Cretaceous to the Middle Jurassic, around 140–168 million years ago. That suggestion is consistent with the close relationship of the Formicidae to the Vespidae (+ Scoliidae) and the phylogenetic position of Cretaceous Vespidae (Carpenter and Rasnitsyn 1990; Brothers and Carpenter 1993). The oldest reliable dated ant fossils (ca. 100 million years old) are early Cretaceous, including both the Aneuretinae (*Gerontoformica* and *Burmomyrma*) and the Sphecomyrminae (e.g., *Sphecomyrma, Cretomyrma, Baikurus* and *Dlusskyidris*) (Grimaldi and Engel 2005; Moreau et al. 2006; and references therein). The oldest known forms of Formicoidea are from ca. 100 million years ago, and the oldest definitive Formicidae (*Sphecomyrma freyi*) are known from approximately 20 million years later; only primitive taxa occur throughout the first 50 million years (pre-Tertiary) of the history of fossil ants (Grimaldi, Agosti, and Carpenter 1997). The adaptive radiation that propelled ants to dominance (ca. 11,800 living species) must have taken place at the beginning of the Tertiary period, because ants are highly represented in Oligocene and Miocene deposits (Grimaldi, Agosti, and Carpenter 1997; Grimaldi and Engel 2005; Moreau et al. 2006).

Even though ants and angiosperms have different histories, both groups, one way or another, became prominent somewhere between late in the Cretaceous and early in the Tertiary, providing for a long and joint evolutionary history (Wilson and Hölldobler 2005; Moreau et al. 2006; see below). It would be interesting to analyze whether through this long history the different types of interactions (e.g., grazing, predation, and mutualism) are widely distributed throughout the different ant and plant groups, or whether certain interactions evolved associated with certain ant and/or plant groups.

Phylogenetic Associations

The most comprehensive and widely cited study on ant phylogeny (Baroni-Urbani, Bolton, and Ward 1992) has supplied the cladistic framework for the discussion of the positions of the oldest definitive worker ants (*Sphecomyrma* and *Brownimecia*) and a reanalysis of the slightly revised data, which provided many newer, still unstable, cladograms (Grimaldi, Agosti, and Carpenter 1997). The cladogram proposed by Grimaldi, Agosti, and Carpenter (1997) differed in some fundamental ways from that of Baroni-Urbani and co-workers (1992). In general, the former authors stress that their analysis raises caution concerning the robustness of the proposed phylogeny of ants, caused by the addition and deletion of taxa as well as the reinterpretation of some of the character states, which had great influence on the topology of the different trees. Thus, the addition or deletion of taxa always resulted in different topologies, which the authors suggest is a clear sign of insufficient data (Grimaldi, Agosti, and Carpenter 1997). The most up-to-date versions of the phylogeny of most major lineages of ants (subfamilies and some tribes) have recently been published by Grimaldi and Engel (2005) (fig. 1.2) and by Moreau et al. (2006).

There are 16 ant subfamilies, the four largest of which include some of the most significant genera that also exhibit important interactions with plants (based on Grimaldi and Engel 2005; but see Moreau et al. 2006):

Dolichoderinae: *Azteca, Dolichoderus, Dorymyrmex, Forelius, Iridomyrmex, Leptomyrmex,* and *Tapinoma.*

Formicinae: *Brachymyrmex, Camponotus, Formica, Lasius, Oecophylla,* and *Paratrechina.*

Myrmicinae: *Aphaenogaster, Crematogaster, Leptothorax, Messor, Monomorium, Myrmica, Pheidole, Pogonomyrmex, Solenopsis, Tetramorium, Wasmannia,* and the tribe Attini (e.g., *Acromyrmex* and *Atta*).

"Ponerinae": *Ectatomma, Pachycondyla,* and *Paraponera.* This paraphyletic group comprises some of the most basal grades of ants (e.g., Amblyoponini and Paraponerini), many of them with a constriction between the first and second gaster segment, but they are well known for their powerful stings (Ward and Brady 2003; Grimaldi and Engel 2005).

Phylogenetic analyses of the Formicidae are important for evolutionary studies; however, the understanding of ant phylogeny still requires serious attention (but see Moreau et al. 2006). Due to the fragility of ants, the

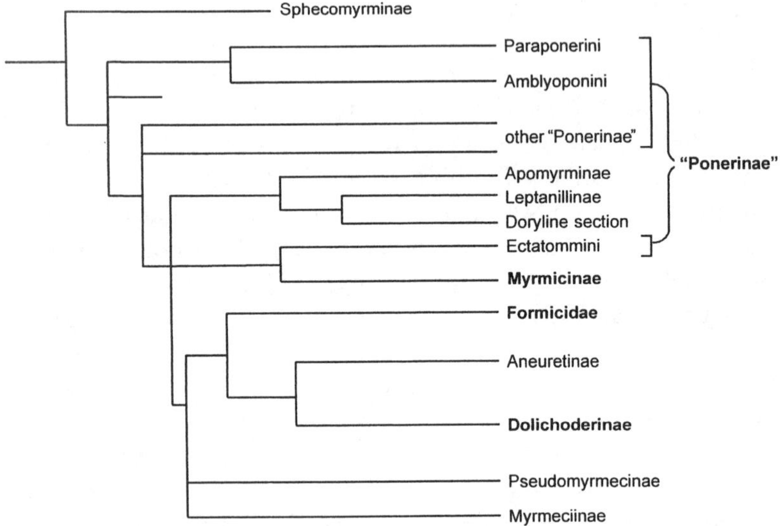

FIGURE 1.2. Cladogram of the ants (Hymenoptera: Formicidae), modified from Grimaldi and Engel 2005; see also Moreau et al. 2006. Shown in boldface are the four most common, abundant, and widespread extant ant subfamilies, according to Wilson and Hölldobler 2005.

discovery of finely preserved amber fossils has allowed retention of many more characters than have compression fossils. This is evident in the less ambiguous placement of the New Jersey amber fossils of primitive ants, as compared to the Eocene compression fossil taxa (Grimaldi, Agosti, and Carpenter 1997). Placement of the finest fossils will depend on stable phylogenies, which are subject to the discovery of fossils with unpredicted combinations of characters; thus, all robust phylogenies will depend on the discovery of new characters and close scrutiny of the homologies of known characters (Grimaldi, Agosti, and Carpenter 1997; Grimaldi and Agosti 2000; Grimaldi and Engel 2005). Currently, the collective fossil record of ants indicates that it was not until 70–80 million years after their origin (ca. 140 mya) that ants evolved into the dominant organisms we know today (Grimaldi and Engel 2005; Moreau et al. 2006).

Efforts to infer angiosperm phylogeny are undermined by similar problems, because the phylogenetic trees inferred are not completely congruent in the interrelationships among the major lineages, and nearly all trees suffer from areas of poor resolution and/or weak support (e.g., APG 1998, 2003; Soltis, Soltis, and Chase 1999; Qiu et al. 1999; Soltis et al. 2000; Graham and Olmstead 2000; Davies et al. 2004; Schneider et al. 2004). However, such efforts have served to establish the general structure of the major lineages of flowering plants (Soltis, Soltis, and Chase 1999). Angiosperm phylogeny is not totally assessed, especially the relationships among the earlier angiosperm groups. Thus, several phylogenies or a combination of them could be useful (e.g., APG 1998, 2003; Davies et al. 2004). However, our goal here is to establish that a number of angiosperm groups have interacted with ants at different times along their evolution and that the interaction has had multiple origins. The analysis by the Angiosperm Phylogeny Group (1998, 2003) serves such purposes (fig. 1.3).

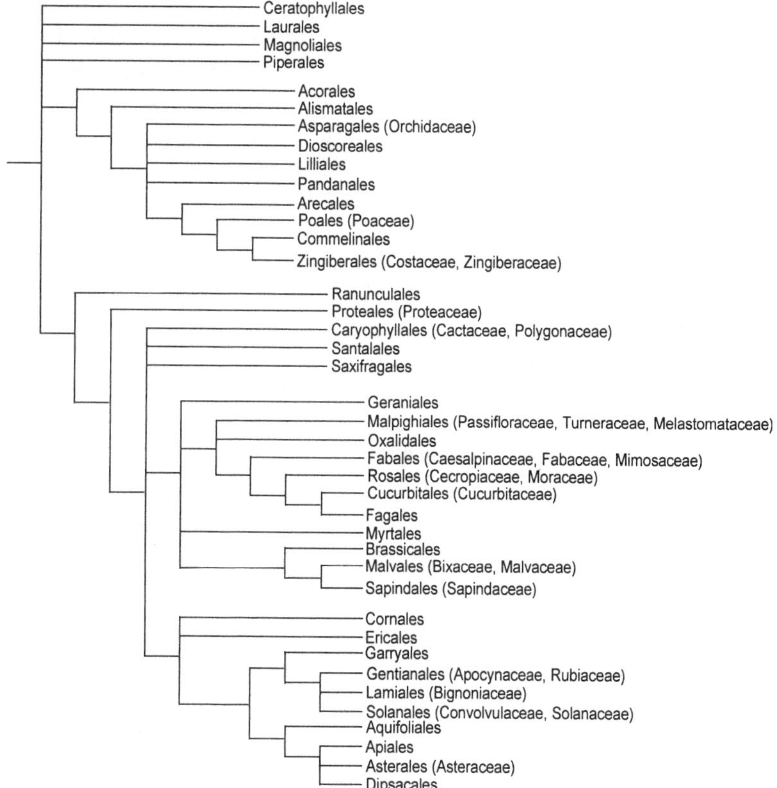

FIGURE 1.3. Phylogenetic interrelationships of the orders of angiosperms, with some relevant plant families that include species producing ant-attractants (e.g., nectaries, food bodies, domatia) and that participate in many of the ant-plant interactions analyzed in this book. Chart based on APG 1998. See also Soltis, Soltis, and Chase 1999; Qiu et al. 1999; Soltis et al. 2000; Graham and Olmstead 2000; Davies et al. 2004.

The study of the phylogenetic history should be an important tool in explaining ant-plant associations because evolutionary history leaves a significant imprint on the composition of the component communities that must exert selection on plant characteristics, and vice versa (Futuyma and Mitter 1997). Some particular examples of phylogenetic or historical studies are shedding new light on the evolutionary ecology of ant-plant interactions: for instance, the association between leaf-cutting ants (Attini) and their Basidiomycetes fungi (Chapela et al. 1994; Hinkle et al. 1994), the analysis of the evolution of mutualism between *Leonardoxa* (Caesalpiniaceae) and its associated ant species (McKey 1989, 1991; Brouat, Mckey and Douzery 2004), the analysis of the Pseudomyrmecinae associated with domatia-bearing plants (Ward 1991; Davidson and McKey 1993), the molecular phylogeny of *Crematogaster* (*Decacrema*) ants and the colonization of *Macaranga* trees (Feldhaar, Fiala, Gadau, et al. 2003), and the comparison between major geographic regions (e.g., Africa and the neotropics [McKey and Davidson 1993]). The use of phylogenetic analyses should be encouraged in future studies of ant-plant interactions.

A Brief History of the Associations

Even though fossil evidence is extremely scarce, the evolutionary history of ant-plant interactions may have developed as early as the mid-Cretaceous. The current understanding of the joint evolutionary history of the two groups has been summarized by Wilson and Hölldobler (2005) in three key events, especially if we follow the four most diverse, abundant, and geographically widespread living subfamilies, the Ponerinae, the Myrmicinae, the Formicinae, and the Dolichoderinae (fig. 1.4).

Ant diversification and ant-angiosperm interactions are probably based in part on a range of previously evolved trophic adaptations that allowed ants to enter into new food-based ecological guilds provided by flowering plants during the late Mesozoic and early Tertiary. Seed plants in general, rather than angiosperms in particular, may have provided the stage for the evolutionary history of ant-angiosperm interactions, because most ant-angiosperm interactions are based on or related to nonreproductive parts of plants (leaves, roots, and stems) (Labandeira and Sepkoski 1993). This may suggest that early ant-plant interactions involved more plant groups than just angiosperms (Beattie 1985). Ants

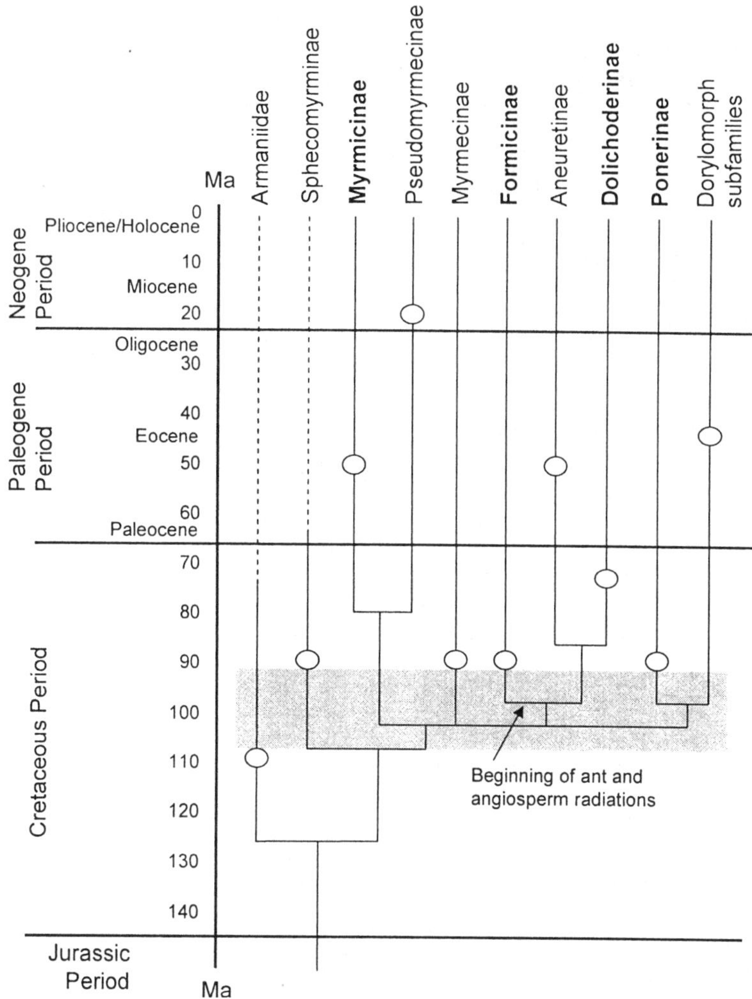

FIGURE 1.4. The evolutionary history of ants. The most important ant subfamilies are shown in boldface (see also Grimaldi and Engel 2005). The circles represent the oldest known fossil record for each subfamily. Modified from Wilson and Hölldobler 2005.

interact with living preangiosperm groups (e.g., ferns) and with angiosperms. Even though modern ferns promote ant activity on their surface with foliar nectaries and rhizomes riddled with tunnels that function as domatia for ants, or produce spores especially adapted for dispersal by ants, the majority of ant-plant interactions are indeed with angiosperms.

After the first evolutionary radiation of ants in the Mesozoic (e.g., the Sphecomyrminae), the Ponerinae emerged as the dominant subfamily during the Middle Cretaceous, with an initial radiation in forest ground litter and soil, coincident with the rise of the angiosperms (Wilson and Hölldobler 2005; Moreau et al. 2006). The other dominant ant subfamilies were also present but with lesser importance (fig. 1.4). Toward the end of the ponerine expansion (Early Eocene), the myrmicines began to radiate, surpassing the ponerines in biomass and diversity. There were diet changes (e.g., adding seeds and elaiosomes) that helped the expansion into deserts and dry grasslands. Especially important were the symbioses with caterpillars of honeydew-secreting butterflies and, particularly, with honeydew-producing hemipterans such as scale insects and treehoppers in tropical and warm-temperate vegetation, aphids in cool-temperate vegetation, and mealybugs underground everywhere. Again, the four most important ant groups were present (Early Eocene), although the importance had now shifted from the Ponerinae to the Myrmicinae (Wilson and Hölldobler 2005).

The evolution of the dolichoderines and the formicines probably occurred later, in the Early to Middle Eocene (fig. 1.4). These groups were presumably less successful than the ponerines and myrmicines in the forest litter environment but more successful at creating symbioses with honeydew-producing hemipterans. However, they were able to occupy environments less available to predators, including cool-temperate climates and tropical forest canopies (see chapter 9). The radiation of the dolichoderines and the formicines (and to some extent that of the myrmicines) was due to a change in diet toward liquid food sources such as honeydew and nectar, a shift aided by the rising dominance of angiosperms over much of the land environment. That expansion began in the Cretaceous and culminated in the Paleocene and Eocene (Wilson and Hölldobler 2005; Moreau et al. 2006). Angiosperm dominance also favored the expansion of honeydew-producing hemipterans and lepidopterans. Thus the expansion of the ant lineages helped by dietary changes led them away from dependence on predation and scavenging, upward into the canopy, and outward into more xeric environments. The four ant groups were present (Middle Eocene), but importance had now shifted to the Dolichoderinae and the Formicinae, which is the current scenario (based on Wilson and Hölldobler 2005). Interestingly, other groups, such as ferns (especially polypod ferns, which comprise >80% of living fern species), also diversified in the shadow of angiosperms, suggesting perhaps an ecological opportunistic response to the diversification of angiosperms, as these came to dominate terrestrial ecosystems (Schneider et al. 2004).

Despite the prevalence and importance of ant-plant interactions, not much is known about their joint evolutionary history (Beattie 1985; Pemberton 1992; but see Wilson and Hölldobler 2005; Moreau et al. 2006), and the fossil evidence is extremely scarce (C. C. Labandeira, pers. comm.). Fossil evidence for ant-plant interactions is very fragmented, dating the association between the two groups in the Oligocene, based on an extrafloral-nectaried *Populus crassa* leaf (Pemberton 1992). The fossil site (Florissan, Colorado, USA) also harbored 32 ant species (Carpenter 1930), which were the most abundant of the fossil insects, as well as predatory and parasitic insects whose modern relatives visit extrafloral nectaries (EFNs). Interestingly, 10 of the 32 ant species belong to five extant genera (*Camponotus, Formica, Iridomyrmex, Lasius,* and *Pheidole*) that have species acting as ant-guard mutualists of plants bearing EFNs. This finding alone, plus the fact that EFNs are visited by a variety of insect groups (e.g., wasps, bees, flies, beetles, and mites) and spiders besides ants, suggests that ant-plant interactions and the relative effect of these groups on the origin and evolution of nectaries and on plant fitness need to be comparatively analyzed using a multispecies approach (e.g., Cuautle and Rico-Gray 2003; Heil and McKey 2003; Oliveira and Del-Claro 2005).

Ants involved in mutualistic interactions, or in interactions whose outcome varies according to a cost-benefit system, creating a continuum between mutualism and antagonism (e.g., pollination, seed dispersal, plant protection, and ant-fed plants), evolved throughout several subfamilies (e.g., Myrmicinae, Pseudomyrmecinae, Formicinae, Dolichoderinae, Ponerinae). That ant species involved in mutualistic associations with plants belong to a variety of subfamilies in the Formicidae agrees well with plant phylogeny. Extrafloral and circumfloral nectaries (the main resource associated with ant-plant mutualisms), as well as other plant rewards (e.g., food bodies) or domatia, have developed across many families and at different times along angiosperm phylogeny (fig. 1.3). This is not surprising, since most ant-plant associations are largely fortuitous, nonspecialized, and facultative, such that specialization between particular ant and plant species is rare, suggesting that only occasionally does selection favor obligate interactions. In the case of systems in which ants protect plants, selective benefits should accrue to plants that attract a broad array of ants; the greater the diversity of ants, the greater the variety of plant enemies they are likely to remove, and the greater the probability that in any given habitat, season, or time of day, some species will forage on the plants (Beattie 1985).

Ant-plant interactions are suitable systems in which to study processes operating at the community level, because of the parallels between the characteristics and features of the two communities (Andersen 1991a, 1993, 1997b; López, Serrano, and Acosta 1994) and also because in some cases plant species number is the best predictor of ant species number (Morrison 1998). Most studies of ant-plant associations have analyzed interactions between small subsets of ants and plants within communities (e.g., Rico-Gray and Thien 1989b; Deslippe and Savolainen 1995). Most ant-plant interactions, however, are highly facultative, and the diversity of ants involved in these interactions can vary considerably even over short geographic ranges, so a landscape view should also be approached (Bronstein 1995). Our understanding of the ecology and evolution of ant-plant interactions and their effects on community organization requires a better understanding of how the diversity of ants and their use of plants varies across an array of regions, gradients, etc. (chapter 10).

Coevolution and Interspecific Interactions

A considerable portion of this book deals with interspecific interactions and coevolution. These two concepts are addressed here to make sure the reader is aware of what we mean. The definition that best describes a coevolutionary process is that of the "geographic mosaic of coevolution" (Thompson 1994, 2005; see below). Although there are no current examples of an ant-plant coevolutionary process using this approach, some research has been done with it in mind (see chapter 10), and we hope that this view will be commonly used in future studies.

Ant-plant interactions are interspecific interactions, and as such they need to be treated in an organization-wise manner, which is more coherent than the classic disconnected views commonly used (studying categories such as leaf-cutter ants, ant-guards, seed-harvesting ants, etc.). Antagonistic and mutualistic interactions are intimately associated. Indeed, the existence of most mutualisms is strongly associated with antagonistic associations (Thompson 1982), so much so that some authors consider that virtually all mutualisms are consumer-resource interactions (Holland et al. 2005). The flow of the chapters in this volume is based on the view that antagonism and mutualism are strongly associated. This association is also strongly emphasized along the evolution of ant-plant interactions, where antagonistic interactions were common at the beginning (ponerines and myrmicines) and then, through the appearance of new plant structures (e.g., elaiosomes and nectaries) and a change of feeding habits (e.g., the use of honeydew), mutualistic interactions rapidly evolved (formicines and dolichoderines) (see above).

* * *

Coevolution can be defined as the reciprocal evolutionary change between interacting species, in which both of them exhibit specific evolutionary changes as an outcome of the interaction; the process relates to the partial coordination of two gene pools that do not mix, and thus it is an interspecific process (Thompson 1982, 1994). Although coevolution is important in the evolution of species, and several authors have offered a variety of definitions (e.g., Janzen 1980; Roughgarden 1979; Schemske 1983; Kiester, Lande, and Schemske 1984; Ricklefs 1984), the term has on many occasions been incorrectly used or defined. The most coherent and clear definition, which integrates the whole of the coevolutionary process, is that of the geographic mosaic theory of coevolution (Thompson 1994, 1997, 2005).

Most species are confronted with the ecological and evolutionary challenge of interacting, evolving, and sometimes coevolving, with at least several other species. Understanding how species meet this challenge is one of the most fundamental problems in ecology and evolutionary biology, because it links three major issues in the evolution of biodiversity (Thompson 1997, 1999b): (1) the diversity of species, (2) the diversity of specialization found among species, and (3) the diversity of ways in which coevolution binds these taxa into biological communities. That is, this issue ties together species biodiversity and interaction biodiversity. Often the challenge has been sidetracked by simply using the term *diffuse coevolution* for all the varied ways in which groups of species evolve together (e.g., Strauss, Sahli, and Conner 2005). In diffuse coevolution, authors lump together a wide range of complex interactions that are highly dynamic and often involve more than one pair of species. When faced with so many conflicting selection pressures, one is likely to assume that pairwise coevolution may not proceed and therefore that coevolution is uncommon or, at best, diffuse (Thompson 1998). If this were true, the discovery of constraints on the trajectories of coevolution would become a basis for arguing that reciprocal selection is rare or all but impossible, and it would be easy to begin to think about coevolution as a rare and ecologically unimportant process (Thompson 1999c). However, diffuse coevolution is a catchall for many forms of reciprocal evolutionary change in interacting species, and the term, fortunately, is gradually being replaced with more testable coevolutionary hypotheses that confront the actual ecological, geographic, and phylogenetic structure of interactions (Thompson 1994, 1997, 2005).

The geographic mosaic theory of coevolution suggests that much of the dynamics of coevolution involving pairs or groups of species often occurs at a geographic scale above the level of local populations and below the level of the fixed traits of interacting species. It is based upon three ecological observations (Thompson 1994, 1997, 2005): (1) species are groups of genetically differentiated populations, (2) outcomes of interactions vary among communities, and (3) interacting species differ in their geographic ranges. These observations, accepted as assumptions, suggest a three-part hypothesis regarding the coevolutionary process: (1) there is a selection mosaic among populations, favoring different evolutionary trajectories to interactions in different populations; (2) there are coevolutionary hotspots, which are the subset of communities in which much of the coevolutionary change occurs; and (3) there is a continual population remixing of the range of coevolving traits, resulting from the selection mosaic, coevolutionary hotspots, gene flow, random genetic drift, and local extinction of populations (Nuismer, Thompson, and Gomulkiewicz 1999). This process, in turn, suggests three predictions about ecological patterns: populations will differ in the traits shaped by an interaction, traits of interacting species will be well-matched in some communities and mismatched in others, and there will be few species-level coevolved traits, because few traits will be globally favored (Thompson 1994, 1997).

Spatial structuring has been recognized for a long time as an important factor influencing demographic and genetic patterns in natural populations (Wright 1943; Levins 1969, 1970; Burdon and Thrall 1999). However, the geographic mosaic theory of coevolution clearly states that different spatial and temporal scales shape the coevolutionary process, and many of the implications of these ideas have now been carried into the broad picture of coevolutionary studies (Burdon and Thrall 1999; Kraaijeveld and Godfray 1999; Lively 1999). The studies of coevolution in interactions have tended to focus on single populations or species, neglecting the complex interactions that are likely to occur between the individual populations of species that occupy a wide geographic range, each of which experiences different biotic and abiotic selection pressures (Thompson 1999b; Benkman 1999; Kraaijeveld and Godfray 1999). If the interaction is to affect the fitness of individuals and favor reciprocal evolutionary change, then coevolution demands some degree of reciprocal specialization by interacting species. The geographic mosaic theory of coevolution is based on the idea that structured populations of interacting species experience local differences in the intensity of selection they impose on each other, which can lead to a geographic patchwork for traits involved in the interaction (Thompson 1994, 1997, 1999a, 1999d). Finally, variation in the ecological outcomes of interactions is the raw material for coevolution. That raw material is molded in different forms of interaction and in different environments. Because interactions between ants and plants occur almost everywhere in terrestrial habitats, they should become a very good model for exploring how the raw material of variation in ecological outcome has been molded in very different ways in different habitats as ants and plants have diversified.

* * *

Interspecific interactions, or interactions between individuals of different species, are one of the most important processes influencing the patterns of adaptation and variation of species (Thompson 1982, 1994, 1999b, 2005; Futuyma and Slatkin 1983) and of community organization and stability (Bondini and Giavelli 1989). In other words, the history of evolution and biological diversity is basically a history of the evolution of interspecific interactions. Like species, interactions also evolve and multiply; they are links between the histories of species and also shape their future evolution (Thompson 1982, 1999b; Futuyma and Slatkin 1983). Organisms in nature are not isolated, and to survive and reproduce, most have evolved in ways that require them to use a combination of their own genetic information and that of other species. However, not much is known about how genomes of separate species become intermeshed in the process of coevolution (see above), or reciprocal evolutionary change through natural selection (Thompson 1999b).

Furthermore, one species does not interact with only one other species. A species is rather immersed in an evolutionary unit of interaction that is the group of species within which selection acting on one of the pairs of species significantly affects selection in the other species (Thompson 1982; Kaiser 1998; Bronstein and Barbosa 2002). For example, the leaf-cutting-ant system is a unit of interactions encompassing several species with interactions producing different outcomes: (1) the ants and their basidiomycetes fungus are a mutualistic pair of species; (2) the ants and the plants they forage on are an antagonistic pair of species; (3) a microfungus (*Escovopsis* sp.) and the ants interact in an antagonistic manner; and (4) the ants and a bacteria (*Streptomyces* sp.), which inhibits growth of the microfungus and increases basidiomycetes biomass, interact in a mutualistic manner (Wilkinson 1999; Currie, Mueller, and Malloch 1999; Currie 2001; see chapter 2).

Another way of assessing or demonstrating the effect of interspecific interactions encompassing more than two species is, for example, to consider the relative strengths of top-down and bottom-up forces in a community (trophic interactions). In a study using the ant-plant *Piper cenocladum* (Piperaceae) in a tropical forest community in Costa Rica, enhanced plant resources (e.g., nutrients and light) had a direct positive effect on plant biomass, but there was no evidence of an indirect effect of plant biomass on herbivores or predators (i.e., cascading through the herbivores), whereas ant activity had indirect effects on plant biomass by decreasing herbivory on the plants irrespective of resource enrichment (Letourneau and Dyer 1998a; Dyer and Letourneau 1999a).

Interspecific interactions are based on an entirely selfish cost-benefit system, which depends on the relative gain to loss in fitness produced by the interaction, so we should expect a continuum from antagonism to mutualism. Moreover, interspecific interactions change in time and space. For instance, a species may be antagonistic to plants in one stage of its life cycle (leaf-chewing caterpillars) and mutualistic in another stage (pollinating butterflies). A population or species may also be antagonistic to plants in one portion of its distribution or habitat, while a population of the same species in another portion of its distribution may be mutualistic with plants. For example, the ant *Formica neorufibarbis gelida* visits the flowers of three plant species, creating a very interesting system of mutualistic and antagonistic interactions: these ants pollinate the gynodioecious *Paronychia pulvinata* (Caryophyllaceae), are herbivores of the gynodioecious *Eritrichum arentioides* (Boraginaceae), and appear to have little effect on the hermaphroditic *Oreoxis alpina* (Apiaceae) (Puterbaugh 1998).

The types of interspecific interactions can be defined on the basis of whether the net effect or outcome of the interaction is an increase or decrease in fitness or has no effect (is neutral) for each interacting species. For example, although the ant *Oecophylla longinoda* enhances the fitness of ant-occupied *Ficus capensis* (Moraceae) individuals by reducing herbivory or competition with other plants, fruit removal is reduced in trees with large ant colonies because nocturnal ant presence on fruits decreases the visitation rate of seed-dispersing bats (Thomas 1988). The net effect of *O. longipoda* ants should be evaluated in order to assess whether the interaction between the ants and *F. capensis* is mutualistic or antagonistic.

* * *

Basically, two types of interactions are considered throughout this book: antagonistic and mutualistic. Although other terms, such as *commensalism, neutralism,* and *amensalism,* are common in the literature, they are not within the scope of this book. The way the chapters are organized is based on the intimate association between antagonism and mutualism. There are several possible ways that antagonism and mutualism are related.

For instance, one way to assess the intimate association between antagonism and mutualism is by considering that virtually all mutualisms are consumer-resource interactions, as Holland et al. explain. For example, ants feed on plant secretions; they are herbivores that have been co-opted to benefit the plant. One could compare mutualisms involving ants to predation or herbivory involving ants and try to assess the association between antagonism and mutualism. Moreover, these authors suggest that the gathering of evidence on the differences between mutualisms and antagonisms is interesting and worth discussing, but future studies should consider that probably the difference has to do with the nature of the few forms of interaction that have been studied to date, rather than with their mutualistic versus antagonistic outcomes (Holland et al. 2005).

On the contrary, some authors consider that, although they are related, a separation should be made between antagonism and mutualism (Thompson 1982, 1994, 2005). For example, some forms of herbivory are true grazing across multiple hosts, and other forms are true parasitism, such that one form or the other may not look like herbivory. The forms of selection are inherently different for interactions involving multiple interactions between free-living species as opposed to intimate symbiotic relationships. In general, we will follow this view.

We begin (chapter 2) with the two clear examples of antagonistic ant-plant interactions (grazing, by leaf-cutter ants; and predation, by seed-harvesting ants). Chapter 3 discusses mutualism from antagonism (primary seed dispersal), and chapter 4 considers the related mutualism from opportunism (secondary seed dispersal). We follow with other types of mutualism from antagonism (pollination, and the defense of plants by ants mediated either by direct or by indirect interactions, chapters 5–7). In chapter 8 we present a mutualistic system between ants and plants that is not evolutionarily based on an antagonistic interaction, but rather on the life-history characteristics of the plants (nutrition of plants by ant mutualists). The following two topics are canopy-dwelling ants in chapter 9 and variation in ant-plant interactions in chapter 10. Canopy-dwelling ants have special significance because this area of study is just now being developed; the matter of variation in ant-plant interactions is one on which we want to encourage more geographically oriented research. Both topics are very important to our understanding of ant evolution, ants' great diversity, and their association with angiosperms.

We turn to the interface between ant-plant interactions and agroecosystems in chapter 11. Even though ants in agriculture may seem disconnected from the focus of the rest of the book, this research area is currently of utmost importance. Because natural, pristine biological communities are practically nonexistent (especially in the tropics), our knowledge needs to expand into communities of secondary origin (especially human-based ones) to address real ecological questions pertaining to them. Current research shows that information gathered on defense of plants by ants, for example, can be used in man-made systems. In chapter 12 we provide an overview and discuss future perspectives on ant-plant interactions.

CHAPTER TWO

Antagonistic Interactions

Leaf-Cutting and Seed-Harvesting Ants

In antagonistic interactions, the fitness of individuals of one of the interacting species increases, while the fitness of individuals of other interacting species decreases as a result of the interaction. Basically, antagonistic interactions occur between species because living organisms are concentrated packages of energy and nutrients (trophic interactions) and because resources are limited (competition) (Thompson 1982). Antagonistic interactions are usually defined in terms of discrete categories such as parasitism, predation, and competition (Ricklefs 1984; Futuyma 1986; Begon, Harper, and Townsend 1990) or are based on the kinds of items eaten (carnivory or herbivory), with certain terms being used with different meanings (e.g., predation for grazing or herbivory in Harper 1977; Begon, Harper, and Townsed 1990). However, when the ways organisms feed on other species are considered, certain evolutionary patterns may result, and although these categories are no more discrete than other concepts, like population or community, they provide a useful tool for comparative studies. Some evolutionary patterns may result from the ways organisms feed on other species. These differ greatly in how they attack their victims, including whether they kill them, how long they remain to feed on a single victim before killing it or leaving it, and how many victims they feed upon during their lifetimes (Thompson 1982, 1994). Thus, antagonistic interactions can be divided into four basic types: parasitism, grazing, predation, and competition.

Ant activities such as collecting extrafloral nectar and food bodies (e.g., grazing) are, in fact, antagonistic. From the plant's point of view, however, protective mutualisms exploit the fact that ants can act simultaneously as herbivores and carnivores (i.e., plant protection usually results from predation or attack by ants on non-ant herbivores; see chapter 6). Ants and plants are associated basically in two categories of antagonistic interactions, then: grazing (leaf-cutting ants) and predation (seed-harvesting ants). Interestingly, these two types of antagonistic interactions are considered to exert relevant effects on terrestrial plant succession (Davidson 1993). In the case of herbivory, palatable species, either individually or as a functional group, can be regarded as keystone resources whose fluctuating abundances are likely to influence communities of both producers and consumer species on a local and landscape level (Vasconcelos and Cherrett 1997; Farji-Brener 2001; Peñaloza and Farji-Brener 2003). However, granivores often behave as keystone predators and may inhibit succession in a diversity of plant communities (Davidson 1993).

On one hand, even though the grazing effect of leaf-cutter ants on plants is straightforward (loss of tissue), in reality it is a very complex system involving a number of different organisms and outcomes (Wilkinson 1999; Currie, Mueller, and Malloch 1999, Currie 2001), which are practically impossible to separate. On the other hand, seed harvesting is a straightforward interaction in which ants harvest and kill seeds (predation), although the impact of that death on plant population dynamics, on the evolution of seed dispersal and seed defense, and on ant-population dynamics is not easily predicted (e.g., Brown et al. 1986). Furthermore, we always have to consider that harvester ants may have a dual role, acting as both seed predators and dispersers of seeds (Retana, Picó, and Rodrigo 2004). In this chapter we review leaf-cutter and seed-harvesting ant systems, analyzing the interactions and their effects on the community.

Leaf-Cutting Ants (Grazing)

Members of the Myrmicinae tribe Attini exhibit the complex habit of culturing and eating fungi (Weber 1966, 1972). The ants are a monophyletic and morphologically distinct group limited to the American continent, with most of the 12 genera and more than 200 species showing

FIGURE 2.1. Leaf-cutting ant (*Atta,* Attini) transporting a leaf fragment to the colony.

a neotropical distribution (Mexico and Central and South America) (Hölldobler and Wilson 1990; Stradling 1991; Chapela et al. 1994; North, Jackson, and Howse 1997; Wilkinson 1999). In contrast, the attine fungi are polyphyletic, most belonging to the Basidiomycetes family Lapiotaceae (Chapela et al. 1994; Hinkle et al. 1994; Mueller, Rehner, and Schultz 1998). From the time of its origin around 50 million years ago, the ant-fungus mutualistic association has resulted in the evolution of unique behavioral (e.g., elaborate manuring regimes that maximize fungal harvest, and elimination of competing "weed" fungi mechanically), physiological, and anatomical modifications (e.g., secretion of antibiotic "herbicides" to control weed molds; and the loss of digestive enzymes, with fungal enzymes relied on instead to produce low-molecular-weight absorbable nutrients) in the ants, which distinguish the Attini from other Formicidae. Corresponding morphological and biochemical modifications are seen in at least some of the fungal associates (Hölldobler and Wilson 1990; Chapela et al. 1994; Hinkle et al. 1994; Mueller, Rehner, and Schultz 1998). Leaf-cutting (or fungus-growing) ants (fig. 2.1) have succeeded at domesticating multiple cultivars (varieties produced under cultivation) and are capable of switching to novel cultivars. A single ant species farms a diversity of cultivars that are shared occasionally between distantly related ant species, probably by lateral transfer between ant colonies (Diamond 1998; Mueller, Rehner, and Schultz 1998).

The Ants and the Fungus

In this mutualistic association, the ants benefit because the fungus provides enzymes that break down plant tissue and detoxify certain plant secondary compounds that may have insecticidal properties, thus enabling the ants to make use of plant material that would otherwise not be available to them; the fungus may also provide the ants with essential chemicals (Powell and Stradling 1991; Kacelnik 1993; North, Jackson, and Howse 1997; and references therein). In exchange, the fungus is cultured in an environment that is kept virtually free of competition from other microorganisms by constant tending and application of antibiotic compounds that the ants produce (North, Jackson, and Howse 1997; Wilkinson 1999; and references therein). The ants also provide substrate, dispersion, and recycling of some of their enzymes (Hölldobler and Wilson 1990; Kacelnik 1993). The fungus serves as the only larval food and probably as an important portion of the adult diet for primitive attines, although workers of those species that cut leaves from living plants also drink sap directly (Littledyke and Cherret 1976; Stradling 1991; Kacelnik 1993). Latex intake (e.g., from Euphorbiaceae [Rico-Gray, Palacios-Rios, et al. 1998]), however, may be quite harmful for the ants (Powell and Stradling 1991).

Although the interaction between the ants and the fungus is considered to be mutualistic, the fungus is not a passive partner in the relationship and may manipulate the ants to provide it with substrate and antimicrobial defense. Experimental evidence has suggested that the fungus can select its substrate by controlling the foraging behavior of the ants by means of chemical feedback mechanisms (see North, Jackson, and Howse 1997).

The mutualistic interaction between leaf-cutting ants and Basidiomycetes fungi plays a central role in neotropical communities because they influence both the grazing and the detritus chains, generating a complex system that involves many other species and thereby generating a wealth of additional interactions. This system of ecological interactions has been investigated through quantitative modeling, but the complexity of the system makes for only simplified representations of reality (Giavelli and Bodini 1990). The obligate mutualistic association of Attini ants and their fungus is based on the antagonistic interaction between the ants and the plants they graze. Grazers are channeled genetically over evolutionary time into requiring a mixed diet (Thompson 1982), and leaf-cutting ants are no exception. Ants are tied to their nest site, so monophagous grazing would impose severe restrictions on their diet, particularly in tropical habitats with high plant species richness, relatively few individuals per species, and large quantities of toxic plant-defense chemicals. Leaf-cutting ants are highly successful herbivores because they are able to graze on a wide variety of plants (fig. 2.2), so that the harvested plant material can be used as substrate for the polyphagous fungi on which they feed (North, Jackson, and Howse 1997; Wirth et al. 2003). For instance, the use of a high diversity of plant sources was reported by Cherrett (1968, 1972) for *Atta cephalotes* in a Guyanese rain forest, where this ant exploited 30%–50% of the accessible plants in a period of 10 weeks. Throughout one year of observation in the rain forest of Barro Colorado Island, Panama, workers from one colony of *Atta colombica* harvested material from 126 plant species, mostly trees and shrubs (64%), and lianas (30%), and more rarely epiphytes and hemiepiphytes and herbs (Wirth et al. 1997; Wirth et al. 2003). Fungi culture has conferred a

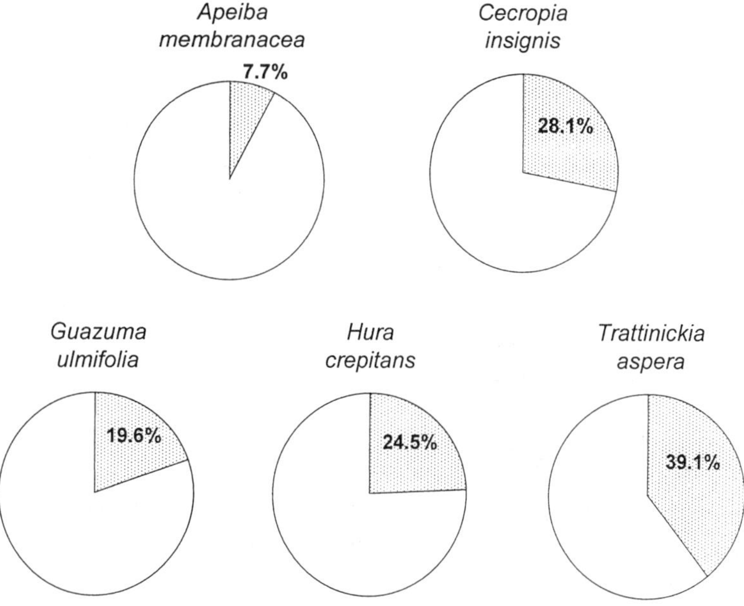

FIGURE 2.2. Estimated annual foliage losses (total leaf area of an individual tree crown per year) of individual trees from five species due to leaf-cutting by one colony of *Atta colombica* in the tropical rain forest on Barro Colorado Island, Panama. Modified from Wirth et al. 2003.

major advantage to *Atta* species inhabiting neotropical forests because it allows them to overcome a wide range of plant chemical defenses and thus be "ecologically polyphagous" (these ants are really monophagous) in the midst of a highly diverse flora (Stradling 1991).

The Ants and the Plants

The range of plant species used as fungal substrate by attines is partly determined by the richness of the habitat and partly by the nutritional value, moisture content, secondary compound content, and physical characteristics of leaf tissues (Powell and Stradling 1991). Attines living in savannas generally use a relatively narrow range of Poaceae, while those inhabiting tropical rain forests graze on a relatively wider spectrum of plant species (Buckley 1982a; Stradling 1991). In the cerrado savanna of Brazil, however, the so-called lower attines (see Hölldobler and Wilson 1990) may use material from as many as 53 species from 28 families as substrate for fungus culturing (Leal and Oliveira 1998, 2000). Plant selection by leaf-cutting ants is related to the presence of ant- and/or fungus-repellent substances in leaf tissues (Hubbell, Wiemer, and Adejare 1983; Hubbell, Howard, and Wiemer 1984; North, Jackson, and Howse 1997), the amount of nutritional contents (e.g., high nutritional contents are preferred despite high tannin concentrations) (Nichols-Orians 1991a, 1991b), and leaf toughness (the softer, younger leaves are preferred over the tougher, older leaves) (Nichols-Orians and Schultz 1989). Another factor affecting foraging by leaf-cutting ants (e.g., *Acromyrmex* spp.) is the presence of endophytic fungi (e.g., *Neotyphodium* spp.), which interact in a mutualistic way with their host grasses by deterring herbivores and pathogens via the production of alkaloidal mycotoxins (Tibbets and Faeth 1999).

The foraging behavior of leaf-cutting ants varies with changes in the environment. In a seasonal tropical forest in Costa Rica, foraging by *Atta colombica* and *A. cephalotes* is predominantly nocturnal during the dry season and diurnal during the rainy season, probably due to changes in precipitation patterns, temperature, or vegetation cover, or even hostile interactions with other ants (Rockwood 1975; Wirth et al. 1997). The largest quantities of new leaves and flowers were grazed as they became available at the beginning of the rainy and dry seasons, in correlation with peaks in the number of plant species producing them, and although mature leaves were always present, the ants used a limited number during the rainy season (Rockwood 1975). Shepherd found that adjacent colonies of *A. colombica* in Antioquia, Colombia, one located in an older and the other in a younger secondary forest, exhibited differences in grazing. The former colony used 130 species and had a dietary diversity of 30.7 equally used species in a cyclical annual pattern of gradually changing diet. The latter colony used 103 species and had a dietary diversity of only 11.8 equally used species, and grazing of this colony was governed by the availability of a small number of high-quality resource species (Shepherd 1985). In a thorough study of vegetation harvested by *Atta colombica* on Barro Colorado Island, Wirth et al. (1997) estimated annual herbivory of green leaves as 3,855 m^2 foliage for one colony (fig. 2.2). Total dry weight of biomass harvested was higher in the dry season because most material collected during the wet season consisted of green leaves, while in the dry season more than 50% of the biomass collected was nongreen plant material such as fruits, seeds, and flowers (Wirth et al. 1997). A seasonal pattern in the use of fungal substrate was also reported for seven genera of lower attines in the Brazilian cerrado (Leal and Oliveira 1998, 2000).

Grazing activities of leaf-cutting ants are also affected by changes in the structure of the woody plant community during the regeneration process (fig. 2.3; Nichols-Orians 1991b; Vasconcelos and Cherrett 1997). In the Brazilian Amazonia the number of plants attacked by *Atta laevigata* was independent of the number available for attack, and plant resistance against ant grazing was not affected by previous damage (Vasconcelos 1997). However, successional changes in the composition of the woody plant community did significantly affect the composition of the leaf diet of the ant, and these variations were strongly correlated with variations in the abundance of *Bellucina imperialis.* As this species increased in abundance, its relative contribution to the ants' diet also increased, suggesting that variations in the diversity of the ants' diet were related to the abundance of preferred species (Vasconcelos 1997). Furthermore, Farji-Brener has suggested an association between the abundance of *Atta* nests and plant succession in tropical systems. The high abundance of pioneer plants (more palatable to ants) in secondary or disturbed forests could explain the high abundance of *Atta* nests in these systems (Farji-Brener 2001). For example, the presence of gaps as foraging sources (where pioneers are more frequent) is more important in primary forests (where they are scarce) than in secondary forests (where they are more abundant) (Peñaloza and Farji-Brener 2003). Farji-Brener

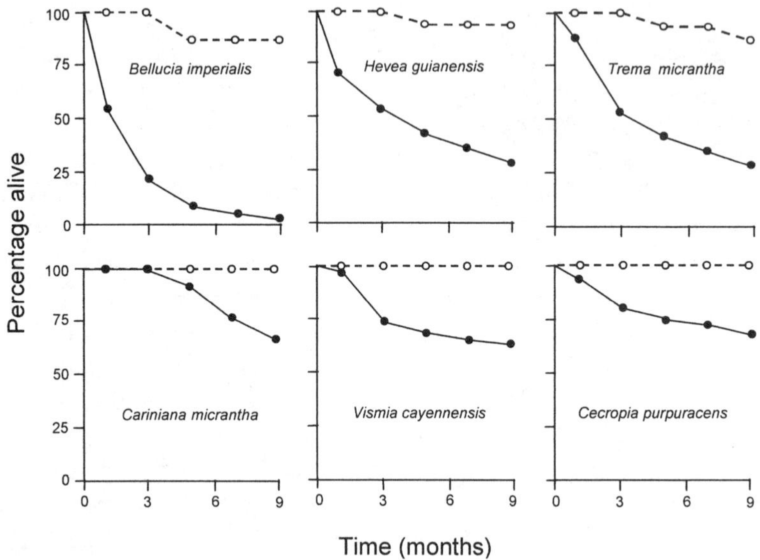

FIGURE 2.3. Effect of leaf-cutting activity by *Atta laevigata* on seedling survival in the rain forest of central Amazonia. Survivorship curves of transplanted seedlings are shown for seedlings protected (broken lines) and unprotected (continuous lines) from leaf-cutters. For each species there was an initial number of 20 protected and 80 unprotected seedlings. Modified from Vasconcelos and Cherrett 1997.

(2005) has also shown for a tropical rain forest of Costa Rica that the effect of abandoned nests on plant assemblage composition is low because between-site variation is more important than variation due to nest effect.

Leaf-cutting ant nests have been considered a major disturbance in tropical soils (Alvarado, Berish, and Peralta 1981), because the ants collect leaves over large areas and concentrate them, enriching the soil at specific sites (Haines 1975, 1988) and inducing particular patterns of associated vegetation (Fowler 1977; Farji-Brener and Silva 1995b; and references therein). Thus, it has been suggested that selective grazing, leaf concentration, and soil enrichment by leaf-cutting ants may be important factors determining patterns of succession in early successional habitats (Cherrett 1989; Nichols-Orians 1991b; Farji-Brener and Silva 1995b; Vasconcelos 1997; Vasconcelos and Cherrett 1997). The location (e.g., *Atta cephalotes* in subterranean chambers, *Atta colombica* on the soil surface near the nest) of the nutrient-rich organic refuse produced by a leaf-cutting ant colony is potentially important in determining patterns of plant recolonization in abandoned or dead ant nests. For example, research done in Costa Rica and Panama shows that soils from *A. cephalotes* nests (exhibiting internal refuse) did not differ from adjacent soils in abundance of fine roots and seed diversity, whereas organic refuse from *A. colombica* nests (exhibiting external refuse) was less diverse in seed composition but had a greater abundance of fine roots than adjacent soils; thus, the relative abundance of these *Atta* species may influence the structure and/or composition of tropical forests (Farji-Brener and Medina 2000). In Brazilian eastern Amazonia, leaf-cutting activity by *A. sexdens* has recently been shown to have an important effect on forest succession on abandoned land by altering the physical and chemical properties of soil, root distribution, and tree growth. The deep soil beneath mature nests at different depths presents a low resistance to penetration and is rich in Ca and Mg when compared to non-nest soil, and such properties are accompanied by increased root biomass beneath *A. sexdens* nests and may favor seedling growth (Moutinho, Nepstad, and Davidson 2003). Thus, leaf-cutter ants can be considered "ecosystem engineers." The effect and potential consequences of large *Atta* nests in tropical rain forests have been summarized in Farji-Brener and Illes 2000.

The effect of leaf-cutting ant nests is not limited to tropical systems. Leaf-cutting ant nests (mainly in the genus *Acromyrmex*) also affect temperate systems, where it has been demonstrated that soil is enriched, vegetation patterns are modified, and nests function as refuges for rare plant species, especially during seasons of high hydric stress (Farji-Brener and Ghermandi 2000, 2004).

Leaf-cutting ants may graze on plants over 100 m from their colony, and a single foraging trip may require several hours (Howard 1991 and references therein). The time and energy required to locate and harvest rare or distant plants of high quality may, at times, make it more efficient for ants to graze more readily available resources of lower quality. Analysis of this potential trade-off between resource quality and cost is central to understanding overall colony foraging strategies. However, differing requirements of the partners in the mutualistic association (ants and fungi) may result in conflicts over resource quality (Howard 1991). For example, Powell and Stradling have suggested that their results provide no support for the optimal foraging hypothesis and that, on the contrary, the analysis revealed a strong influence of high protein and low phenol content of leaves upon foraging preferences, and the importance of plant tissue pH for the fungus. In this complex association, plants acceptable to the grazers may contain antifungal compounds that the ants cannot detect. Substrate selection by higher attines, although constrained by the presence of substances directly toxic to the grazers, includes the collection of material that supports little or no fungal growth. The repeated switching between a wide variety of plant species in nature will dilute and minimize any deleterious effects due to the collection of substrates unsuitable for fungal growth and at the same time obviate the necessity to search on foot for a restricted number of target species in a diverse and complex habitat (Powell and Stradling 1991). Observations on the foraging activities of *Atta texana* in the semiarid Zapotitlán Valley in Mexico show that these ants use a wide variety of plant species (and humans' food leftovers) and that a given colony does not graze on a particular plant individual for more than 10 days, rarely eliminating the colony's food source (Rico-Gray, Palacios-Rios, et al. 1998; unpubl. obs.). The temporal pattern of leaf harvest by *Atta colombica* on Barro Colorado Island indicates that ants switch among plant species in harvesting periods of a few days to several months and resume the use of individual species at different times throughout the year (Wirth et al. 2003).

In this ant-fungus-plant system, the fitness of the ants depends on the growth of the fungi, which in turn depends on the supply of substrate, so that the foraging behavior of the ants is likely to be adapted for efficient substrate supply, both in quality and in quantity (Kacelnik 1993). Ant workers are central-place foragers, since they deliver resources to the colony's garden, and they may be expected to maximize the rate of resource delivery to the central place. However, individual workers may not be optimal foragers; the maximizing agent in this case is the colony and not necessarily each individual worker (Kacelnik 1993; Burd 1996a, 1996b; see below).

The size of leaf fragments cut and carried by leaf-cutting ants affects the time and energy costs of providing substrate to the colony's fungal gardens. Burd (1996a) estimated the costs and benefits for individual workers of *Atta colombica* and concluded that the load masses needed to maximize the rate or energetic efficiency of individual foraging are greater than the average fragment masses actually carried by *A. colombica* foragers. His analysis of foraging rates of *A. cephalotes* suggested that fragments carried were also below rate-maximizing size. Thus, individual rate or efficiency maximization appears not to be the "strategic" basis of foraging behavior in these leaf-cutting ants. Fragment size might be constrained by handling requirements, but little is known about this aspect of leaf cutting. Short absolute return times to the nest (and therefore lightweight loads) might be favored to reduce moisture loss from fragments, to reduce exposure time to parasitoid attack, or to enhance information transfer to nest mates. An alternative possibility is that small loads are rate maximizing, but at the level of the colony rather than the individual worker (Burd 1996a, 1996b). Burd suggests that group foraging success might be enhanced if individual workers reduce their cutting time, thereby taking smaller fragments individually but allowing a greater number of their nest mates to obtain fragments per unit of time. Finally, load selection in leaf-cutting ants is unlikely to be understood by considering foragers as isolated agents, and individual maximization models fail to explain fragment selection by *Atta colombica* and *A. cephalotes* (Burd 1996a); both of these models use a group measure of foraging success (Burd 1996b). In the past two decades, a number of studies have addressed the issue of load size and food plant selection by attine ants, and a rich discussion on the multiple aspects affecting harvest choice can be found in the specific literature (Nichols-Orians and Schultz 1990; Nichols-Orians 1991a, 1991b; Roces and Lighton 1995; Folgarait et al. 1996; Burd 2000; Röschard and Roces 2002).

Some of the grazing activity of fungus-growing ants can have a positive effect on plants when, instead of removing leaves or parts of leaves, they remove seeds and fruits from the forest floor (fig. 2.4; Leal and

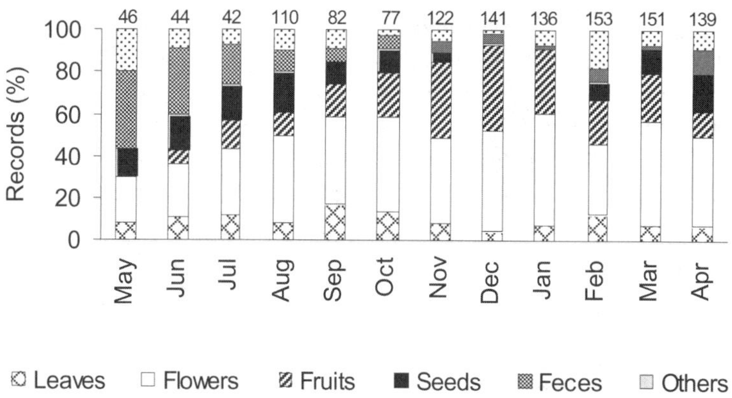

FIGURE 2.4. Material collected by attine ants during 12 months in an area of cerrado savanna at Mogi-Guaçu, southeastern Brazil. Values above bars refer to total number of records per month and include the following genera: *Cyphomyrmex, Mycetarotes, Mycocepurus, Myrmicocrypta, Sericomyrmex,* and *Trachymyrmex*. Modified from Leal and Oliveira 2000.

Oliveira 1998, 2000). Indeed, it has been demonstrated that, on the one hand, *Mycocepurus goeldii* facilitates germination in *Hymenaea courbaril* (Caesalpiniaceae) (Oliveira et al. 1995), that *Atta laevigata* aids to the establishment success of *Tapiria velutinifolia* (Anacardiaceae) (Farji-Brener and Silva 1996), and that *A. colombica* aids in seed dispersal of *Miconia argentea* (Melastomataceae) (Dalling and Wirth 1998; see chapter 4). On the other hand, the activity of attines like *Atta cephalotes* and *Trachymyrmex* sp. (Nascimento and Proctor 1996) and *Sericomyrmex aztecus* (Kaspari 1993) in removing the seeds of *Peltogyne gracilipes* (Caesalpiniaceae) and *Miconia affinis* (Melastomataceae), respectively, are unlikely to have any dispersal function since the seeds are cut up and used as fungus substrate (i.e., predation).

The Leaf-Cutting Ant-Plant System and Its Effect on the Plant Community

Studies on fungus-growing ants encompass a wide variety of topics that range from the ant-fungus association, nest structure, social behavior (e.g., Jaffe and Vilela 1989; Hölldobler and Wilson 1990; Roces and Hölldobler 1995, 1996), and interactions with other ant species (Vasconcelos and Casimiro 1997), to the effect of leaf-cutting ants on vegetation structure, succession (Nichols-Orians 1991b; Farji-Brener and Silva 1995b; Vasconcelos and Cherrett 1997; Farji-Brener and Medina 2000; Farji-Brener 2005), and soil characteristics (Alvarado, Berish, and Peralta 1981; Farji-Brener and Silva 1995a; Farji-Brener and Medina 2000; Moutinho, Nepstad, and Davidson 2003). Their foraging strategies (Giavelli and Bodini 1990; Howard 1991; Kacelnik 1993; Burd 1996a, 1996b) and activity (Rockwood 1975; Shepherd 1985; Nichols-Orians and Schultz 1989; Nichols-Orians 1991b; Wirth et al. 1997) have also been studied, as well as their effect on seed dispersal and germination (Oliveira et al. 1995; Farji-Brener and Silva 1996; Nascimento and Proctor 1996; Dalling and Wirth 1998; Leal and Oliveira 1998). Other researchers have studied the way they cope with plant defense mechanisms (Hubbell, Wiemer, and Adejare 1983; Hubbell, Howard, and Wiemer 1984; Powell and Stradling 1991; North, Jackson, and Howse 1997).

The understanding of the foraging activities of leaf-cutting ants can prove crucial for some neotropical regions. Although lower attines have very small nest populations and rarely cut leaves, using a variety of materials as substrate for their fungus (fig. 2.4), higher attines or "true" leaf-cutting ants (*Atta, Acromyrmex*), due to the amount of leaves they graze, are considered serious pest species in certain regions (Giavelli and Bondini 1990; Stradling 1991; North, Jackson, and Howse 1997). For example, leaf-cutting ants remove between 12% and 17% of leaf production in neotropical forests alone, while a colony of grass-cutting ants may remove, depending on the species, between 30 and 250 kg of dry matter per year (Hölldobler and Wilson 1990 and references therein). The ability to control the foraging activity of leaf-cutting ants through their own fungus offers exciting new prospects for the biological control of leaf-cutting ants in the neotropics. It may be possible to prevent the damaging effect of leaf-cutting ants on crops and plantations with pesticides that mimic the semiochemical produced by the fungus; secondary compounds of crop plants could be associated with fungal toxins in baits (North, Jackson, and Howse 1997). Furthermore, the study of leaf-cutting ants could be important for finding new antimicrobial compounds from their mandibular glands (North, Jackson, and Howse 1997; Wilkinson 1999). Interestingly, it has been argued that leaf-cutting ant debris could be used as a short-term defense against the ants' own foraging activities. Because these debris exhibit repellent properties against leaf-cutting ants, it could serve as a possible mechanism of herbivory control that is nontoxic to humans and economically feasible (Farji-Brener and Sasal 2003 and references therein).

The fungus-growing ant system is also a good example of the continuum between antagonism and mutualism and of a unit of interaction encompassing several species that interact to produce different outcomes (Thompson 1982). The ants and their basidiomycetes fungus are a mutualistic pair of species; the ants and the plants they graze on are an antagonistic pair of species; the microfungus *Escovopsis* sp. and the ants interact antagonistically; and the ants and the bacteria *Streptomyces* sp., which inhibits growth of the microfungus and increases basidiomycetes biomass, interact mutualistically (Wilkinson 1999; Currie, Mueller, and Malloch 1999; Currie 2001).

Seed-Harvesting Ants (Predation)

Like grazers, many predators are channeled genetically over evolutionary time into requiring a mixed diet (Thompson 1982), and seed-harvesting ants are no exception. Because ants are tied to their nest site, monophagous harvesting would severely restrict their diet, particularly because seed availability is highly unpredictable in dry habitats (where seed-harvesting ants are more abundant) and because they have to cope with the presence of defensive seed traits (Rios-Casanova, Valiente-Banuet, and Rico-Gray 2004, 2006). Seed-harvesting ant species regularly forage for seed on the ground (rarely on the plants) or in frugivore defecations; seeds are then taken to the nest to be eaten, stored, or fed to the larvae (Whitford 1976; Buckley 1982a; Andersen 1991a; MacMahon, Mull, and Crist 2000). The plant may benefit from this ant activity if seeds are discarded on the way to the nest, abandoned in shallow underground granaries, or dumped in ant refuse piles where they are protected from fire and/or from other predators and can then germinate (Levey and Byrne 1993).

The Ants

Harvester ants represent a broad assemblage of many different evolutionary lines (more than 100 species worldwide), mainly within the subfamily Myrmicinae (particularly the genera *Messor, Monomorium, Pheidole,* and *Pogonomyrmex*), but some also in the Ponerinae and Formicinae (Hölldobler and Wilson 1990; Andersen 1991b; Taber 1998) (fig. 2.5). This wide range of ant species harvest a great variety of seeds in different habitats (Carroll and Janzen 1973; Buckley 1982a; Hölldobler and Wilson 1990; Moutinho 1991). Seed harvesting by ants is practically restricted to arid environments (deserts and dry grasslands) in warm temperate and tropical regions around the world (Buckley 1982a; Rios-Casanova, Valiente-Banuet, and Rico-Gray 2006). However, in Australia seed harvesting by ants occurs throughout the continent and is prominent in practically all major vegetation associations (Andersen 1991b).

The transition from antagonism to mutualism is clear when we combine the studies on seed predation and seed dispersal (Thompson 1982; Levey and Byrne 1993; Rodgerson 1998; Retana, Picó, and Rodrigo 2004). Indeed, elements of both antagonism and mutualism appear in some of the interactions reported in chapters 3 and 4 (seed dispersal) because for some dispersal systems it is often difficult to separate one outcome from the other, although it is clear that certain cases are "mostly predation" and others are "mostly dispersal" (e.g., Boyd 1996).

Seed harvesting by ants has a major relevance in the sclerophyllous vegetation of southern Australia, where the highest concentrations of

FIGURE 2.5. A worker of a *Pogonomyrmex* harvester species transporting a seed (Poaceae).

myrmecochorous species (i.e., plants whose seeds are adapted for dispersal by ants) occur. In this region dispersal is predominantly accomplished by unspecialized species of *Pheidole* and *Rhytidoponera,* which are also the major seed predators of other plant species (Andersen 1988, 1991a; Westoby, Hughes, and Rice 1991; and references therein). The relationship between antagonism (seed predation) and mutualism (seed dispersal) in Australia may be based on the availability of unspecialized seed-collecting ants that contribute to the prevalence of myrmecochory (Andersen 1991b). In North America, however, the evolution of specialized seed structures for dispersal may be precluded by the assemblage of seed-harvesting ants whose workers are much larger than those ants normally associated with elaiosome-mediated seed dispersal (large workers would consume the elaiosome and the seed; see Rissing 1986).

Seeds as Food for Harvester Ants

The use of seeds as food by ants is not surprising because seeds are rich in lipids and proteins and have a high nutritional value (Janzen 1971). However, seeds can be hard to find and ants may have to overcome the seeds' physical and chemical defenses. The degree of commitment of harvester ants to a seed diet varies from generalist omnivores (e.g., *Atopomyrmex mocquerysi* of Africa) to specialist granivores (e.g., *Monomorium*

(= *Chelaner*) *whiteri* of Australia or *Messor* (= *Veromessor*) *pergandei* of North America) (Hölldobler and Wilson 1990; Andersen 1991a). Seed foraging by granivorous ants involves a feedback between seed selection and seed availability; while most harvester ants eat a wide variety of seeds under natural conditions, seed choice is largely determined by size (fig. 2.6), morphology, and availability of the seeds (Andersen 1991a).

In a study of seed preferences by the harvester ant *Pogonomyrmex occidentalis*, it was found that seed availability varied seasonally and that seed preference by ants was correlated with the seasonal availability of preferred species, but not with unpreferred seeds (Crist and MacMahon 1992). Ants removed 9%–26% of the potentially viable seed pool each year and as much as 100% of available preferred species. From the soil seed pool, ants preferentially harvested small seeds (fig. 2.6), with foraging activities concentrated near the nest, where seed densities were low. In controlled preference experiments, however, ants were unselective near nests but preferred large seeds with higher energy content in trials 10 m from nests, indicating that preferences measured under experimen-

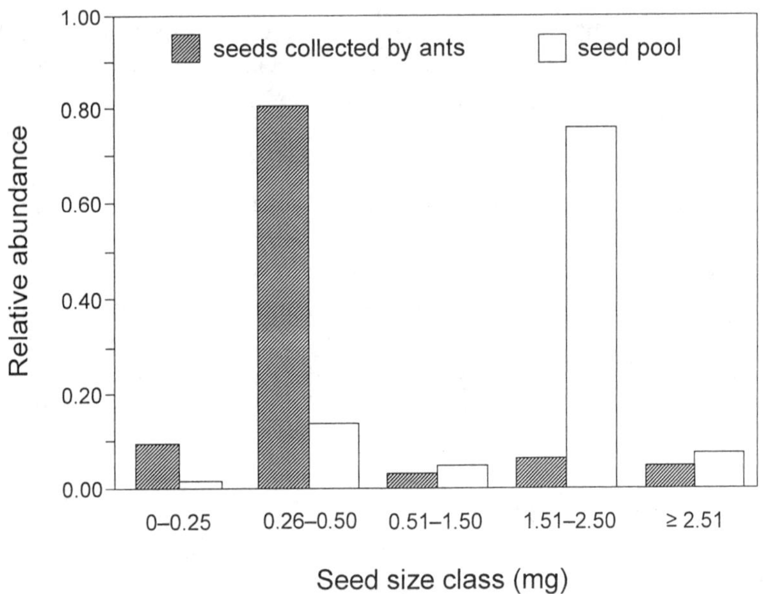

FIGURE 2.6. Size distributions of seeds available on the soil surface (N = 3,793) and of those returned by *Pogonomyrmex occidentalis* harvester ants (N = 3,410), during two years in a shrub-steppe ecosystem in Wyoming (USA). Modified from Crist and MacMahon 1992.

tal conditions may not correspond to natural diets (Crist and MacMahon 1992). Moreover, seed removal rates from both low- and high-density seed patches were higher near foraging trails than away from them, suggesting that differences in removal rates by colonies of *P. occidentalis* are due in part to a higher probability of discovery of patches located near trails (Mull and MacMahon 1997).

The spatial and temporal foraging patterns of *Pogonomyrmex* spp. and their important direct and indirect effects on community structure and ecosystem functioning have recently been summarized by MacMahon, Mull, and Crist. These authors consider that harvester ants directly influence community structure by exerting effects on plants and other taxa and on ecosystem functioning. Their direct effects on plants include removal and consumption of seeds and other materials, storage and rejection of seeds, and plant clipping for nest maintenance. Harvester ants may also directly affect other granivore taxa (e.g., rodents), predators (e.g., snakes, birds, and lizards), and myrmecophiles that live in or on their nests and associated soils (e.g., beetles, orthopterans, termites, hemipterans, collembollans, thysanurans, diplurans, millipeds, spiders, and mites). Harvester ants have direct influences on ecosystem processes through their effects on food-web structure, energy flows, nutrient transport, and soil modification; alternatively, they may have important disturbance effects (MacMahon, Mull, and Crist 2000 and references therein).

Effect of Seed Harvester Ants on Plant Communities

Arid environments are characterized by a great spatial and temporal variation in seed availability; thus few plant species can produce a profitable crop of seeds during any given year. However, seed-harvesting ants cope with the fluctuations in their food supply by storing seeds, so as to be relatively independent of plant production (Brown, Reichman, and Davidson 1979; Buckley 1982a; Davison 1982). In the case of seed defenses, grass seeds are often preferred, probably because they are relatively low in toxins (Carroll and Janzen 1973; Buckley 1982a; Davison 1982) and/or easy to carry (Buckley 1982a), but a wide variety of other species are also utilized (O'Dowd and Hay 1980; Buckley 1982a and references therein). However, the presence of endophytic fungi in grass seeds may alter ant foraging activities and decrease seed removal by some desert seed-harvesting ant species due to the production of alkaloid metabolites (Knoch, Faeth, and Arnott 1993). In general, many seeds are protected by physical traits, and those adapted for ant dispersal are relatively strong and suffer low levels of predation compared to relatively weak seeds that are destroyed at much higher rates (Rodgerson 1998).

Because plant-based food resources change seasonally (Rico-Gray 1989, 1993; Rico-Gray, García-Franco, et al.1998), ants undergo diet changes over time (Rico-Gray and Sternberg 1991). Food resources are patchily distributed for most animals, regulating their feeding behavior, their population dynamics, and ultimately their evolution (Bronstein 1995). The responses of animals to patch structure on different spatial scales influences how processes such as foraging and species interactions vary spatially (e.g., Azcárate and Peco 2003) or translate across different levels of organization. In this context, the scale effects of vegetation can be important determinants of forager movement and seed removal by ants (Crist and Wiens 1994; Mull and MacMahon 1997). Indeed, broad-scale differences in vegetation structure (i.e., soil characteristics and grazing intensity by herbivores) have been shown to affect forager movements and seed harvesting by ants, whereas the distance of patches from nests and fine-scale factors of seed patches (i.e., seed species and cover type) seem less important in determining among-colony variation in harvesting activity (Crist and Wiens 1994). Although individual ant movements can partially explain differences in seed removal among fine-scale cover types, processes operating at the population level, such as predation and grazing intensity, largely contribute to the variation of seed removal among ant colonies (see review by MacMahon, Mull, and Crist 2000).

Seed-harvesting ants can exert severe and significant effects on plant communities through at least two different processes (Brown and Human 1997). Direct effects result from selective harvesting of seed species that can affect the relative and absolute abundance of flowering plants, especially in deserts, grasslands, and other xeric habitats where these ants are most abundant (Buckley 1982a; Hölldobler and Wilson 1990; Andersen 1991b). Indirect effects include the fact that their nest mounds provide enriched edaphic environments that can support plant communities that differ from the surrounding vegetation (King 1976, 1977a, 1977b, 1977c) and the ants' interactions with rodents that in turn affect the vegetation.

High seed removal rates by ants have been recorded in different studies carried out in a variety of Australian regions and habitats (Buckley 1982a; Andersen 1991b; and references therein). For instance, it was estimated that ants collected 4,000 seeds m^{-2} during an autumn bout of harvesting (Briese 1982), and up to 5,000 seeds h^{-1} were recorded passing into a single nest (Smith and Atherton 1944). It is clear that the volume of seeds removed by ants is at times quite considerable, but the impact on plant populations has been poorly studied. Some authors consider that the activity of harvester ants tips the balance of competition among some plant species, promotes equilibrium in others, rearranges the local distributions of surviving species, or reduces seed density and the vegetative mass of plants (MacMahon, Mull, and Crist 2000; Beattie and Hughes 2002; and included references). However, in an ant-exclusion experiment in California, no changes in the plant community were detected when comparisons were done between plots with and without the activity of the harvester ant *Messor andrei* (Brown and Human 1997; but see Andersen 1991b; Brown, Reichman, and Davidson 1979). Moreover, Andersen (1991b) suggests that the impact of seed losses will depend more on how many seeds remain, and where, rather than on how many are lost, and he points out that losses should be placed in the context of the overall dynamics of the plant population. The latter is directly related to the availability of safe sites for germination and establishment (Louda and Potvin 1995; Kelly and Dyer 2002).

In the case of many stable populations of long-lived perennials, seedling recruitment is negligible in the absence of disturbance, such that recruitment is limited by the availability of safe sites rather than by seed supply (Andersen 1991b). For instance, although seedling density of *Eucalyptus baxteri* increased by a factor of 15 after removal of ants, complete seedling mortality within a year resulted in no ultimate impact on recruitment caused by ants (Andersen 1988, 1991b). Similarly, Laman (1996b) has shown that even though *Pheidole* sp. ants are more associated with *Ficus stupenda* and *F. punctata* (Moraceae) than with any other canopy plant in a rain forest in Borneo and readily harvest the seeds of these strangler fig trees, the number of *Ficus* seedlings found in the canopy is considerably lower (239 to 10) when the ants are absent. Because no *Ficus* seedlings are found in the vicinity of *Pheidole* sp. ant nests, it seems that the few seeds the ants drop, abandon, or discard form the basis for recruitment in these plant species in an environment with very limited suitable sites for seed germination and establishment (Laman 1996b). Similar results were obtained by Levey and Byrne (1993), studying *Pheidole* spp. in a Costa Rican rain forest, where the few seeds of *Miconia nervosa* and *M. centrodesma* (Melastomataceae) placed by the ants in refuse piles form the basis of plant recruitment patterns in these shrub species.

In the case of ephemerals, however, seed removal by predators has a direct effect on plant populations because these species are often limited by seed supply (Andersen 1991b). For example, annual plants were 50% denser after two seasons in plots where ants had been excluded, than in plots with ants (Brown, Reichman, and Davidson 1979).

Seed-harvesting ant nest mounds are sources of spatial heterogeneity in soil biota and soil chemistry and can affect the distribution and productivity of plant species in a community (Wight and Nichols 1966; Rissing 1986). The soil mounds of *Pogonomyrmex barbatus* support 30 times higher densities of microarthropods and 5 times higher densities of protozoa than surrounding control soils (Wagner, Brown, and Gordon 1997). Moreover, the soils of *P. barbatus* nests are marginally more acidic and have higher concentrations of nitrate, ammonium, phosphorus, and potassium than control soils (Wagner, Brown, and Gordon 1997). In addition, soil temperature can be significantly higher in nest mounds than in surrounding soils (Brown and Human 1997).

As expected, the spatial heterogeneity generated by the peculiar properties of nest mounds can have a strong influence on plant communities, with variable effects on individual plant species (Wagner 1997). For instance, significantly more grasses and fewer forbs were present on mounds of the seed-harvesting ant *Messor andrei,* although at least one forb (*Lepidium nitidum*) produced twice as many seeds when it grew on nest mounds (Brown and Human 1997). Significantly more plant species of nutrient-rich soils were present on nest mounds of *Messor capensis* than on neighboring soils; some of the species were taller, whereas others produced longer inter-nodes, or even more seeds, than conspecifics growing away from the mounds (Dean and Yeaton 1993). When growing on nest mounds of the seed-harvesting ants *Messor* (= *Veromessor*) *pergandei* and *Pogonomyrmex rugosus* in the Mohave Desert, individuals of *Schismus arabicus* and *Plantago insularis* experience a 15.6- and 6.5-fold increase in fruit or seed production, respectively, when compared to nearby controls; and for the former plant species seed mass is also augmented (Rissing 1986).

* * *

In summary, due to the large numbers of seeds removed by ants and the often intense interspecific competition for seeds among ants, granivory and seed-harvesting have been considered to be important interactions structuring plant communities, particularly in desert regions where many plant populations exist solely as seeds for long periods and where many granivorous taxa occur in peak abundance (e.g., Brown 1975; Davidson 1977a, 1978; Brown, Reichman, and Davidson 1979; Rissing 1986). Although seed harvesting by ants is basically considered predation, seed dispersal can be simultaneously effected when ants discard seeds along the trail or in ant nest refuse piles, where increased growth has been reported. Dispersal and enhanced survival of such seeds may represent a relatively primitive form of ant-dispersal devoid of seed morphological specialization such as a rich elaiosome (see chapters 3 and 4).

The relationship between antagonism (seed predation) and mutualism (seed dispersal), as discussed in chapter 1, may be based on the availability of unspecialized seed-collecting ants that contribute to the prevalence of myrmecochory, as in Australia (Andersen 1991b); the evolution of specialized seed structures for dispersal may also have been precluded by the assemblage of seed-harvesting ants whose workers are significantly larger than those ants normally associated with elaiosome-mediated seed dispersal (i.e., large worker size may allow consumption of elaiosome *and* seed), as in North America (Rissing 1986). Finally, mutualisms may be more common among the more social species within a taxon, and the richness of social behaviors within a species may be in part an evolutionary result of mutualisms that allow species to spend less time foraging for food (Thompson 1982). Indeed, contrary to the variety of mutualistic interactions between ants and plants, antagonistic interactions between ants and plants fall into two groups only: leaf-cutting ants (grazing) and seed-harvesting ants (predation). Moreover, based on the amount of leaves and seeds removed by the ants, the impact of these interactions on the plant community can be at times quite severe and relatively easy to estimate. Antagonistic interactions may appear more clear-cut to the human mind due to severe damage caused by leaf-cutter and seed-harvesting ants to agriculture (and removed seeds or damage to leaves can be quantified or estimated). In contrast, it is frequently difficult to assess the impact of mutualisms on the interacting species, and how they ripple (Thompson 1982) throughout the community. While we are prone to admire the beauty inherent in some mutualistic interactions (e.g., pollination), we do not fully comprehend their impact.

CHAPTER THREE

Mutualism from Antagonism

Ants as Primary Seed-Dispersers

Interspecific interactions are based on an entirely selfish cost-benefit system, which depends on the relative gain as compared to loss in fitness produced by the interaction (e.g., Willmer and Stone 1997; Bronstein 1994a). As explained in chapter 1, antagonistic and mutualistic interactions are related in many ways. It can even happen that over evolutionary time certain antagonistic interactions exhibit a shift in outcome so that the interacting species benefit from the interaction. A change in outcome from antagonism to mutualism is most likely in interactions that are inevitable within the lifetimes of individuals and may have their evolutionary origin in the defense reactions of species (Thompson 1982). If it is unlikely that individuals can avoid a specific antagonistic interaction, then selection will favor individuals that have traits causing the interaction to have at least a less negative effect on them. Ants and plants are involved in two interactions of this kind: seed and fruit dispersal (chapters 3 and 4) and pollination (chapter 5).

Seed dispersal by animals, whether vertebrate or invertebrate, usually involves a reward (fruit, aril, or elaiosome) for the animal. Combined research on seed predation and seed dispersal frequently makes the transition from antagonism to mutualism more evident; the two merge in the studies of predator or parasite satiation and seed dispersal (Thompson 1982; Rodgerson 1998). For example, in interactions involving nutlets or cones as the reward, the dispersal unit itself (seeds) is eaten by the mutualist (predation in part), and the mutualism is based on the subset of seeds not eaten; in interactions involving fleshy fruits, arils, or elaiosomes as the reward, a separate resource is offered to the mutualist (Howe and Smallwood 1982; Thompson 1981a, 1982; Boyd 1996).

Seed dispersal is a process that has profound consequences for populations: (1) ecologically, because dispersal influences the dynamics and persistence of populations, the distribution and abundance of species, and community structure, and is a key process determining the spatial structure of plant populations; and (2) evolutionarily, because dispersal determines the level of gene flow between populations and affects processes like local adaptation, speciation, and the evolution of life-history traits (Beattie 1978; Peakall and Beattie 1995; Dieckmann, O'Hara, and Welsser 1999; Kalisz et al. 1999; Nathan and Muller-Landau 2000). Nathan and Muller-Landau point out that the patterns of seed dispersal vary among plant species, populations, and individuals at different distances from parents, different microsites, and different times. The final outcome of seed dispersal depends on how the process operates at the various stages of the plant life cycle, including the adult plant, seed production, seeds on plants, seed predation, seed dispersal per se, seeds in transient soil seed banks, secondary dispersal and dormancy, seeds in persistent soil seed banks, germination and growth, seedlings, and plant growth (Nathan and Muller-Landau 2000). Here, seed dispersal will mean only the departure of a diaspore (e.g., fruit or seed) from the parent plant (Howe and Smallwood 1982).

In this chapter we review the general characteristics of the reward offered by the plants to the ants (the elaiosome). We offer some general concepts on the selective advantage to plants of seed dispersal by ants that has been associated with a variety of major benefits or hypotheses, followed by examples provided by research done in different regions of the world. We close by discussing the importance of multispecific studies in the analysis of the importance of seed dispersal by ants. Recent studies have demonstrated that a guild of ant species visiting a plant can simultaneously have different effects on plant fitness. Certain ant species in the guild may protect the plant against herbivores while others may disperse the seeds; furthermore, a single ant species may not be simultaneously a good defender and a good dispersal agent (Cuautle and Rico-Gray 2003; Cuautle, Rico-Gray, and Díaz-Castelazo 2005; Dutra, Freitas, and Oliveira 2006).

The Reward: Elaiosomes

Myrmecochory has led to the development of special anatomical, biochemical, and phenological adaptations in the plant, enhancing ant attraction and thus increasing the effectiveness of seed dispersion (Gorb and Gorb 2003). The elaiosome is the ant-attractive portion of the diaspore (fig. 3.1), and it has been shown that intact diaspores and elaiosomes are removed by ants but seeds without elaiosomes are not (O'Dowd and Hay 1980; Beattie 1985), or are removed at much lower rates (Leal 2003). For example, when organisms other than ants (e.g., beetles) remove the elaiosome, the seed lacking elaiosome will no longer attract ants and thus will not be dispersed (Ohara and Higashi 1987).

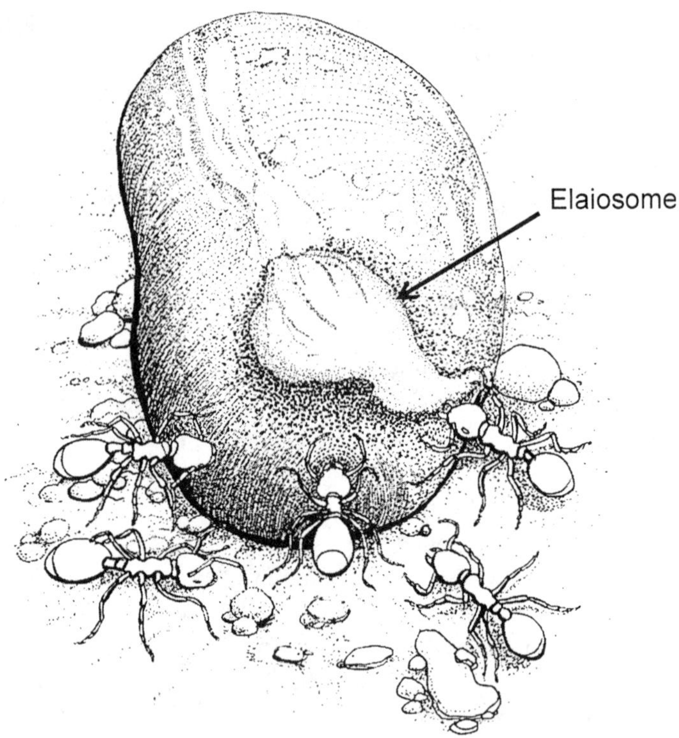

FIGURE 3.1. Workers of *Pheidole* sp. gather around an elaiosome-bearing seed of *Croton priscus* (Euphorbiaceae) on the floor of a Brazilian forest. The ants use the lipid-rich elaiosome as a handle to remove the seed; seedlings of *C. priscus* are often found on refuse piles left by foraging ants near the nest. After Passos and Ferreira 1996.

TABLE 3.1 **Examples of analysis of lipids and fatty acids in elaiosomes in *Datura discolor* and *Viola* spp.**

Seed traits	Plant species		
	Datura discolor	*Viola selkrikii*	*Viola verecunda*
Weight of seed (mg)	9.9 mg per unit	0.61 ± 0.12 mg	0.42 ± 0.03 mg
Weight of elaiosome (mg)	0.65 mg per unit	0.02 ± 0.004 mg	0.006 ± 0.0004 mg
Proportion of elaiosome to diaspore (%)	—	3.3 ± 0.12	1.5 ± 0.02
Proportion of seeds with elaiosome removed by ants	73% in 15 min	ND	ND
Proportion of seeds without elaiosome removed by ants	Seeds remained in depots for >15 days	ND	ND
Proportion of lipid to elaiosome (%)	Not tested	26.7 ± 0.2	40.0 ± 0.01
Fatty acid (% proportion to total):			
Palmitic acid	24.8	21.8	19.5
Palmitoleic acid	—	5.5	4.4
Stearic acid	14.8	7.0	4.9
Oleic acid	0.0	30.1	33.3
Linoleic acid	25.7	27.3	25.3
Linolenic acid	34.7	8.3	12.3

Sources: O'Dowd and Hay 1980; Ohkawara and Higashi 1994.
Note: ND = no data.

Elaiosomes contain ant attractants, particularly the abundant oleic and linoleic acids (in diglycerid and free fatty acid fractions), that stimulate ants to carry the entire seed or diaspore back to the nest (Brew, O'Dowd, and Rae 1989; Lanza, Schmitt, and Awad 1992; and references therein) (table 3.1). The elaiosome is removed without damaging the seeds, which are then discarded either below the parent plant, in an abandoned gallery, or in a refuse pile close to a nest entrance (O'Dowd and Hay 1980; Buckley 1982; Beattie 1985; Keeler 1989; Westoby, Hughes, and Rice 1991).

Elaiosomes vary in chemical composition, but they are usually rich in lipids, with sympatric species differing in amino acid content and lipid composition (Beattie 1985; Brew, O'Dowd, and Rae 1989; Lanza, Schmitt, and Awad 1992). For example, the elaiosomes of *Trillium* species (*T. erectum, T. grandiflorum, T. undulatum*) contain protein, triglycerides, diglycerides, monoglycerides, and free fatty acids (Lanza, Schmitt, and Awad 1992). Ants obtain from elaiosomes food for larvae (Beattie 1985). Moreover, access to elaiosomes can significantly influence the dynamics of an ant colony by shifting the mass and numerical investment ratio in colony reproductive output toward female bias but not affecting the number of workers or queen size (Morales and Heithaus 1998). It has been suggested that the nutrient composition of elaiosomes may provide the underlying selective advantage for ants in seed dispersal and that specific compounds may manipulate ant behavior; ants are more prone to select and disperse those seeds and thus maximize seed dispersal (Brew, O'Dowd, and Rae 1989; Boyd 1996). More recently, elaiosomes were also shown to have a quantitative effect on development of a larval life cycle because larvae that accumulated more radio-label from elaiosomes tended to develop into virgin queens, whereas other female larvae developed into workers (Bono and Heithaus 2002).

The nutritional constituents of elaiosomes should be of significant importance to ants so that they are attracted to use elaiosomes as an alternate food (in place of insect prey and/or the embryo of seeds) and thus effect seed dispersal. However, Hughes, Westoby, and Jurado have demonstrated that the chemical composition and the behavior releaser compound (see below) in elaiosomes have converged only with the insect prey of ants, not with seed composition. For example, the fatty acid composition of elaiosomes is more like those of insects than those of seeds, and the levels of palmitic, palmitoleic, stearic, and oleic acids in elaiosomes and insects are particularly similar (Hughes, Westoby, and Jurado 1994 and references therein). Furthermore, the diglyceride 1,2-diolein, which is the main ant attractant in elaiosomes (Marshall, Beattie, and Bollenbacher 1979; Skidmore and Heithaus 1988; Brew, O'Dowd, and Rae 1989; Lanza, Schmitt, and Awad 1992), is also an important component of insect haemolymph (Hughes, Westoby, and Jurado 1994). The attraction of carnivorous and omnivorous ant species may be an important adaptive advantage for the development of elaiosomes (Carroll and Janzen 1973). Thus the development of elaiosomes simultaneously transferred the reward away from the seed itself, lessening the cost of the interaction, and recruited carnivorous and omnivorous ant species as potential seed-dispersers (see also chapter 4). Furthermore, this relationship between the chemical composition of elaiosomes and insects is noted in the capitula of eggs of many stick insect species (order Phasmotodea). The eggs are similar to seeds in size, shape, color, and texture, and the capitulum resembles an elaiosome in function and appearance (Hughes and Westoby 1992a). Ants carry the eggs to the nest, use the capitula for food, and bury the eggs; phasmatid nymphs emerge and suffer reduced rates of parasitism by wasps, although there are alternative explanations (see Hughes and Westoby 1992a).

Elaiosomes also vary in size, and ants in many instances preferentially disperse diaspores with the greatest elaiosome/diaspore mass ratio, although plants produce seeds of different sizes, allowing for a variety of ants to effect their dispersal (Gorb and Gorb 1995). Furthermore, ant body size predicts dispersal distance of ant-adapted seeds. However, the link between ant body size and seed dispersal distance, combined with the dominance of invaded communities by typically small ants, predicts the disruption of native ant-seed dispersal mutualisms in invaded habitats (Ness et al. 2004). It has been shown experimentally that workers of *Myrmica punctiventris* removed seeds of large and small elaiosome/diaspore mass ratio, probably responding to a complex of diaspore characters of the *Trillium* species used (*T. erectum, T. grandiflorum, T. undulatum*), rather than to size alone (Gunther and Lanza 1989). Elaiosomes also vary intraspecifically. Workers of *Myrmica ruginodis* removed first the diaspores of *Hepatica nobilis* (Ranunculaceae) with the largest elaiosome (and largest achene); however, Mark and Olesen found that elaiosome size was more important to removal than achene size or the elaiosome/achene size ratio. Thus, if ant dispersal increases plant fitness, elaiosome size and hence diaspore size would be expected to increase over time; however, such directional selection mediated by the ants is probably counterbalanced by the plant. Seed predators and a negative trade-off between number and size of seeds would, among other factors, select for smaller diaspore size, i.e., counteract the effect of the ants' preference for larger elaiosomes (Mark and Olesen 1996). However, producing seeds of different sizes and different elaiosome/diaspore mass ratio would allow for diaspores to be dispersed by a wider variety of ants, and not specifically by those ants capable of removing large diaspores. The differential response of ant species to such characteristics as seed arrangement (solitary or clumps) and elaiosome/seed ratios demonstrates the way in which ant behavior may have been an important selective force in the evolution and maintenance of myrmecochory (Hughes and Westoby 1992b; Gorb and Gorb 1995).

Seed Dispersal by Ants

Ants are the most likely example to be the major selective force toward seed dispersal by arthropods, because they take food back to the nest, moving diaspores away from the parent plant and placing the seed in an appropriate germination site: below ground, away from predators, and probably in enriched soil (Culver and Beattie 1978; Thompson 1981a).

Seed dispersal by ants is considered a relatively recently derived mode of dispersal in most taxa studied (Berg 1966, 1972; Thompson 1981a; Higashi and Ito 1991; also discussed in Horvitz 1991). For example, the ant-dispersed species in the genus *Trillium* (Liliaceae) (Berg 1958), or those in the ant-dispersed Marantaceae (Horvitz 1991), are derived from bird-dispersed species. The seeds of over 3,000 plant species in more than 90 genera are adapted for seed dispersal by ants (see appendixes 3.1–3.3). The main component of the myrmecochorous syndrome is the presence of specialized fat bodies (elaiosomes) on the diaspores for attracting ants (Gorb and Gorb 2003). Furthermore, part of this myrmecochorous group of plants is composed by diplochorous plant species (plants that use ballistic dispersal in addition to ant dispersal) (Nakanishi 1994).

Many species in four subfamilies of the Formicidae (Formicinae, Myrmicinae, Ponerinae, and Dolichoderinae) gather seeds (Beattie 1985). Ants known as harvester or granivore ants store seeds in underground granaries and consume them during the winter or the dry season. These ants are granivores, and the net outcome of the interactions is usually predation (see chapter 2). Other ants gather seeds and fruits that present elaiosomes, which represent the reward offered by the plant to ants in exchange for seed transportation. Regardless of origin (e.g., elaiosome, aril, caruncle, funiculus, or pericarp), and because these structures are similar both morphologically (van der Pijl 1982) and chemically (Hughes, Westoby, and Johnson 1993), from here on, reward equals elaiosome. For example, the fleshy aril on the seeds of *Turnera ulmifolia* (Turneraceae) (fig. 3.2) are dispersed by different ant species (*Camponotus planatus, C. atriceps, Forelius analis, Paratrechina longicornis, Monomorium cyaneum, Pheidole* spp. and *Dolichoderus lutosus*) (Cuautle, Rico-Gray, and Díaz-Castelazo 2005). Finally, some characteristics of spores of several myrmecophytic fern genera (*Solanopteris, Lecanopteris, Drynaria, Selliguea*) have been associated with transport by the ant *Iridomyrmex cordatus* (Dolichoderinae) (Tryon 1985).

The selective advantage to plants of seed dispersal by ants has been associated with a variety of major benefits or hypotheses (e.g., Howe and Smallwood 1982; Beattie 1985; Andersen 1988; Westoby, Hughes, and Rice 1991; Giladi 2006). These advantages (described below) could be general (points 1–6) or could be seen as comparative (point 7): the advantage of using ants versus other dispersal agents (see also Giladi 2006).

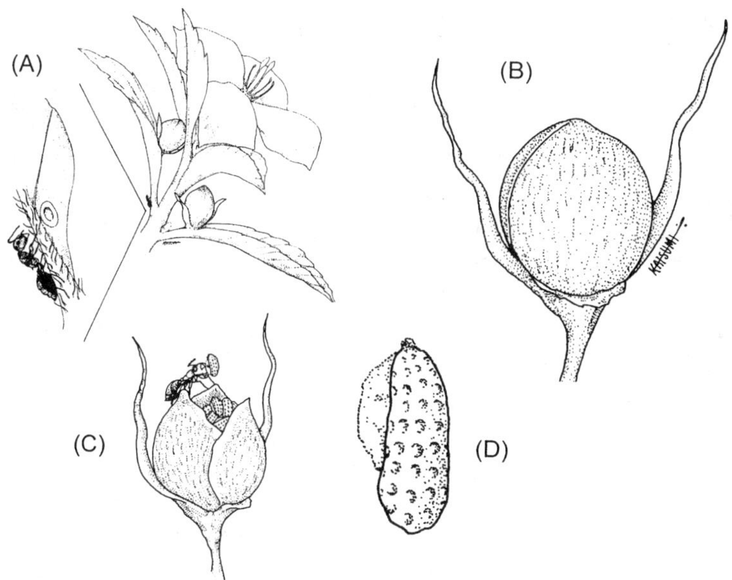

FIGURE 3.2. (A) Branch showing the position of flowers and fruits of *Turnera ulmifolia* (Turneraceae), and a detail of *Crematogaster brevispinosa* (ca. 2.5 mm) visiting an extrafloral nectary. (B) Closed fruit. (C) *C. brevispinosa* workers removing *T. ulmifolia* seeds. (D) Seed showing the attached fleshy aril used as food by the ants. Modified from Rico-Gray 2001.

1. *Avoidance of predators.* Seeds released from the parent plant are quickly taken by ants to their nest, where ants remove the elaiosome and end up losing interest in the seed, which in the nest is beyond the reach of predators (e.g., small mammals, birds, or other seed-eating ants). If ants do not discover released seeds in a few hours, these will be eaten by seed predators (O'Dowd and Hay 1980; Heithaus 1981; Smith, Forman, and Boyd 1989; Nakanishi 1994; Ruhren and Dudash 1996; Pizo and Oliveira 1998).
2. *Avoidance of competitors.* Interspecific competition for germination and for appropriate microsites for seedling growth is substantially reduced by ant dispersal (Westoby et al. 1982; Bond, Yeaton, and Stock 1991). Of three sympatric *Carex,* the ant-dispersed *C. pedunculata,* which did not grow well in the presence of *C. platyphylla* and *C. plantiginea,* benefited when ants took their seeds to microsites (nests) that neither of the two other species could exploit (Handel 1976, 1978; Boyd 1996).
3. *Precision placement.* Seeds taken by ants into their nests remain safe from incineration during forest fires. Seeds of more than 3,000 species in the dry sclerophyllous shrublands of Australia (Berg 1975) and in the fynbos in the

Cape Province of South Africa (Milewski and Bond 1982; Bond and Slingsby 1983) are protected from fires when deposited by ants in their nests. This system has a strong relationship with forest regeneration after fire because many of the seeds deposited by ants in their nests need the high temperatures of fire to trigger germination. Seeds benefit from ant manipulation if the average depth of burial is such that the seeds receive sufficient heat to germinate, without being killed (Majer 1982b). This kind of system may collapse when a new ant species invades and displaces the dominant native ants. The invasion of *Linepithema* (*Iridomyrmex*) *humile* in Cape fynbos may cause the extinction of many endemic Proteaceae. *L. humile* does not remove seeds fast enough or to the appropriate distance, and they do not store them in nests below the soil. Consequently, seeds are eaten by vertebrate and invertebrate predators and suffer the effect of fire (Bond and Slingsby 1984). More recently, Christian and Stanton (2004) have shown that the net effect of seed dispersal by ants on plant populations in fynbos will depend on how temporal fluctuations in fire regimes influence cost-benefit thresholds related to seed dormancy and seed mortality.

4. *Decrease in seed mortality with increased dispersal distance.* If seedling mortality is inversely related to dispersal distance, then the transportation of seeds by ants across a significant distance may be the selective advantage of myrmecochory. In diplochorous plants (i.e., plants with a two-stage dispersal system) seeds ejected explosively away from the parent plant are then discovered by ants and removed. Those seeds carried away beyond the sphere of influence of adult plants may favor selection for ant manipulation (Westoby and Rice 1981; Boyd 1996; Kalisz et al. 1999). This advantage may be critical for ant-dispersed species that form large clones (Pudlo, Beattie, and Culver 1980).
5. *Dispersal distance per se.* It is assumed that ants remove seeds to the nearest nest, and thus the shape of the dispersal curve generated by myrmecochory will clearly be influenced by the density and dispersion of ant nests. Andersen suggested that in Australia dispersal distance is a potential benefit from myrmecochory. Even though mean dispersal distance was just 2.1 m at a site in southeastern Australia, the dispersal curve (influenced by nest density and dispersion, population size, and territoriality of seed-dispersing ants) was characterized by a narrow peak and a long tail. This is optimal when safe sites for seedling establishment are rare, as is typically the case for Australian myrmecochores in the absence of fire, independent of any targeting of seeds to ant nests (Andersen 1988). Due to variation in the composition of ant assemblages, however, dispersal curves generated by ants can vary markedly between natural and disturbed habitats. Andersen and Morrison (1998)

showed that, because widely foraging ant species are more common in disturbed areas of northern Australia, mean dispersal distances at these sites were nearly twice as high as at natural sites (3.9 m as against 2.2 m).

6. *Deposition in a nutrient-rich environment.* The soil of ant nests is usually significantly different, both chemically and physically, from surrounding soils (Wagner 1997) and may increase the heterogeneity of the plant community. These differences may be on a number of factors: temperature, porosity, moisture, pH, organic content, minerals, and diversity and abundance of microorganisms (Beattie and Culver 1977, 1982, 1983; Beattie 1985; King 1976, 1977a, 1977b, 1977c; Smith et al. 1989; Woodell and King 1991; McGinley, Dhillion, and Neumann 1994). However, this hypothesis was rejected for the Australian sclerophyll vegetation (Rice and Westoby 1986) and for Protaceae in the Cape fynbos (Bond and Stock 1989), where seedlings emerge in nutrient-poor sites if seeds are dispersed by ants, whereas seeds that are not dispersed by ants reach germination sites richer in nutrients.
7. *Myrmecochory associated with low-nutrient soils.* Myrmecochores in Australia account for much of the flora in sclerophyll vegetation, which is found on low-nutrient soils; other vegetation associations grow on richer soils, according to Westoby, Hughes, and Rice. These authors suggest that the higher level of adaptation for dispersal by ants on the infertile soils is at the expense of fleshy fruits for dispersal by vertebrates, which account for a larger percentage of plant species on the fertile soils. Biomass costs for the ant-dispersal syndrome are much lower than for the vertebrate dispersal syndrome; therefore, adaptations for ant dispersal might be present in large part because they are cheap and can achieve a distance that is effective for smaller plants (Westoby, Hughes, and Rice 1991). In particular, the dispersal spectra should be analyzed in relation to potassium availability in soils, rather than to soil fertility in general, which will help to explain observed correlations between dispersal mode and soil fertility (Hughes, Westoby, and Johnson 1993; see chapter 4).

Certain of the above points may seem closely related: for instance, points 1 and 4 or points 3, 5, and 6 seem to be variations of the safe-site theme. In general, the role of ants as dispersal agents focuses on the advantages of seeds being placed in appropriate sites for protection against fires and predators, or to avoid competitors, or for the likelihood of germination and seedling development. No one of the above hypotheses alone, however, can explain the selective advantage of seed dispersal by ants. Moreover, points 1–7 are not exclusive, and their relative importance tends to vary depending on the region or habitat (e.g., Smith, Forman, and Boyd 1989; Giladi 2006). Multiple hypotheses were supported in most studies that tested more than one hypothesis, suggesting that the various selective advantages conferred from myrmecochory are seldom exclusive (Giladi 2006).

As discussed later in this chapter, the role of ants while visiting a plant may not be solely as seed-dispersers. Recent studies have shown that an assemblage of ants visiting extrafloral nectaries of a single plant species may have different effects on the plant: while certain ant species are good at deterring herbivores, others are good as seed-dispersers (fig. 3.1) (Cuautle and Rico-Gray 2003; Cuautle, Rico-Gray, and Díaz-Castelazo 2005). Other factors that could be considered are, for instance, Does seed dispersal vary depending on the plant's life form? and Are there upper limits to seed size? The hypotheses presented above are basically the result of research done in the northeastern United States, Australia, and South Africa; we really need to have more results from other regions of the world (e.g., the tropics) to prepare a more general discussion and to move toward constructing a unified theory on seed dispersal by ants.

Myrmecochory: Distribution and Significance Worldwide

Seed dispersal by ants, or myrmecochory, has been studied in a variety of plant species and habitats. Of the approximately 3,000 plant species whose seeds are ant-dispersed, 51.7% are from Australia (>1,500 species [Berg 1975]) and 44.8% from South Africa (>1,300 species [Bond and Slingsby 1983]). Only 1.7% belong to temperate areas (eastern North America, Japan, the Mediterranean region), and 0.2% represent the neotropics. Ants also secondarily disperse seeds from fleshy fruits, which otherwise would be vertebrate-dispersed seeds (chapter 4).

Ants disperse more than 30% of the spring-flowering herbaceous species in the deciduous forests of eastern North America (Beattie and Lyons 1975; Culver and Beattie 1978; Beattie and Culver 1981; Handel 1976, 1978; Pudlo, Beattie, and Culver 1980; Handel, Fisch, and Schatz 1981; appendix 3.1). Myrmecochory is particularly common in sclerophyll vegetation on low-nutrient soils of Australia (Berg 1975; Andersen 1988, 1991b; Westoby, Hughes, and Rice 1991) and the South African fynbos (Bond and Slingsby 1983; Bond, Yeaton, and Stock 1991; appendix 3.2) and probably also in Mediterranean shrublands (Aronne and Wilcock 1994a; Wolff and Debussche 1999; appendix 3.1). To date a relatively reduced number of examples of myrmecochory are available from the neotropics (e.g., Horvitz and Beattie 1980; Horvitz 1981, 1991; Cuautle, Rico-Gray, and Díaz-Castelazo 2005; appendix 3.3; see also chapter 4).

Northeastern North America

Ants disperse (via myrmecochory, including diplochory) more than 30% of the spring-flowering herbaceous species in the temperate broadleaf deciduous forest biome of eastern North America (e.g., Gates 1943; Beattie and Lyons 1975; Culver and Beattie 1978; Beattie and Culver 1981; Handel 1976, 1978; Pudlo, Beattie, and Culver 1980; Handel, Fisch, and Schatz 1981; Smith, Forman, and Boyd 1989; Smith et al. 1989; Ruhren and Dudash 1996; Kalisz et al. 1999). Moreover, ants disperse plant species in the same genera (Ohara and Higashi 1987; Higashi and Ito 1991; Ohkawara 1995; Ohkawara, Higashi, and Ohara 1996), as well as in other genera from other temperate habitats, such as Japan (Nakanishi 1994) and Europe (Gorb and Gorb 1995, 2003).

More than describing the dispersal process, or listing which elaiosome-bearing plants are dispersed by which ant species (appendix 3.1) in temperate forests (particularly in northeastern North America), we would like to highlight the association between myrmecochory and ornithochory and the evolution toward a seasonal distribution of types of seed dispersal (myrmecochory vs. ornithochory). This evolution is ecologically an important event in community organization in the temperate forests of northeastern North America (Thompson 1981a) and probably in other formations as well (e.g., neotropical forests [Horvitz 1991] and Mediterranean vegetation [Aronne and Wilcock 1994a; chapter 4]).

In deciduous forests of eastern North America, there is a pronounced seasonal separation of the early-maturing (spring) ant-dispersed seeds and the late-maturing (late summer and fall) bird-dispersed seeds, even within single families (e.g., the Liliaceae) (Thompson 1981a). This seasonal dichotomy seems to arise from a series of habitat, plant, and/or disperser characteristics. The pattern of availability of ants and birds differs. The rates of seed removal by ants are highest early in the reproductive season of plants; later (in July and August) the availability of food for ants changes when a wider array of food resources becomes available (e.g., insects), and the rate of seed removal by ants becomes lower (Culver and Beattie 1978). Changes in the availability of food resources for ants, resulting in changes in their diet, have also been reported for tropical systems (Smythe 1982; Rico-Gray 1989, 1993; Rico-Gray and Sternberg 1991). The availability of birds, however, is greatest during late summer and fall migrations, when they need energy sources, and surely this exerts a strong selection pressure for plant flowering and fruiting in summer and fall (but see Herrera 1984). If plants produced fleshy fruits early in the season, the low rates of removal would allow a high probability of destruction of fruits by invertebrates. Thus, according to Thompson, low availability of avian frugivores and high probability of fruit damage by invertebrates will tend to select against plants with early-maturing fleshy fruits. The herbaceous ephemerals flowering early in the season, before the forest canopy closes and the summer flora has fully developed, are of short stature. When the summer flora fully grows, the ephemerals become inconspicuous, which makes them rather invisible to dispersal agents like birds, which are attracted to colorful displays. In summary, the evolutionary alternative for early ephemerals is to use ants as dispersal agents, and the availability of birds late in the season will tend to select for plants with fleshy fruits (Thompson 1981a).

Mediterranean Region

Even though the reproductive characteristics of the shrubs of the Mediterranean region are under the selective pressures of limited soil nutrients and water availability (Aronne and Wilcock 1992, 1994a, 1994b), like the sclerophyll vegetation in Australia or the fynbos of South Africa (Westoby, Hughes, and Rice 1991), few examples of seed dispersal by ants are known from the Mediterranean region (appendix 3.1). This is true probably either because this system has been seen as unimportant (review in Aronne and Wilcock 1994a) or because most species have been reported as bird-dispersed (e.g., Herrera 1982, 1984; Jordano 1982, 1989). Recent studies in the region (e.g., Italy, France, Spain), however, emphasize two distinct seed-dispersal systems, myrmecochory (sensu stricto) involving elaiosome-bearing seeds, and dyszoochory, accomplished by harvester ants (considered mostly as predation in this book), in which case seeds either are abandoned on the way to the nest or forgotten in underground granaries, or they are mistakenly rejected intact on the refuse pile (Baiges, Espadaler, and Blanché 1991; Aronne and Wilcock 1994a; Espadaler and Gómez 1996; Wolff and Debussche 1999; Garrido et al. 2002).

The analysis of the content of nests of *Messor minor* in southern Italy showed the existence of seeds from scrub species, as well as those from Fabaceae and Poaceae, suggesting that ant dispersal may be much more widespread in the Mediterranean environments than has been realized (Aronne and Wilcock 1994a). Many of these seeds came from fleshy-fruited shrub species (e.g., *Rhamnus alaternus, Myrtus communis, Smilax aspera, Pistacia lentiscus,* and *Phillyrea latifolia*) previously reported as bird-dispersed (Aronne and Wilcock 1994a). However, *R. alaternus* and *M. communis* do have elaiosomes. In general for the Mediterranean region, ants gather seeds and bring them back to the nest; while some seeds are discarded damaged (dyszoochory/predation), a large number of seeds are left undamaged on the ant mound (Aronne and Wilcock 1994a; Wolff and Debussche 1999), which once again shows the fine line between seed predation and seed dispersal (between antagonism and mutualism). Moreover, this pattern can change with the dynamic of the plant community. In a mosaic of different-aged abandoned vineyards in France, it was shown that the proportion of dyszoochorous plants is higher during early succession of the vegetation, as well as that of granivorous ants; a mature oak forest (*Quercus pubescens*) exhibited the lowest proportion of dyszoochorous plants, and no granivorous ants were present (Wolff and Debussche 1999).

The studies on the evolution of seed dispersal in the Mediterranean region have mostly concentrated on seed dispersal by birds (Jordano 1982, 1989; Herrera 1984), probably because this dispersal mode is more frequent. However, the seed dispersal system of both *R. alaternus* and *M. communis* involves a variety of dispersal mechanisms, including birds, mammals and ants. It seems that changes in the vegetation since the Tertiary (probably a tropical type of vegetation) have also conveyed a shift in plant reproductive systems, from predominantly dioecious shrubs bearing few, large seeds (e.g., *Osyris alba*) to a more recent group of multiseeded, dry-fruited hermaphrodites (e.g., Cistaceae, Fabaceae), where ant dispersal seems to be more widespread (Wilcock and de Almeida 1988; Aronne and Wilcock 1992, 1994a, 1994b; Aronne, Wilcock, and Pizzolongo 1993).

It is considered that ant dispersal in the Mediterranean region is the result of both a rapid shift from fleshy fruits to dry fruits, which conveys a shift from bird-dispersed to ant-dispersed systems (Aronne and Wilcock 1994a), and a change during vegetation succession from mostly dyszoochory (high abundance of granivory ants) early in succession to myrmecochory (increase in elaiosome-bearing plants, absence of granivorous ants) in the mature forest (Wolff and Debussche 1999).

Australia and South Africa

Two of the regions where seed dispersal by ants is prominent, and where this mechanism has been extensively studied, are the low-nutrient soils of the fynbos shrublands in the Cape of South Africa (e.g., Bond and Slingsby 1983; Bond, Yeaton, and Stock 1991), and those of the sclerophyll vegetation of Australia (e.g., Berg 1975; Drake 1981; Andersen 1988, 1991b; Westoby, Hughes, and Rice 1991). A few examples are presented in appendix 3.2; however, full lists of myrmecochorous plants for Australia and South Africa are offered in Berg (1975) and Bond and Slingsby (1983), respectively. These two regions exhibit remarkable convergence in the evolution of myrmecochory across diverse lineages. Moreover, several of the hypotheses mentioned above assessing the selective advantage to plants of dispersal of seeds and fruits by ants are based on studies accomplished in these regions.

AUSTRALIA. Seed dispersal in the sclerophyll vegetation of Australia (both myrmecochory and diplochory) is present in at least 1,500 plant species in 87 genera and 24 families (65% only in the Dilleniaceae, Fabaceae, Mimosaceae, Rhamnaceae, and Rutaceae) (Berg 1975; Andersen 1988, 1991b; Westoby, Hughes, and Rice 1991; Hughes and Westoby 1992c). Myrmecochory is very common in the Australian arid zone, and most plants in the Australian *Acacia*-temperate and arid zones are ant-dispersed (Davidson and Morton 1981b, 1984). Moreover, 30%–50% of the flora in Australia is ant-dispersed (Westoby, Hughes, and Rice 1991).

Most of the above plants are shrubs whose seeds have low water content and firm, waxy, and persistent elaiosomes. Plants release diaspores singly, many times ballistically (diplochory), and therefore they are found by ants as single items (Westoby, Hughes, and Rice 1991). Removal rates by ants are high in summer (a few hours) and slightly slower in winter (a few days) since few seeds would be available naturally in winter; removal by other organisms appears insignificant (Hughes and Westoby 1990; Westoby, Hughes, and Rice 1991).

Two features of ant dispersal in Australia are relevant: (1) dispersed seeds are protected from fires when deposited by ants in their nests, and (2) most plants adapted for myrmecochory are found on low-nutrient soils. However, these features are confounded by fire and phylogenetic differences. There seems to be a strong relationship between ant dispersal and forest regeneration after fire, but there is not much data to support this, although it is known that many of the seeds deposited by ants in their nests need the high temperatures of fire to trigger germination. Seeds benefit from ant manipulation if the average depth of burial is such that the seeds receive sufficient heat to germinate without being killed (Majer 1982b). However, in environments elsewhere where fires also occur, myrmecochory is very rare (Bond, Yeaton, and Stock 1991). The higher level of adaptation for dispersal by ants on the infertile soils, though, is at the expense of fleshy fruit for dispersal by vertebrates, which accounts for a larger percentage of plant species on the fertile soils (Westoby, Hughes, and Rice 1991; Hughes, Westoby, and Johnson 1993). Since biomass costs for the ant-dispersal syndrome are much less than for the vertebrate dispersal syndrome, adaptations for ant dispersal might be present in large part because they are cheap and can achieve a distance that is effective for smaller plants (Westoby et al. 1982; Westoby, Hughes, and Rice 1991).

SOUTH AFRICA. Myrmecochory in the Cape flora has been estimated in plants of at least 78 genera in 29 families (5 endemic to the Cape) of both monocotyledons and dicotyledons, but many of these still need to be tested in the field (Slingsby and Bond 1981; Milewski and Bond 1982; Bond and Slingsby 1983). These plants include tall, treelike shrubs, short shrubs, and perennial herbs; myrmecochory is rare in geophytes and absent from annuals (Bond and Slingsby 1983, 1984; Bond, Yeaton, and Stock 1991). Myrmecochory is essentially a characteristic of the fynbos shrublands on nutrient-poor soils and is rare in adjacent vegetation (e.g., the Karoo). Moreover, genera that cross edaphic boundaries have myrmecochorous members in fynbos and wind-dispersed members in adjacent shrublands in arid climates or on nutrient-rich soils (Bond and Slingsby 1983; Bond, Yeaton, and Stock 1991). Elaiosomes are mostly firm and durable rather than soft and ephemeral, and, as in other regions, the use of two dispersal mechanisms (diplochory and myrmecochory) is also common (Bond and Slingsby 1983). However, despite the richness of myrmecochorous plants, the number of ant species involved is quite low. Bond, Yeaton, and Stock report that although the dominant dispersers for most of the fynbos (middle and lower mountain slopes and coastal fynbos) are two species of *Anoplolepis* (*A. custodiens* and

A. steingroeveri), other ants disperse seeds as well, particularly in areas where *Anoplolepis* is absent. Unlike the plants, many of the ants have widespread distributions beyond the boundaries of the fynbos (Bond, Yeaton, and Stock 1991).

Myrmecochory in the fynbos has been studied mostly in the Proteaceae, in particular, the woody shrubs with relatively large achenes and an elaiosome formed by the pericarp, in the genera *Leucospermum* (46 species) and *Mimetes* (14 species) (Bond, Yeaton, and Stock 1991). Smaller-seeded species, with diaspores less attractive to ants, have been less intensively studied, and thus they may differ from the Proteaceae in the ecology of the interaction. Some of the ant species associated with the latter are *Tetramorium quadrispinosum, Meranoplus peringueyi, Rhoptromyrmex* spp., and *Pheidole capensis* (Bond and Slingsby 1983). Unfortunately, the invasion of the fynbos by the ant *Linepithema* (*Iridomyrmex*) *humile* threatens to dramatically disrupt the mutualistic association between the Proteaceae and the native ant species, so much that some of these plant species may become extinct. *L. humile* does not remove seeds fast enough and moves them to short distances, and these ants do not store seeds in nests below the soil; consequently, seeds are eaten by vertebrate and invertebrate predators and suffer the effect of fire (Bond and Slingsby 1984). Christian (2001) has shown that presence of *L. humile* shifts the composition of the fynbos plant community and causes a disproportionate decline in the densities of large-seeded plants, probably because many of the remaining native ants were less effective dispersers of large seeds. In California (USA), displacement of harvester ants (*Pogonomyrmex subnitidus*) by *L. humile* appears to decrease the dispersal of *Dendromecon rigida* (Papaveraceae) seeds and may be increasing the loss of seeds due to predation (Carney, Byerley, and Holway 2003).

The Neotropics

There are even fewer studies of seed dispersal by ants in the tropics (appendix 3.3), and, like studies of northeastern North America and in particular the Mediterranean region, they emphasize the close association between seed dispersal by birds and seed dispersal by ants (Horvitz 1991; Horvitz and LeCorff 1993). Seed dispersal by ants in the neotropics ranges from dispersal of elaiosome-bearing seeds (e.g., Cuautle, Rico-Gray, and Díaz-Castelazo 2005) to secondary seed dispersal of seeds from fleshy fruits. Another significant aspect of seed dispersal by ants in the tropics is their effect on the postdispersal fate of nonmyrmecochorous seeds. Ants may remove and/or prey on seeds that reach the forest floor either spontaneously or because they are dropped by vertebrate frugivores (i.e., rearranging vertebrate-generated seed shadows); they may also function as secondary seed-dispersers by removing seeds from vertebrate droppings (chapter 4).

Seed removal in some tropical systems can be higher than that reported for North American temperate forests (Horvitz and Schemske 1986b). An interesting feature of myrmecochorous systems in the neotropics is the presence and activity of large forest-dwelling, carnivorous, ponerine ants (e.g., *Odontomachus, Pachycondyla:* Ponerinae) (e.g., Horvitz and Beattie 1980; Horvitz 1981; appendix 3.3). These ants pick up single seeds and quickly disappear in the leaf litter; the seeds may be carried up to 10 m, and the reward is removed and then consumed by the workers and larvae in the nest (Horvitz and Beattie 1980; Horvitz 1981; Horvitz and Schemske 1986b). However, as Horvitz points out, even if a seed is not carried to the nest, it may benefit from being abandoned in a safe spot (usually under the leaf litter) where appropriate microconditions may favor seedling establishment. Added benefits of this system may be increased germination rates and reduced infestation by fungi. For example, the seeds of *Calathea microcephala* (Horvitz 1981) exhibit a significant increase in germination rates following the removal of the reward by the ants. The effect of ponerine ants varies among sites and plant species involved, and there is also a general marked spatial variation in ant-seed interactions, which is probably due to variation in the relative abundance of ant species (Horvitz 1981; Horvitz and Schemske 1986b; LeCorff and Horvitz 1995).

Ponerine ants are not the only ant species attracted to seeds in the neotropics, and the effect of nonponerines can vary widely. For instance, the important dispersers of the two elaiosome-bearing species *Croton priscus* (Passos and Ferreira 1996) and *Calathea micans* (LeCorff and Horvitz 1995) are *Atta sexdens* and *Pheidole* sp., and *Aphaenogaster araneoides* (all Myrmicinae), respectively. The effect of other ants (e.g., *Pheidole, Solenopsis,* and *Wasmannia*), which usually recruit many nest mates to the seed and consume the aril on the spot without effecting seed dispersal (i.e., displace seeds away from the parent plant) (Horvitz and Schemske 1986b), may be considered an interference with the seed-dispersal system. Aril or pulp removal by minute ants, however, can benefit the seed by reducing fungal infection in the humid leaf litter (Oliveira et al. 1995; Pizo and Oliveira 1998). Indeed, a multispecific approach to the study of ant-plant interactions has shown that different ant species in a guild of ants visiting a plant species may have different roles (e.g., defense and dispersal), which can only be assessed by studying the whole process (Cuautle, Rico-Gray, and Díaz-Castelazo 2005).

Research done with the tropical coastal shrub *Turnera ulmifolia* (Turneraceae) has shown that when a multispecies approach is used, it becomes clear that different groups of organisms with which a plant can interact (e.g., herbivores, pollinators, seed predators, and seed-dispersers) combine to exert upon it an influence that is seldom additive (Juenger and Bergelson 1998; Herrera 2000; Cuautle and Rico-Gray 2003). For example, at least 25 ant species interact with *T. ulmifolia;* 14 are associated with extrafloral nectaries (EFNs) and seeds, 8 only with EFNs, and 3 only with seeds. The most common are *Camponotus planatus* (interacting predominantly with EFNs), *C. atriceps* (EFNs and seeds), *Forelius analis* and *Paratrechina longicornis* (EFNs and seeds), *Monomorium cyaneum* (EFNs and elaiosomes), *Pheidole* spp. (predominantly seeds), and *Dolichoderus lutosus* (EFNs). All ant species present forage for the nectar produced by the extrafloral nectaries. However, certain ant species (plus wasps) effect defense (mediated by extrafloral nectaries [Cuautle and Rico-Gray 2003]), while other species effect seed dispersal (mediated by elaiosome-bearing seeds [Cuautle, Rico-Gray, and Díaz-Castelazo 2005]), and no one ant species is good at providing both services to the plant (table 3.2). A multispecific approach also helped to explain why *T. ulmifolia* sustains tiny ants that do not fulfill a defense function but play a vital role in seed dispersal, and to clarify that the presence of extrafloral nectaries could influence seed dispersion by attracting a certain ant community. Dutra, Freitas, and Oliveira have recently described a similar case involving two ant attractants (pearl bodies and fleshy fruits) on the part of the neotropical nettle *Urera baccifera* (Urticaceae) in a Brazilian forest. The plant is visited by 22 ant species that forage on leaves for pearl bodies and on fruiting branches, where they collect fleshy fruits. Large ant species act as primary seed-dispersers by removing entire fruits to their nests. Except for fungus-growing ants (*Acromyrmex* and *Atta*) that are only seen at fruits, all ants exploit both pearl bodies and fruits and attack lepidopteran larvae on leaves of *U. baccifera*. Field experiments revealed that herbivores are more abundant on ant-excluded than on ant-visited shrubs of *U. baccifera* and that

TABLE 3.2 **Functions performed by different ants when associated with *Turnera ulmifolia* in Mexican coastal dunes**

	Role in plant defense	
Role in seed dispersal	Defends	Does not defend
Good disperser	?	*Forelius analis* *Pheidole* spp.
Average disperser	*Camponotus atriceps*	*Paratrechina longicornis* *Solenopsis geminata*
Poor disperser	?	*Monomorium cyaneum*

Sources: Modified from Cuautle and Rico-Gray 2003; Cuautle, Rico-Gray and Díaz-Castelazo 2005.

ant-excluded plants have significantly faster leaf abscission rates than ant-visited plants. This facultative system is unique in that herbivore deterrence caused by pearl-body- and fruit-harvesting ants can also add to leaf longevity (Dutra, Freitas, and Oliveira 2006).

Conclusion

Besides the undisputed benefits that myrmecochory offers to plants and the description of the process of myrmecochory in a variety of habitats, the origin, evolution, and chemical composition of the reward offered to ants by plants, the elaiosome, is probably one of the most important aspects of myrmecochory. As is the case for most ant-plant interactions, which are nonspecific, there is usually a specialized structure mediating the interaction (e.g., food bodies, extrafloral nectaries, and domatia). Seed dispersal by ants is no exception; it is a nonspecific system mediated by a specialized structure, the elaiosome. The development of the elaiosome and its peculiar chemical composition, which evolved many times among a variety of plant families and from a variety of seed structures, is central to the development of myrmecochory (Gorb and Gorb 2003). The latter has allowed plants to attract to their advantage ants that would otherwise be seed or insect predators. More attention should be given to this key aspect of myrmecochory and also to the effect of other insects (e.g., ground beetles [Ohara and Higashi 1987; Higashi and Ito 1991; Ohkawara 1995; Ohkawara, Higashi, and Ohara 1996]) that are attracted to the nutritive reward and may disrupt the ant-plant mutualism, and to the effect of insects such as wasps that feed on elaiosomes but do effect seed dispersal (Jules 1996). The relationship between seed dispersal by birds and by ants should emphasize the origin and type of reward and the environmental factors responsible for the shift in dispersal agents (Horvitz and Le Corff 1993; Horvitz et al. 2002).

Future research should stress the study of the "quality" component of ants as dispersers (Hughes and Westoby 1990, 1992b, 1992c; Gorb and Gorb 2003), including the importance of ant behavior in seed dispersion and the spatial relationship between ants and seedlings of myrmecochorous plant species (Davidson and Morton, 1981a; Gorb, Gorb, and Punttila 2000; Cuautle, Rico-Gray, and Díaz-Castelazo 2005; Giladi 2006). Although there is an impression of asymmetry in the levels of adaptation among the ants and the plants, the presence of particular colony organizations and morphological and behavioral traits is clearly required for an ant species to be considered an effective disperser. For example, ant behavior determines the number of seeds removed, dispersal distances, the number of damaged seeds, and seed germination, and may also determine seedling establishment in *Turnera ulmifolia* (Cuautle, Rico-Gray, and Díaz-Castelazo 2005).

APPENDIX 3.1 **Examples of seed dispersal by ants in temperate broadleaf deciduous forest biomes (eastern North America, Japan, and Europe)**

Plant species	Growth form	Associated ant genera (number of species)	Source
Aristolochiaceae			
Asarum europaeum	herb	*Formica* (1)	Gorb and Gorb 1995
Berberidaceae			
Jeffersonia diphylla	herb	Genera not provided	Smith et al. 1989
Vancouveria chrysantha	herb	*Formica* (3), *Lasius* (1), *Liometopum* (1)	Berg 1972
V. hexandra	herb	*Formica* (3), *Lasius* (1), *Liometopum* (1)	Berg 1972
V. planipetala	herb	*Formica* (3), *Lasius* (1), *Liometopum* (1)	Berg 1972
Brassicaceae			
Lobularia maritima[a]	herb	*Messor* (1)	Retana, Picó, and Rodrigo 2004
Cyperaceae			
Carex pedunculata	herb	*Aphaenogaster* (1)	Handel 1976, 1978
Euphorbiaceae			
Euphorbia nicacensis	herb	*Camponotus* (1), *Cataglyphis* (1), *Crematogaster* (1), *Formica* (1), *Messor* (1), *Pheidole* (1), *Tapinoma* (1)	Wolff and Debussche 1999
Ricinus communis	shrub	*Prenolepis* (1)	Gates 1943
Fumariaceae			
Dicentra spectabilis	herb	Genera not provided	Galen 1983
Liliaceae			
Erythronium americanum	herb	*Aphaenogaster* (1)	Ruhren and Dudash 1996
E. japonicum	herb	*Lasius* (1), *Myrmica* (1), *Paratrechina* (1), *Stenamma* (1)	Ohkawara, Higashi, and Ohara 1996
Trillium erectum	herb	*Myrmica* (1)	Gunther and Lanza 1989
T. grandiflorum	herb	*Camponotus* (1), *Formica* (3), *Lasius* (1), *Myrmica* (5)	Gates 1943; Gunther and Lanza 1989; Kalisz et al. 1999

(*continued*)

APPENDIX 3.1 **(continued)**

Plant species	Growth form	Associated ant genera (number of species)	Source
T. nivale	herb	Genera not provided	Smith, Forman, and Boyd 1989
T. undulatum	herb	*Myrmica* (1)	Gunther and Lanza 1989
Myrtaceae			
Myrtus communis	shrub	*Messor* (1)	Aronne and Wilcock 1994a
Papaveraceae			
Chelidonium majus	herb	*Formica* (1)	Galen 1983; Gorb and Gorb 1995
Corydalis (7 spp.)	herb	Genera not provided	Nakanishi 1994
Dendromecon rigida[b]	tree	*Dorymyrmex* (1), *Formica* (3), *Linepithema* (1), *Liometopum* (1), *Monomorium* (1), *Pogonomyrmex* (1)	Berg 1966; Carney, Byerley and Holway 2003
Sanguinaria canadensis	herb	*Aphaenogaster* (4), *Camponotus* (3), *Crematogaster* (1), *Formica* (3), *Lasius* (2), *Myrmica* (2), *Prenolepis* (1), *Solenopsis* (1), *Stenamma* (1)	Gates 1943; Pudlo, Beattie, and Culver 1980; Ness 2004
Ranunculaceae			
Helleborus foetidus	herb	*Aphenogaster* (2), *Cataglyphis* (1), *Crematogaster* (2), *Formica* (4), *Lasius* (1), *Leptothorax* (1), *Messor* (1), *Myrmica* (3), *Pheidole* (1), *Plagiolepis* (1), *Tapinoma* (2), *Tetramorium* (1)	Garrido et al. 2002, Fedriani et al. 2004
Hepatica acutiloba	herb	Genera not provided	Smith, Forman, and Boyd 1989
H. americana	herb	*Pogonomyrmex* (1)	Skidmore and Heithaus 1988
H. nobilis	herb	*Myrmica* (1)	Mark and Olesen 1996
Rhamnaceae			
Rhamnus alternus	shrub	*Messor* (1)	Aronne and Wilcock 1994a
Solanaceae			
Datura discolor[b]	herb	*Pogonomyrmex* (1), *Messor* (1, = *Veromessor*)	O'Dowd and Hay 1980

Sterculiaceae			
Fremontodendron decumbens[b]	shrub	*Messor* (1)	Boyd 1996
Violaceae			
Asarum canadense	herb	Genera not provided	Smith et al. 1989
Viola alba	herb	*Componotus* (1), *Cataglyphis* (1), *Crematogaster* (1), *Formica* (1), *Messor* (1), *Pheidole* (1), *Tapinoma* (1)	Wolff and Debussche 1999
V. blanda	herb	*Aphaenogaster* (3), *Formica* (2), *Lasius* (1), *Leptothorax* (2), *Myrmica* (2), *Tapinoma* (1)	Beattie and Lyons 1975; Culver and Beattie 1978; Beattie 1978
V. cucullata	herb	*Lasius* (1), *Prenolepis* (1)	Gates 1943
V. hirta	herb	*Formica* (3), *Lasius* (1), *Myrmica* (2)	Culver and Beattie 1980; Gorb and Gorb 1995
V. matutina	herb	*Formica* (1)	Gorb and Gorb 1995
V. mirabilis	herb	*Formica* (1)	Gorb and Gorb 1995
V. odorata	herb	*Aphenogaster* (1), *Formica* (3), *Lasius* (1), *Myrmica* (3), *Tapinoma* (1)	Beattie and Lyons 1975; Marshall, Beattie, and Bollenbacher 1979; Culver and Beattie 1980
V. papilionacea	herb	*Aphaenogaster* (3), *Formica* (2), *Lasius* (1), *Leptothorax* (2), *Myrmica* (2), *Tapinoma* (1)	Beattie and Lyons 1975; Culver and Beattie 1978; Beattie 1978
V. pedata	herb	*Aphaenogaster* (3), *Formica* (2), *Lasius* (1), *Leptothorax* (2), *Myrmica* (2), *Tapinoma* (1)	Beattie and Lyons 1975; Culver and Beattie 1978; Beattie 1978
V. pensylvanica	herb	*Aphaenogaster* (3), *Formica* (1), *Lasius* (1), *Leptothorax* (2), *Myrmica* (1), *Tapinoma* (1)	Culver and Beattie 1978; Beattie 1978
V. rostrata	herb	*Aphaenogaster* (3), *Formica* (2), *Lasius* (1), *Leptothorax* (2), *Myrmica* (2), *Tapinoma* (1)	Beattie and Lyons 1975; Culver and Beattie 1978; Beattie 1978
V. selkirkii	herb	*Lasius* (1), *Myrmica* (1)	Ohkawara and Higashi 1994
V. striata	herb	*Aphaenogaster* (3), *Formica* (2), *Lasius* (1), *Leptothorax* (2), *Myrmica* (2), *Tapinoma* (1)	Beattie and Lyons 1975; Culver and Beattie 1978; Beattie 1978
V. triloba	herb	*Aphaenogaster* (3), *Formica* (1), *Lasius* (1), *Leptothorax* (2), *Myrmica* (1), *Tapinoma* (1)	Culver and Beattie 1978; Beattie 1978
V. verecunda	herb	*Lasius* (1), *Myrmica* (1)	Ohkawara and Higashi 1994

Note: Rewards to ants by different plant species include elaiosomes, arils, caruncles, and funiculi.

[a]This plant species offers no reward to ants.

[b]These species are from California, USA.

APPENDIX 3.2 **Examples of seed dispersal by ants in Australia and South Africa**

Plant species	Growth form	Associated ant genera (number of species)	Source
Chenopodiaceae			
Dissocarpus biflorus biflorus	herb	*Rhytidoponera* (1)	Davidson and Morton 1981b
Sclerolaena convexula	herb	*Melophorus* (1), *Pheidole* (1)	Davidson and Morton 1981b
S. diacantha	herb	*Iridomyrmex* (1), *Pheidole* (1), *Rhytidoponera* (2)	Davidson and Morton 1981b
Dilleniaceae			
Hibbertia ovata	shrub	*Aphaenogaster* (1), *Pheidole* (1), *Rhytidoponera* (1)	Hughes and Westoby 1992b
Fabaceae			
Acacia (13 spp.) (Mimosaceae)	shrub/tree	*Iridomyrmex* (>1), *Meranoplus* (?), *Pheidole* (?), *Rhytidoponera* (>1)	Davidson and Morton 1984
A. terminalis	shrub	*Iridomyrmex* (6), *Meranoplus* (1), *Monomorium* (1), *Pheidole* (3), *Rhytidoponera* (1)	Hughes and Westoby 1992b, 1992c
Dillwynia retorta (Fabaceae)	shrub	*Iridomyrmex* (6), *Meranoplus* (1), *Monomorium* (1), *Pheidole* (3), *Rhytidoponera* (1)	Hughes and Westoby 1992b, 1992c
Proteaceae			
Leucospermum conocarpodendron	tree	*Anoplolepis* (2), *Crematogaster* (1), *Pheidole* (1)	Bond and Slingsby 1983; Bond, Yeaton, and Stock 1991
Mimetis cucullatus	shrub	*Anoplolepis* (2), *Crematogaster* (1), *Linepithema* (1, = *Iridomyrmex*), *Meranoplus* (1), *Rhoptromyrmex* (>1), *Pheidole* (1), *Tetramorium* (1)	Bond and Slingsby 1983, 1984

Note: See also Berg 1975; Westoby et al. 1982; Bond and Slingsby 1983; Bond, Yeaton, and Stock 1991.

APPENDIX 3.3 **Examples of seed dispersal by ants in the neotropics**

Plant species	Growth form	Associated ant genera (number of species)	Source
Euphorbiaceae			
Croton priscus	shrub	*Atta* (1), *Camponotus* (3), *Ectatomma* (1), *Pachycondyla* (1), *Pheidole* (3), *Pogonomyrmex* (1, = *Ephebomyrmex*), *Solenopsis* (1)	Passos and Ferreira 1996
Gesneriaceae			
Chrysothemis friedrichsthaliana	herb	*Azteca* (2), *Pheidole* (1), *Paratrechina* (1)	Lu and Mesler 1981
Marantaceae			
Calathea micans	herb	*Aphaenogaster* (1), *Ectatomma* (3), *Hypoponera* (1), *Odontomachus* (3), *Pachycondyla* (4), *Pheidole* (1), *Trachymyrmex* (1)	Horvitz 1991; LeCorff and Horvitz 1995
C. microcephala	herb	*Odontomachus* (2), *Pachycondyla* (2), *Paratrechina* (>1), *Solenopsis* (1), *Wasmannia* (1), *unidentified Ponerinae* (1)	Horvitz and Beattie 1980; Horvitz 1981
C. ovandensis	herb	*Odontomachus* (2), *Pachycondyla* (2), *Pheidole* (1), *Solenopsis* (1), *Wasmannia* (1)	Horvitz and Beattie 1980; Horvitz and Schemske 1986a, 1986b
Turneraceae			
Turnera ulmifolia	shrub	*Camponotus* (1), *Crematogaster* (1)	Cuautle, Rico-Gray, and Díaz-Castelazo 2005

Note: For ant-garden myrmecochores, see chapter 8.

CHAPTER FOUR

Mutualism from Opportunism

Ants as Secondary Seed-Dispersers

Studies on seed dispersal of tropical species have traditionally focused on fruit consumption and seed deposition patterns created by primary seed-dispersers (Howe and Smallwood 1982; Estrada and Fleming 1986; Fleming and Estrada 1993). This is not surprising, given that in tropical forests nearly 90% of the trees and shrubs bear fleshy fruits and rely on vertebrate frugivores such as birds, bats, or monkeys for seed dispersal (Frankie, Baker, and Opler 1974; Jordano 1993). More recently, however, the relevance of postdispersal events for seed fate and demography of plant species has been repeatedly emphasized for a number of dispersal systems (Wenny 2001; Wang and Smith 2002; Vander Wall and Longland 2004). Indeed, recent studies have demonstrated that postdispersal events, some of them involving ants as seed vectors, can markedly affect seed fate in numerous plant species from different regions (e.g., Levey and Byrne 1993; Chambers and MacMahon 1994; Andresen 1999; Farji-Brener and Medina 2000; Pizo, Passos, and Oliveira 2005). Although myrmecochory can be an important dispersal strategy for some plant taxa in neotropical forests (Horvitz and Beattie 1980, Passos and Ferreira 1996), typical myrmecochores are especially common in arid Australia and South Africa and in Mediterranean and temperate areas (chapter 3).

In tropical habitats, fleshy fruits present a plethora of sizes, shapes, and colors, and the chemical composition of the edible portion also varies widely (Corlett 1996; Forget and Hammond 2005). Fallen diaspores (i.e., any seed, fruit, or infructescence that acts as the plant's unit of dispersal) constitute a large proportion of the litter on the floor of tropical forests (Jordano 1993). They can reach the ground spontaneously (van der Pijl 1982), be dropped by frugivores (Howe 1980; Laman 1996b), or arrive in vertebrate feces (Kaspari 1993). In the rain forest at La Selva Biological Station in Costa Rica, seed rain is estimated as 49 seeds m^{-2} mo^{-1} (Denslow and Gomez-Dias 1990); and in humid forests of southeastern Brazil, the quantity of fallen fruits can be as much as 400 kg ha^{-1} yr^{-1} (Morellato 1992). Such a huge amount of fruit-fall is evidenced not only at the level of the community, but for individual plant species as well. For instance, Laman (1996a) estimated that more than 50% of the seed crop produced by *Ficus* trees (Moraceae) in the Bornean rain forest falls beneath the parent plant, and Pizo reported that nearly 30% of the diaspores taken by birds in the canopy of *Cabralea canjerana* (Meliaceae) drop to the ground, equal to nearly 8,000 diaspores for some especially fecund trees over the entire fruiting season in the Brazilian Atlantic forest (Pizo 1997).

The remarkable abundance and diversity of tropical ground-dwelling ants make them the most probable organisms to find and consume fleshy fruits or seeds (hereafter *diaspores, sensu* van der Pijl 1982) on the floor of tropical environments. Ant density in tropical rain forests may exceed 800 workers m^{-2} (Hölldobler and Wilson 1990). At the La Selva Biological Station, densities of ant colonies surpass 4 nests m^{-2} (Kaspari 1993), and more than 400 ant species are known to inhabit this 1,500 ha rain-forest reserve (Longino, Coddington, and Colwell 2002). Not surprisingly, ant-seed interactions are widespread in tropical forests. For instance, 886 interactions between ants (41 species) and naturally fallen fleshy diaspores (56 plant species) were recorded in the Brazilian lowland Atlantic rain forest (Pizo and Oliveira 2000), whereas in coastal sandy forests, 562 such interactions comprised 48 ant species and 44 species of plants (Passos and Oliveira 2003) (see fig. 4.1).

Ants affect seed distribution through two general mechanisms: by harvesting edible seeds that may subsequently escape predation and by collecting diaspores to eat an ant-attractive nutritive part (elaiosome, aril, or fleshy pulp) and discarding unharmed viable seeds (van der Pijl 1982; Handel and Beattie 1990). In the past few years, subtle relations involving ants and nonmyrmecochorous fleshy diaspores (i.e., lacking special adaptations for ant dispersal) have been discovered in tropical forests and savannas. Interactions cover a broad range of ant and diaspore sizes

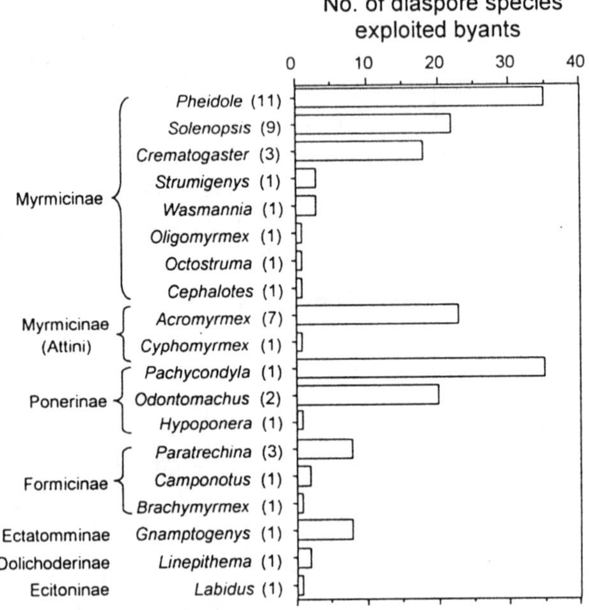

FIGURE 4.1. Ant fauna exploiting fallen fleshy diaspores on the floors of two different coastal forests in Brazil. Ants and diaspores were surveyed during two years of monthly samplings at the two sites. The number of species in each genus is given in parentheses. Data from Pizo and Oliveira 2000; Passos and Oliveira 2003.

and include both untouched freshly fallen diaspores and those previously manipulated and dropped by vertebrate frugivores or even found in their feces. In this regard, it would be interesting to test for similarities between the effect of ants and that of dung beetles (Andresen 1999) on rearranging seed shadows, originated on seeds deposited in vertebrate feces. Recent findings from tropical habitats have shown that ant activity at fallen fleshy diaspores can rearrange the seed shadow generated by vertebrate dispersers, affect seed-bank dynamics and seed germination, and influence patterns of seedling distribution and survival in primarily vertebrate-dispersed species (Roberts and Heithaus 1986; Levey and Byrne 1993; Böhning-Gaese, Gaese, and Rabemanantsoa 1999; Passos and Oliveira 2002). Even leaf-cutter ants (chapter 2), traditionally considered pests, have been shown to positively affect seed biology (Farji-Brener and Silva 1996; Leal and Oliveira 1998; Wirth et al. 2003).

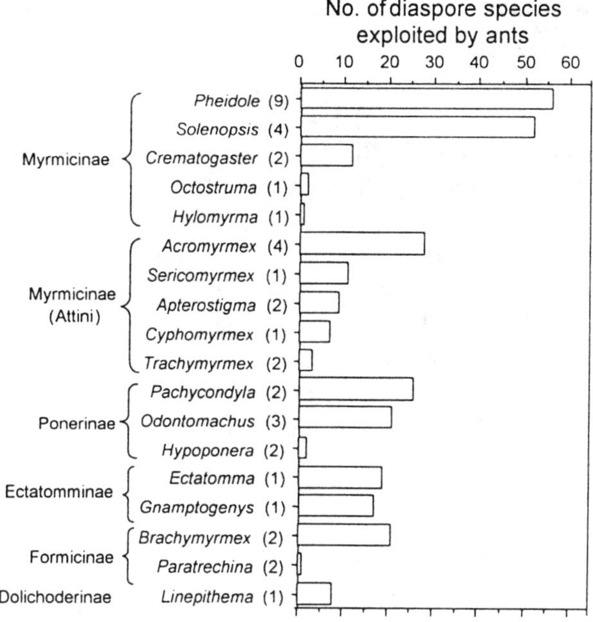

FIGURE 4.1. (continued)

In this chapter we summarize recent findings showing that the use of fallen fleshy diaspores by opportunistic ground-dwelling ants can have relevant effects on seed and seedling biology of primarily vertebrate-dispersed plant species. We provide a general characterization of the plant and ant species involved in these interactions, address the particular attributes of ants and diaspores that mediate the interaction, and discuss the possible consequences of the interaction for the plants. Possible causes underlying the observed patterns are examined and discussed.

The Reward: Fallen Fleshy Diaspores

The diaspores with which ants interact on the floor of tropical forests present a broad range of size and nutrient content in the fleshy portion (either pulp or aril) exploited by the ants (Pizo and Oliveira 2000; Passos and Oliveira 2003). Because lipids are an important food resource for ants, serving a variety of purposes that include nutrition, physiological constituents, and behavioral releasers (see Marshall, Beattie, and Bollenbacher 1979), the outcome of ant-diaspore interactions can be largely determined by the lipid content of the diaspores (see also chapter 3). Indeed, prior to most studies that investigated ant-fruit/seed interactions, Carroll and Janzen (1973, 235) appropriately wrote, "Seeds with an oily covering may be fed on by almost any kind of ant." Their statement is supported, for instance, by the suggestion that the relative amounts of lipids and carbohydrates in fleshy diaspores are a good predictor of the type of ant attending such food resources in tropical forests (Pizo, Passos, and Oliveira 2005). Finally, because ant colonies must take in adequate protein to meet the nutritional requirements of larvae and functional queens (Hölldobler and Wilson 1990), protein-rich diaspores may also represent an important dietary complement for carnivorous ants under certain conditions (Passos and Oliveira 2004).

A great variety of ant and diaspore species interact on the floor of Brazilian coastal rain forests (fig. 4.1). Ants in the subfamily Myrmicinae are by far the most frequent attendants to fallen fleshy diaspores in these areas, followed by the Ponerinae and the Formicinae. Treatment given by the ants to fallen seeds and fruits may depend largely on the size of the ant relative to the size of the diaspore. Ants are seen transporting entire diaspores, tearing pieces off diaspores, and collecting liquids from them. While large ponerine ants such as *Pachycondyla* and *Odontomachus* (1.0–1.5 cm length) are able to individually remove entire diaspores of up to 1 g to their nests, smaller ants such as *Pheidole* and *Crematogaster* usually recruit many workers to consume the diaspore on the spot (fig. 4.2). The piecemeal removal of pulp or aril by small myrmicines usually lasts less than 24 h. Some *Solenopsis* species may cover the diaspore with soil before collecting liquid and solid food from it (Pizo, Passos, and Oliveira 2005). Large fruits (>1 g) are usually consumed on the spot, but if they contain small seeds, these may be removed with bits of pulp attached by large ponerine and attine ants (Pizo and Oliveira 2000; Passos and Oliveira 2003).

The distance of diaspore displacement by ants varies widely. Small diaspores (<0.10 g) can be transported nearly as far as 100 m by leaf-cutting *Atta* spp. (Wirth et al. 2003; Leal and Oliveira 1998), but medium- to large-sized diaspores are either consumed on the spot (fig. 4.2) or carried only a few meters by attines (Oliveira et al. 1995). Large ponerines can move diaspores for 10 m or more (Horvitz 1981; Fourcassié

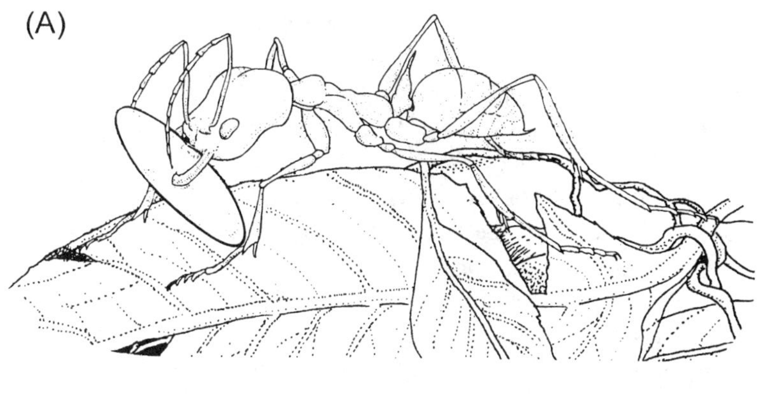

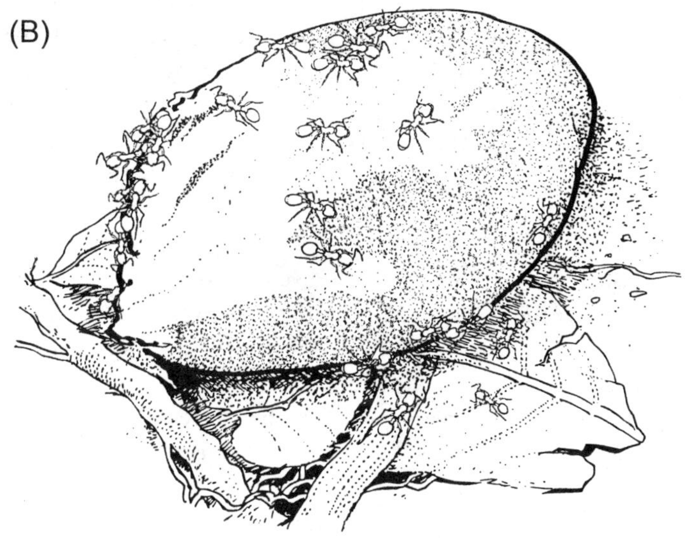

FIGURE 4.2. Ant behavior at fallen fleshy diaspores on the leaf litter of a rain forest in Brazil. (A) *Odontomachus chelifer* carrying a fallen diaspore (containing up to 17 seeds) of *Clusia criuva* (Clusiaceae) to its nest. (B) Recruited workers *Pheidole* sp. removing bits of aril from a seed of *Virola bicuhyba* (Myristicaceae). Both plant species are dispersed primarily by birds in neotropical forests (Galetti, Laps, and Pizo 2000).

and Oliveira 2002), while small ants (<0.5 cm) usually do not carry diaspores beyond 2 m (Pizo and Oliveira, 1999).

Inside the nest, ponerine ants feed fleshy portions of diaspores (either pulp or aril) to larvae and adult nest mates. After a residence time of 2–9 days inside the nest, cleaned intact seeds (i.e., without the fleshy part) are deposited on refuse piles outside the nest with no evidence of seed predation by the ants. Attine ants have also been observed to discard intact seeds of several species in refuse piles outside their nests (Leal and Oliveira 1998, 2000; Dalling and Wirth 1998). In the rain forest at Barro Colorado Island (Panama), Farji-Brener and Medina (2000) found seven times higher densities of viable seeds around *Atta* nests than in adjacent soils. Milesi and Lopez de Casenave (2004) have recently shown that leaf-cutting *Acromyrmex* spp. can, as a by-product of their predation on fruits and seeds, also rearrange seed distribution by leaving viable seeds around their nests in open woodland in Argentina. Other myrmicines (e.g., *Pheidole*) prey upon some of the seeds they collect but also cache intact seeds inside their nests (Levey and Byrne 1993).

Ants also frequently have access to fleshy diaspores when these reach the ground partially undigested inside frugivore defecations (Roberts and Heithaus 1986; Kaspari 1993). As mainly canopy-dwelling insects, ants may additionally exploit fruit matter from vertebrate feces encountered on branches or leaves in tropical forests (Martínez-Mota, Serio-Silva, and Rico-Gray 2004). Because nutrient-rich bits of pulp or aril remain attached to feces-embedded seeds, large ant assemblages can displace seeds from vertebrate defecations (Davidson 1977b; Pizo and Oliveira 1999; Pizo, Guimarães, and Oliveira 2005). Although ants of the subfamily Myrmicinae are the prominent removers of seeds from feces of vertebrate frugivores in neotropical forests (Roberts and Heithaus 1986; Byrne and Levey 1993; Kaspari 1993), ponerine ants may account for 12% of the ant records in connection with feces-embedded lipid-rich arillate seeds of *Clusia criuva* (Clusiaceae) in coastal Brazilian forests (Passos and Oliveira 2002).

Diaspore Attributes and Patterns of Ant Attendance

In true myrmecochory (chapter 3) the outcome of the interaction is affected not only by the size of diaspores but also by the presence of a lipid-rich appendage called the elaiosome (Hughes and Westoby 1992c; Gorb and Gorb 1995; Mark and Olesen 1996). Similarly, recent evidence has shown that the chemical composition of nonmyrmecochorous fleshy diaspores, particularly the lipid content, can also play an important role in the interaction with ants. Several authors have already stressed that lipids, particularly fatty acids, are important mediators of ant-diaspore interactions (e.g., Marshall, Beattie, and Bollenbacher 1979; Skidmore and Heithaus 1988; Brew, O'Dowd, and Rae 1989), and nonmyrmecochorous diaspores have a fatty acid composition remarkably similar to that of elaiosomes in true myrmecochores (Hughes, Westoby, and Jurado 1994; Pizo and Oliveira 2001). Fleshy fruits of tropical forests vary widely in size, shape, color, and chemical composition of the edible portion (e.g., Corlett 1996), and therefore ants in such habitats interact with a broad range of fruits differing in morphology and nutrient content. It has recently been suggested that patterns of ant attendance to fallen diaspores parallel what happens in true myrmecochory and can be largely mediated by the size of the diaspore and the lipid content of the fleshy part as well (Pizo and Oliveira 2001). Ant response to such diaspore features can be relevant for the outcome of the ant-diaspore interaction in tropical forests, with important consequences for seed and seedling biology of primarily vertebrate-dispersed plants (see below).

In a study involving ground-dwelling ants and selected bird-dispersed species representing three discrete diaspore size classes (small: <0.1 g; medium: ≅1 g; large: >3 g), and two extremes relative to the lipid content of their fleshy portions (<8% and >60% of dry mass), Pizo and Oliveira highlighted some general tendencies in the response pattern of ants relative to these diaspore traits on the floor of the Atlantic rain forest (fig. 4.3). Both day and night, more ants attended lipid-rich diaspores than lipid-poor ones, and the lipid content of the fleshy portion of large

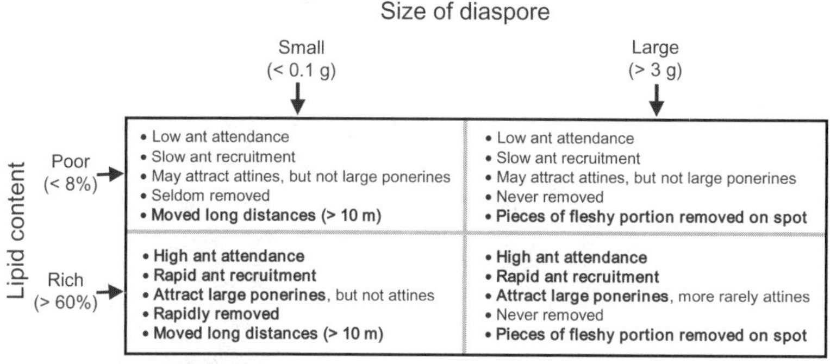

FIGURE 4.3. Summary of the documented response patterns of ground-dwelling ants to fallen fleshy diaspores in the Atlantic forest of Brazil. Ant behaviors are categorized according to the size and lipid content (on a dry-mass basis) of diaspores. Responses that are most likely to benefit diaspores appear in boldface. Modified from Pizo, Passos, and Oliveira 2005.

and medium-sized diaspores positively affected the recruitment rate of ants to newly discovered diaspores. Both removal rate by ants and displacement distance were negatively affected by diaspore size. Based on these general trends, Pizo and Oliveira (2001) predicted that small, lipid-rich diaspores would more likely benefit from interactions with ants on the floor of tropical forests.

The diet of ground-dwelling ants is obviously crucial in determining their preferences for fallen fleshy diaspores. Carnivorous ponerines are more frequently seen on lipid-rich diaspores, whereas fungus-growing attines more commonly attend lipid-poor diaspores (fig. 4.4). Like the elaiosome-bearing seeds of true myrmecochores, lipid-rich diaspores are analogous to insect prey (Carroll and Janzen 1973; Hughes, Westoby, and Jurado 1994), and it is not surprising that carnivorous ponerines would prefer lipid-rich diaspores. Why attine ants, in contrast, exploit lipid-poor diaspores more often than lipid-rich ones should be investigated in greater detail (Beattie 1991).

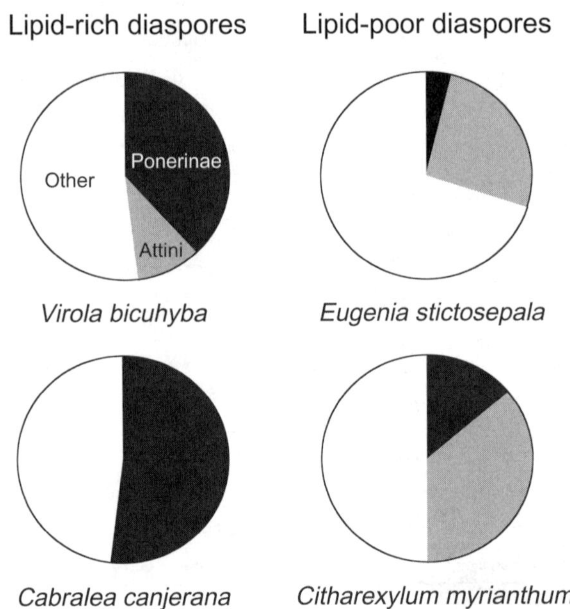

FIGURE 4.4. Percentage of attendance by Ponerinae, Attini, and other ants to lipid-rich (>60% of lipids in aril dry mass) and lipid-poor (<8% of lipids in pulp dry mass) diaspore species placed on the forest floor. Fifty diaspores of each species were used in the censuses. Modified from Pizo and Oliveira 2001.

Ant Effects on Seeds and Seedlings

We have shown that a wide assemblage of ants interact with a large range of fleshy diaspores once these reach the ground in tropical forests and that different ants treat the diaspores in different modes. A central issue for this interaction is whether ants can play any significant role in population recruitment of these primarily vertebrate-dispersed plants (Horvitz and Schemske 1986a). Possible benefits to fallen fleshy diaspores secondarily dispersed by ants are not necessarily different from those received by true myrmecochorous diaspores (see chapter 3); they may include increased germination success, reduced damage to seeds or seedlings, reduced competition, and the deposition of seeds in sites where seedlings find improved conditions for establishment and growth (i.e., "directed dispersal") (see Howe and Smallwood 1982; Wenny 2001).

Unless fruits and seeds dropped on the ground by vertebrate frugivores have their fleshy portion removed, infestation by fungal pathogens may rapidly cause them to rot on the humid leaf litter of tropical habitats (Augspurger 1990). Thus, perhaps the first advantage acquired by fleshy diaspores upon interacting with ants on the floor is that, by removing the pulp or aril from the diaspores, the ants can considerably reduce fungus infestation of seeds (Horvitz 1981; Oliveira et al. 1995; Leal and Oliveira 1998; Pizo and Oliveira 1998; Pizo, Passos, and Oliveira 2005). Removal of fleshy matter from seeds beneath the parent plant may further benefit the viable seed by allowing it to remain available for longer periods to secondary vertebrate dispersers, such as the case of agoutis and guapinol seeds (*Hymenaea courbaril*) (see Hallwacks 1986; Oliveira et al. 1995).

Removal by ants of the fleshy portion of primarily vertebrate-dispersed seeds has been shown to increase by 19% to 63% the germination success of several species (Oliveira et al. 1995; Pizo and Oliveira 2001, Passos and Oliveira 2002). In the cerrado savanna of Brazil, pulp or aril removal by attine ants significantly increased germination success in seven plant species, and in *Psycotria stachyoides* (Rubiaceae) germination was increased even when compared to control seeds whose pulp was removed by hand, indicating that ant-induced mechanical and/or chemical factors facilitate germination (Leal and Oliveira 1998). Finally, ants may also enhance germination by removing undigested portions of diaspores from frugivore defecations, as suggested by the experiments of Passos and Oliveira (2002).

By displacing the seeds from below the parent plants, ground-dwelling ants can further help the seeds escape from density- or distance-oriented seed predators (Janzen 1970; O'Dowd and Hay 1980; Heithaus 1981; Pizo and Oliveira 1998). In Brazilian Atlantic forests, a series of exclosure experiments performed with different plant species to compare removal rates of caged (no access to vertebrates) and uncaged fleshy diaspores (free access to vertebrates and ants) have revealed that (1) ants can remove up to 93% of the diaspores in 24 hours, with small lipid-rich diaspores usually experiencing the highest removal rates, and (2) removal attributed to vertebrate seed predators (estimated by the difference in removal rates between caged and uncaged treatments) increased with diaspore size because large fruits are not transported by ants and thus become available to rodents (Pizo and Oliveira 2001; Passos and Oliveira 2003, 2004). Therefore, if escape from seed predators can benefit nonmyrmecochorous fleshy diaspores as a result of their interaction with ants, this advantage would be greatest for small, lipid-rich diaspores, which are rapidly removed by ants (especially the primarily carnivorous ponerine ants) (see fig. 4.3).

Directed dispersal occurs when a given dispersal agent deposits disproportionate amounts of seeds in suitable locations, generating a two-component process: nonrandom arrival and survival at predictable sites (Howe and Smallwood 1982). To benefit from directed dispersal, a diaspore must have traits that make it more likely to be transported by one type of agent than by others, and/or it must have a morphology that enables it to reach certain locations more often than expected by chance (Venable and Brown 1993). For the case of fleshy diaspores secondarily dispersed by ants, however, the plants need not be adapted specifically for directed dispersal by these insects but, instead, may have multiple dispersal agents that act at different spatial scales (Roberts and Heithaus 1986; Horvitz and Le Corff 1993). Because ant foragers return to a fixed nest location after finding food, cases of directed dispersal by ants have been repeatedly reported, especially in tropical forests. Several studies have shown that dispersal by vertebrate frugivores is often only one stage of a two- or three-phase dispersal process that may involve ants at a final step (Clifford and Monteith 1989; Kaufmann et al. 1991; Böhning-Gaese, Gaese, and Rabemanantsoa 1999). Litter-dwelling ants frequently remove seeds of tropical *Miconia* trees (Melastomataceae) from bird defecations (Kaspari 1993; Byrne and Levey 1993). In Costa Rican rain forests, *Pheidole* ants harvest seeds from frugivore droppings and cache them in their nests in partially decomposed twigs. Although the ants eat most of the harvested seeds, about 6% of the *Miconia* seeds are deposited on refuse piles, where seedlings grow faster and survive better compared to control topsoil (Levey and Byrne 1993). These results illustrate the complexity of the dispersal system of tropical trees and show how seed-harvesting ants can be simultaneously harmful and beneficial toward seeds, destroying most but significantly helping some.

Directed Dispersal of Seeds by Ponerine Ants

Clusia criuva (Clusiaceae) is a common tree growing in a coastal sandy plain forest in Brazil. The plant can produce up to 18,000 diaspores over an entire fruiting season, each consisting of up to 17 seeds enveloped by a red aril that has one of the highest lipid contents (83.4% of dry mass) yet described in the literature (Jordano 1993). Arillate seeds are dispersed by at least 14 bird species, which ingest the whole diaspore and defecate intact seeds. The birds are thus legitimate seed-dispersers and occasionally drop intact diaspores beneath the parent plant (M. A. Pizo, pers. comm.; Galetti, Laps, and Pizo 2000). Once on the ground, either dropped directly from the parent plant or dispersed by birds, *C. criuva* diaspores are exploited by a diverse assemblage of ground-dwelling ants (16 species), but most especially two primarily carnivorous ponerines—*Pachycondyla striata* and *Odontomachus chelifer*—which transport the diaspore to their nests (Passos and Oliveira 2002, fig. 1A). Interestingly, the two large ponerines also interact intensively with another common bird-dispersed tree in the area, *Guapira opposita* (Nyctaginaceae), which produces mature drupes (1 seed each) whose pulp has high protein content (28.4% of dry mass). In a manner similar to what happens with *Clusia,* if fruits of *Guapira* fall spontaneously with the pulp intact or are dropped by birds with bits of pulp attached, *P. striata* and *O. chelifer* rapidly remove them to their nests (Passos and Oliveira 2004). During two years the effects of these two large ponerines on seed fate and seedling performance in *Clusia* and *Guapira* were investigated in a coastal sandy forest of Brazil; the main results are summarized in table 4.1.

Although the ponerines behave similarly toward seeds and seedlings of both species, the plants are not equally affected by the two ant species. For instance, although seedlings of *C. criuva* are more abundant

TABLE 4.1 **Effects of two common ponerine ants, *Pachycondyla striata* and *Odontomachus chelifer*, on seeds and seedlings of *Clusia criuva* and *Guapira opposita* in a Brazilian coastal sandy forest**

Ant behavior and ant effects on seeds and seedlings	*Clusia criuva*		*Guapira opposita*	
	P. striata	*O. chelifer*	*P. striata*	*O. chelifer*
Remove fallen diaspores	Yes	Yes	Yes	Yes
Remove diaspores from bird feces	Yes	Yes	—	?
Feed bits of aril/pulp to larvae in the nest[a]	Yes	Yes	Yes	Yes
Discard intact seeds outside nest	Yes	Yes	Yes	Yes
Removal of fleshy part by ants enhances germination[b]	Yes	Yes	—	—
Increased seedling recruitment near nest[c]	Yes	Yes	—	Yes
Increased seedling survival (1 yr) near nest[c]	Yes	No	—	—
Increased soil nutrients near nest[c,d]	Yes	No	—	Yes
Higher soil penetrability near nest[c]	—	—	—	Yes
Potential herbivore deterrence near nest[c,e]	—	—	—	Yes

Sources: Data from Passos and Oliveira 2002, 2004. Table modified from Pizo, Passos, and Oliveira 2005.
Note: A dash indicates that ant activity and/or ant effects on plants were not evaluated.
[a] Observations from captive colonies.
[b] Effect of cleaning activity by ants was not evaluated for *Guapira,* but pulp removal by hand increased germination in this species.
[c] Compared to random control plots without nests.
[d] *Pachycondyla* nests had increased concentrations of total N, and P in the *Clusia* study. *Odontomachus* nests had increased concentrations of P, K, and Ca in the *Guapira* study.
[e] Evaluated by recording attack rates by ants on dipteran larvae placed on seedlings growing near ant nests and in control plots.

in the vicinity of nests of *P. striata* and *O. chelifer* than in areas without nests, seedling survival in this species is positively affected only around nests of *P. striata,* no effect being detected near *O. chelifer* nests (fig. 4.5). *Guapira* seedlings, in contrast, are more frequent close to *O. chelifer* nests than in random areas without nests (table 4.1). Interspecific variation in the effects of ants on *Clusia* and *Guapira* may have resulted from different soil properties around ant nests, variable nutrient requirements of the seedling species, or unknown temporal factors.

The multistep dispersal process in *Clusia criuva* illustrated in figure 4.6 involves complex interactions between the plant, its guild of bird primary dispersers, and its ant secondary dispersers. The possible sequential events in the life of the seed indicate that ant-seed interactions can markedly affect seed fate in this primarily bird-dispersed species. Of the diaspores produced by *C. criuva* trees, 83% are taken by birds, and seeds reach the ground either in bird feces or as fallen fresh diaspores. Most seeds embedded in feces are removed by ants within 24 h, and

P. striata and *O. chelifer* account for more than 12% of seed removal. Approximately 90% of the fresh diaspores on the ground surface are also removed in 24 h, and ants are responsible for 97.5% of this removal (more than 34% due to *P. striata* and *O. chelifer*). Inside the colony, the two ponerines remove the lipid-rich aril and discard the seeds on refuse piles. Aril removal and separation from bird defecations increase germination success, and directed dispersal to ant nests affects the distribution and development of *C. criuva* seedlings (fig. 4.5).

The interactions of *Clusia* and *Guapira* with *Pachycondyla* and *Odontomachus* ants indicate that, as reported for myrmecochores of xeric environments (chapter 3), directed dispersal provided by the two ponerines results in seed deposition in nutrient-enriched soil close to their nests (table 4.1). Soil enrichment near the nests probably results from the deposition of organic material on adjacent refuse piles, and variation in soil fertility at such microsites should vary with the type of food items

(A) Ant effects on seedling distribution

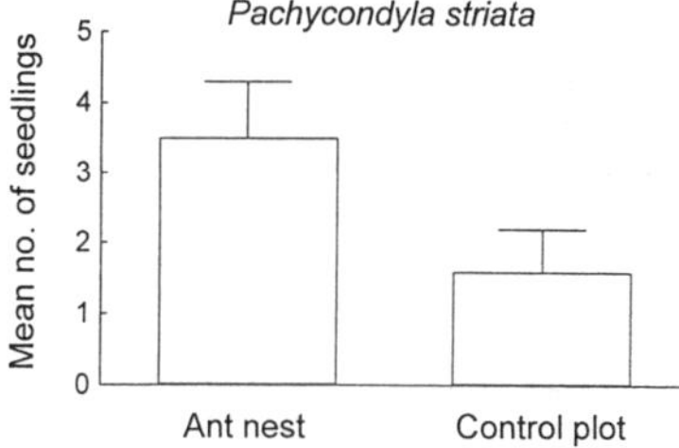

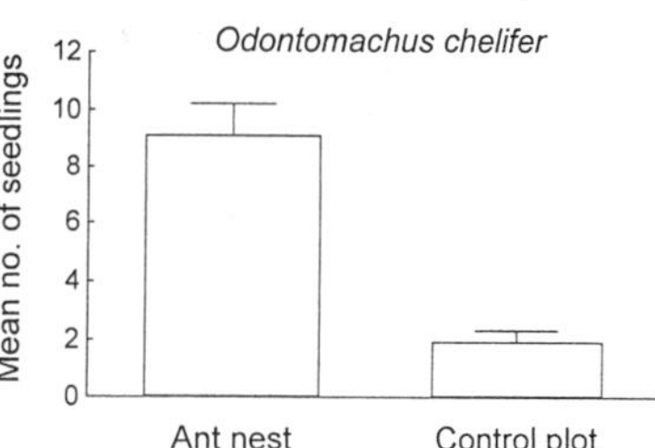

(B) Ant effects on seedling survival over 1 year

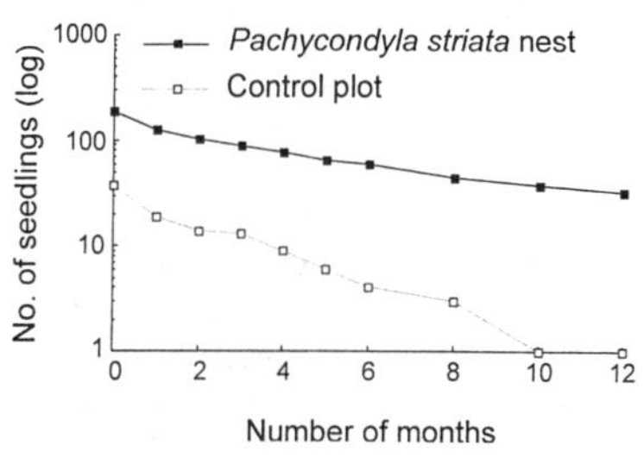

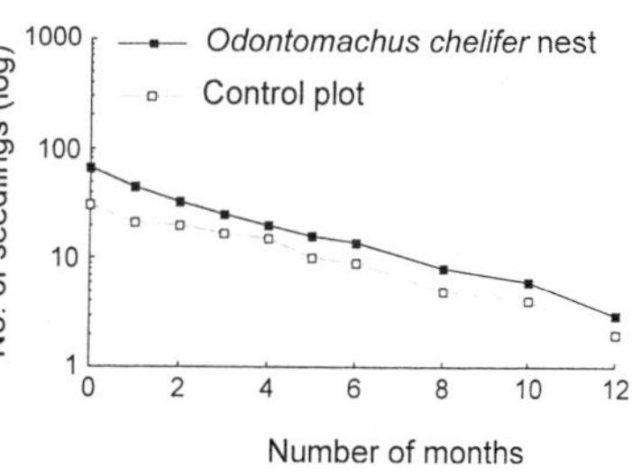

FIGURE 4.5. Effects of two ponerine ants on the distribution and survival of seedlings of *Clusia criuva*, a primarily bird-dispersed tree. (A) Mean number (+1 SE) of seedlings on nests of *Pachycondyla striata* and *Odontomachus chelifer* and in adjacent control plots. (B) Survivorship curves for seedlings growing on nests of *P. striata* and *O. chelifer* and in control plots. Modified from Passos and Oliveira 2002.

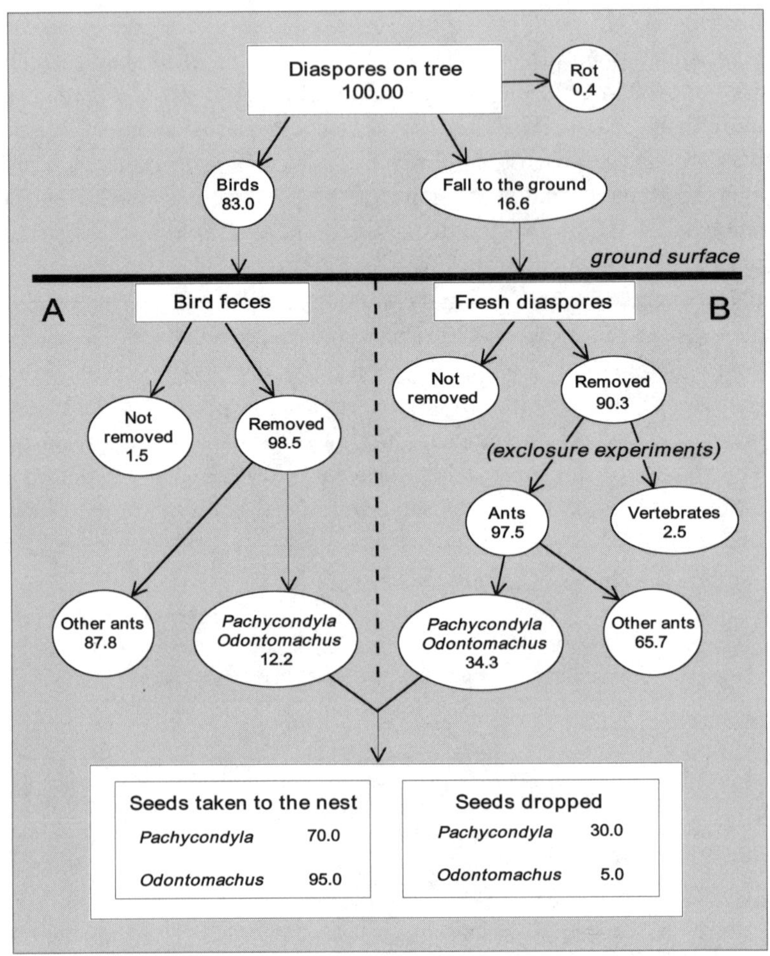

FIGURE 4.6. Diagram of the multistep dispersal process of *Clusia criuva* in a sandy forest in Brazil. The process begins with 100% diaspores produced, of which on average 83.0% are taken by birds, 16.6% fall to the ground, and 0.4% rot. Seeds reach the forest floor embedded in bird feces (A) or as fallen fresh diaspores (B). On average, 98.5% of the seeds in feces are removed by ants in 24 h, and *Pachycondyla striata* and *Odontomachus chelifer* account for about 12% of seed removal. Of the fresh diaspores on the ground, 90.3% are removed in 24 h, and ants are responsible for 97.5% of removal. *Pachycondyla* and *Odontomachus* together remove about 34% of the diaspores. Feces-embedded seeds and diaspores (seeds enveloped by a red aril [see fig. 4.2]) of *Clusia* are taken by *Pachycondyla* to nests in 70%, and by *Odontomachus* in 95% of the records. Modified from Passos and Oliveira 2002.

brought by foragers through time. Similar effects by *Odontomachus* ants on seed fate and seedling establishment have been shown for other primarily vertebrate-dispersed plants in the same area, such as *Anthurium* sp. (Araceae), *Myrcia rostrata* and *Psidium catleyanum* (Myrtaceae) (Passos and Oliveira 2003). Such a spatial association is analogous to epiphyte ant gardens in which the plants grow mainly on arboreal ant nests and benefit from seed dispersal, mineral provisioning, and protection by ants (chapter 8; Davidson 1988; Orivel and Dejean 1998). The behavioral observations of *Pachycondyla* and *Odontomachus* ants toward *Clusia* and *Guapira* diaspores (table 4.1) support previous reports on treatment of seeds by ponerines (Horvitz and Beattie 1980; Horvitz 1981; Hughes and Westoby 1992c; Pizo and Oliveira 1998) and contrast with other ant-seed systems in which ant-induced seed movements are caused by both seed predation and "true" myrmecochory (Handel and Beattie 1990; Levey and Byrne 1993; Garrido et al. 2002).

Ants in the subfamily Ponerinae are predominantly carnivorous (Hölldobler and Wilson 1990), and the lipid-rich fleshy part of diaspores can be regarded as an arthropod-prey mimic complementing the diet of this group of ants (Carroll and Janzen 1973; Hughes, Westoby, and Jurado 1994). Chemicals mediate the behavior of ants toward potential food items, and lipids are known to be the major attractants in the interactions between ants and elaiosome-bearing seeds of true myrmecochores (Marshall, Beattie, and Bollenbacher 1979). The results with *Clusia criuva* and other plant species strongly suggest that the lipid-rich aril of primarily bird-dispersed plants is a key feature in the attraction of ponerine ants on the floor of tropical forests (Pizo and Oliveira 1998, 2001; Passos and Oliveira 2002, 2003). Although the role of proteins as ant-attractants in fleshy diaspores is still to be assessed, the behavior of *Pachycondyla* and *Odontomachus* toward protein-rich *Guapira* fruits suggests that these might complement the protein intake in colonies of primarily carnivorous ponerines (Passos and Oliveira 2004), especially in areas such as sandy forests with low biomass of litter arthropods (Pizo, Passos, and Oliveira 2005).

Prospect

The importance of vertebrates as primary seed-dispersers of plants bearing fleshy fruits has been suggested to considerably exceed that of ants as secondary dispersers (Böhning-Gaese, Gaese, and Rabemanantsoa 1999). The studies reported in this chapter, however, indicate that directed dispersal by ants strongly affects recruitment in primarily vertebrate-dispersed plants. Indeed, although distances of seed displacement by ants may be small, such ant-induced movements were shown to produce nonrandom seedling recruitment patterns in primarily bird-dispersed plants (but see Horvitz and Le Corff 1993 for a discussion on scale of bird and ant dispersal). The associations between ants and fleshy diaspores occur without special adaptations on the part of the plants or the ants. Considering the huge amount of fruit-fall (Jordano 1993) and the abundance and diversity of ground-dwelling ants in tropical habitats (e.g., Longino, Coddington, and Colwell 2002), it is not surprising that many ant species interact opportunistically with nutrient-rich fallen diaspores, with important consequences for seed and seedling biology in these areas (Levey and Byrne 1993; Leal and Oliveira 1998, 2000; Pizo and Oliveira 2000, Passos and Oliveira 2003). Seed cleaning per se by ants (i.e., removal of fleshy matter from seeds) can be enough to render some benefit to the plant (Oliveira et al. 1995). Seed dispersal, germination, and early seedling development are the most critical stages in determining where plants recruit within a landscape (Herrera et al. 1994; Schupp 1995). Ants are dominant ground-dwelling insects (Kaspari 1993), and their nests have specific temperature, moisture, texture, and nutrient characteristics (Hölldobler and Wilson 1990; Farji-Brener and Medina 2000). Thus, directed dispersal by ants likely facilitates significantly seed germination and seedling establishment (Horvitz 1981) and can even potentially render seedlings some protection against herbivores (Davidson and Epstein 1989; Passos and Oliveira 2004). Due to seed clumping around nests, however, additional density-related mortality factors such as seedling competition and pathogen attack should also be considered to affect establishment and survivorship of seedlings (Augspurger 1984; Howe and Schupp 1985; Howe 1989).

CHAPTER FIVE

Mutualism from Antagonism

Ants and Flowers

The evolution of interactions between plants and their pollinators provides some of the clearest examples of change in the outcome of interactions from antagonistic to mutualistic (Thompson 1982). Early insect pollinators of angiosperms fed on pollen, ovules, seeds, and flower parts (Thien 1980; Thien et al. 1985, 1998; Pellmyr and Thien 1986). The vast majority of these interactions were detrimental to the plants, and the closed carpels of angiosperms were probably a defense against these flower visitors (Mulcahy 1979). However, these antagonistic interactions provided a basis on which selection could act, because some flower visitors were less detrimental to flower parts than others and some plants possessed floral traits that caused the interaction to be less detrimental to the plant and, at some point in time, beneficial. The development of floral nectaries reduced even more the cost of the interaction relative to the gain, because it transferred some of the reward away from the plant's reproductive structures themselves and likely lessened the cost of the interaction (Baker and Baker 1979; Thompson 1982).

Although ants are present in most communities and are constant plant visitors and avid sugar collectors, there are few confirmed cases of pollination by ants. The literature can be grouped into the categories of pollination by ants and discouragement of floral visits by ants. It suggests that natural selection has not favored ant pollination in many environments but does favor avoidance of nectar thievery. Nevertheless, ants do

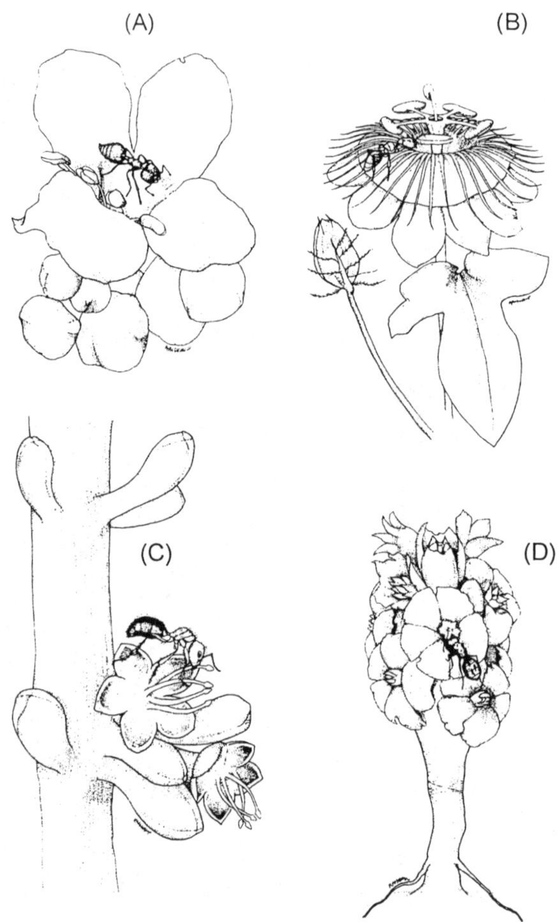

FIGURE 5.1. Ants foraging for floral nectar. (A) *Crematogaster brevispinosa* and *Avicennia germinans.* (B) *Camponotus* sp. and *Passiflora foetida.* (C) *Camponotus planatus* and *Coccoloba uvifera.* (D) *Camponotus sericeiventris* and *Bdallophyton bambusorum.* (D) based on photo by J. G. García-Franco.

visit flowers (fig. 5.1) and other reproductive parts of plants living in lowland dry tropics (Elias and Prance 1978; Haber et al. 1981; Rico-Gray 1980, 1993; Fisher and Zimmerman 1988; Rico-Gray and Thien 1989a, 1989b; García-Franco and Rico-Gray 1997; Rico-Gray, García-Franco, et al. 1998), semiarid areas (Rico-Gray, García-Franco, et al. 1998; Rico-Gray, Palacios-Rios, et al. 1998), Mediterranean regions (Herrera, Herrera, and Espadaler 1984; Gómez and Zamora 1992; Bosch, Retana, and Cerdá 1997), and alpine environments (Puterbaugh 1998).

Pollination by Ants

Hickman (1974) proposed ant pollination as a low-energy system and described ten traits that would allow for predictive inferences concerning ant pollination.

1. Plants must live in hot and dry habitats where ant activity and frequency are high.
2. Nectaries must be readily accessible to a small, short-tongued insect.
3. Plants must be short or prostrate, or flowers must be close to the ground.
4. Populations must be dense or stands must contain few species.
5. Synchronously blooming flowers must be few per plant, since the presence of many attractive flowers will maximize intraplant rather than interplant foraging and pollen transport.
6. If plants are erect, flowers must be sessile or nearly so. If plants are matted, flowers must be on the mat surfaces.
7. Pollen volume per flower must be small to avoid stimulation of self-grooming.
8. Seeds must be few per flower, since each seed requires at least one pollen transfer.
9. Flowers must be small and of minimal visual attraction.
10. Nectar quantity must be small enough to support ants while discouraging visits by insects whose energy demands while they are foraging exceed the available reward.

One of the reasons most often cited for the rarity of ant pollination is the combination of limited foraging areas and the lack of wings on worker ants; in this situation the ants cannot effect gene flow among the plants they service. However, pollen movement by ants is similar to that of many winged insects (e.g., Peakall, Handel, and Beattie 1991; Gómez and Zamora 1992). But research based on the finding that ant secretions inhibit pollen function has produced interesting results. Beattie et al. (1984, 1985) and Hull and Beattie (1988) demonstrated that pollen exposed for brief periods to ants exhibit reduced viability, reduced percentage of germination, and shorter pollen tubes, relative to control pollen. These effects are produced by myrmicacin, a powerful antibiotic secreted by the ants' metapleural glands. Beattie et al. (1984, 1985) suggested that myrmicacin incidentally kills pollen grains, blocking the evolution of ant pollination systems (but see Peakall, Angus, and Beattie 1990). Nevertheless, pollination by ants has been reported, whether anecdotally or by experimental demonstration (e.g., Hickman

1974; Jones 1975; Petersen 1977a, 1977b; Nilsson 1978; Brantjes 1981; Wyatt 1981; Wyatt and Stoneburner 1981; Svensson 1985; Peakall, Beattie, and James 1987; Peakall 1989; Peakall and James 1989; Peakall and Beattie 1989, 1991; Peakall, Angus, and Beattie 1990; Gómez and Zamora 1992; Ramsey 1995; Gómez et al. 1996; Gómez 2000; Schürch, Pfunder, and Roy 2000). Because the number of examples is not large (16) and the systems studied are relatively different among them, broad generalizations about pollination by ants are really precluded. Of the 16 examples of pollination by ants, 9 (56.3%) are from Spain, 3 (18.8%) from the United States, 1 from Switzerland, and 3 (18.8%) from Australia (appendix 5.1). Except for one of these examples, *Leporella fimbriata* (Orchidaceae), pollination is effected by worker ants. In the following paragraphs we summarize some of the clearer demonstrations of pollination by ants.

Hickman reports that *Polygonum cascadense* (Polygonaceae), a small, apparently self-incompatible annual plant in the United States, is regularly cross-pollinated by workers of the ant *Formica argentea,* which actively forages for nectar in dense, monotypic stands of the plant. The perianth shape brings the ant's head in contact with sticky pollen on the anthers if the protandrous flower is in the "male" stage, and with the receptive stigmas if the flower is more mature ("female stage"). Transfer of pollen grains from anther to ant, and from ant to stigma, was observed in the field. Most pollen is carried on the mouthparts and lower parts of the head, but occasionally grains adhere to antennae, legs, and gaster. No ants were observed to engage in grooming behavior while foraging in *P. cascadense* populations, and all *F. argentea* workers that were collected carried pollen. In the absence of ants (greenhouse experiments), seed set is very low (0 to 7%), while seed set is between 85% and 100% in the field (Hickman 1974).

According to Wyatt and Stoneburner, the flowers of *Diamorpha smallii* (Crassulaceae), a succulent, self-incompatible annual, endemic to granite outcrops in the United States, are visited by workers of *Formica schaufussi* and *F. subsericea,* which exhibit similar behavior. The ants move rapidly from flower to flower seeking for nectar, and an average of 15 to 20 ants can be observed at one time. *Formica* ants carry up to 20 of the sticky pollen grains adhering to hairs and indentations primarily on their thoraces. No pollen grains were observed on the antennae of any ant, possibly due to grooming. The distances moved by ants (1–2 m) were strongly skewed to the right and significantly leptokurtic, indicating that pollen dispersal is concentrated within a small area; but substantial numbers of long-distance dispersal events also occur. However, the highly leptokurtic pattern of ant-mediated pollen dispersal is closely comparable with other observations of plant pollination by animals, such as bees, butterflies, and beetles. Finally, ants were not the only flower visitors, and their efficiency was not compared to that of the flying visitors (honeybees, native bees, flies, butterflies, and a pollen-eating beetle) (Wyatt 1981; Wyatt and Stoneburner 1981).

Leporella fimbriata (Orchidaceae) is an Australian self-compatible orchid that propagates vegetatively, creating large clones, and whose pollination is effected by sexually attracted male winged ants (*Myrmecia urens*), which pseudocopulate with the flower (Peakall, Beattie, and James 1987; Peakall 1989). The ants carry pollen masses on their thorax, and the pollen is later deposited on the stigma. Pollinator movements are usually restricted within a clone. However, selfing was not as high as suggested because there is some outcrossing mediated by ants, and there is extensive seed flow (Peakall and James 1989). The distribution of pollinator movements was leptokurtic (3.141 ± SE 4.59 m). Pollination was widespread but variable from site to site and season to season, with a maximum of 70% of all flowers being pollinated. This system does not fall into the categories of pollination by ants proposed by Hickman (1974). However, males of *M. urens* lack the metapleural glands that secrete pollenicidial substances; nevertheless, they have a strong disruptive effect on pollen. Peakall, Angus, and Beattie explain that during pollination the orchid secures the pollen mass to the ant surface by stigmatic secretions, so that normal pollen function, fruit production, and seed set result. It appears that both ant and plant traits are preadaptive, having evolved for functions other than ant pollination (Peakall, Angus, and Beattie 1990).

Microtis parviflora (Orchidaceae) is effectively pollinated by flightless worker ants of the species complex *Iridomyrmex gracilis* in Australia (Peakall and Beattie 1989). Field and laboratory observations and experiments showed that this orchid is self-compatible but not autogamous and that ant pollination results in very high levels of seed set (table 5.1). Because ants had limited access to flowers during ant treatment in the laboratory, it is possible that the higher seed set for field compared to ant treatment results from field flowers being visited by increased numbers of ants over several days. Ants forage persistently for nectar, visiting individual flowers and inflorescences repeatedly. Foraging patterns resulted in high levels of selfing (51% of pollen transfers)

TABLE 5.1 **Seed set and percentage of seed with normal embryos for ant- (*Irydomyrmex gracilis*), field-, and self-pollinated flowers, and untreated flowers of *Microtis parviflora* in Australia**

Pollination treatment	Seed set (range)	% Normal embryos
Ant (*N* = 15)	881 ± 298 (167–1260)	87 ± 10
Field (*N* = 15)	1519 ± 907* (311–2938)	89 ± 9
Self-pollinated by hand (*N* = 10)	803 ± 268 (428–1239)	83 ± 13
No treatment (*N* = 15)	5 ± 10 (0–30)	—

Source: Modified from Peakall and Beattie 1989.
Note: Unpollinated flowers from which pollinia were removed did not set seed (*N* = 10).
*This mean seed (± SD) set is statistically different from the "Ant" and "Self-pollinated" seed sets (*t* tests).

because they were accomplished within a clone, but leptokurtic distributions of pollinator travel distances (mean 12.4 ± 14.9 cm, maximum of 89 cm) suggest that some pollen transfers result in cross-pollination (Peakall and Beattie 1991). The pollen carried by the ants is unaffected by the antibiotic secretions from the ants' metapleural glands. Pollinia are separated from the ants' integument by a short stalk and are always carried on the frons (fig. 5.2), remote from the metapleural glands. Peakall and Beattie (1989, 1991) showed that wingless worker ants can be efficient pollinators and suggested that the presence of metapleural glands is not necessarily inimical to the evolution of ant pollination, particularly if mechanisms such as stalked pollinia prevent direct contact between ants and pollen. Moreover, in the absence of clonality, it is likely that ant foraging would have yielded a mixed mating system similar to those reported for a wide array of insect pollinators (Peakall and Beattie 1991). This was the first study to show that wingless ants can be exclusive pollinators.

Blanfordia grandiflora (Liliaceae) is a self-compatible perennial herb, pollinated by worker ants of an undetermined species of *Iridomyrmex* in Australia. According to Ramsey, seed set is about 17%, which is significantly greater than autonomous self-pollination (<1%) but less than experimental self-pollination (40%); ants did not cross-pollinate flowers. Differences in anther-stigma separation did not affect seed set. The ants have thoracic metapleural glands, suggesting that antibiotic secretions could inhibit pollen function. However, the difference in germination between pollen grains that had contacted the ants' thorax and control pollen was only 6%. Also, the seed set of flowers that were cross-pollinated with pollen that either had been or had not been in contact with ants did not differ significantly (78.3% vs. 76.3%). Pollen loads of ants were small;

individual ants carried about 28 pollen grains, 50% of which were carried on the legs, and hence away from potential metapleural gland secretions, so that secretions only marginally affected pollen function. Flowers were also visited by honeybees and, more rarely, by birds (which are the common pollinators in other areas). Ramsey (1995) suggests that by assuring reproduction, ant-mediated self-pollination may have been an important factor in the evolution of self-compatibility and may have reduced selection for autonomous self-pollination.

Worker ants of *Proformica longiseta* visit and pollinate the flowers of *Hormathophylla spinosa* (Cruciferae), a mass-flowering woody plant in Spain. Thirty-nine species of insects in 18 families visited the flowers of *H. spinosa;* all were winged visitors, except for *P. longiseta,* which made up more than 80% of the total number of insects found on the flowers, as reported by Gómez and Zamora. These authors quantified the

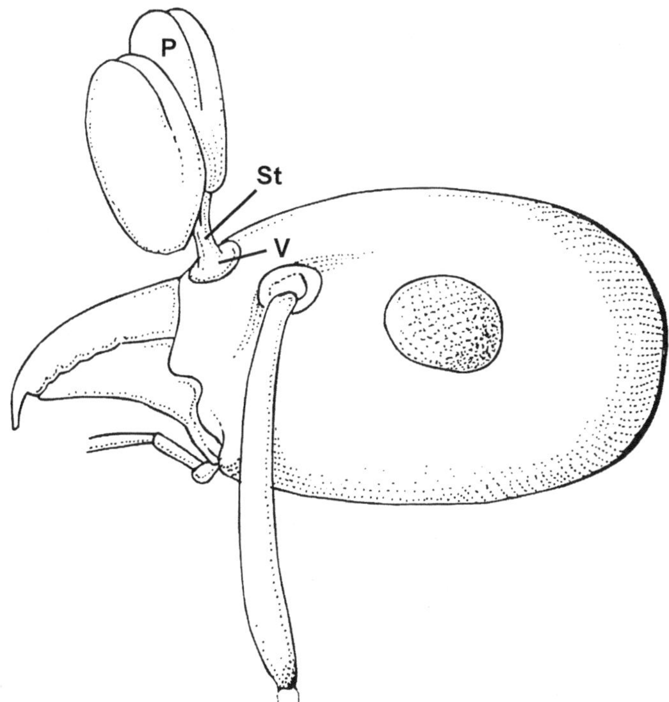

FIGURE 5.2. The mode in which pollinia of *Microtis parviflora* are transported by pollinating workers of *Iridomyrmex gracilis.* P = pollinium; St = stipe; V = viscidium. Modified from Peakall and Beattie 1989.

abundance and foraging behavior of *P. longiseta* and examined the role of ants as true pollinators in comparison with winged flower visitors. All pollinators exhibited similar foraging patterns, with 98% of total movements made between flowers within the same plant. The average distance between flowers of the same plant visited by ants (1.96 ± 1.9 cm) exceeded that of certain winged pollinators visiting the same plant. In fact, no significant differences were found between the distribution of interflower distances of *P. longiseta* and that of all the winged pollinators combined. When foraging for nectar, ants always made contact with the plant's reproductive organs and transferred large numbers of pollen grains. However, pollen exposed to ants for brief periods exhibited significantly reduced percentage germination. The percentage of fruit set in *H. spinosa* mainly depends on the effect of winged insects, whereas the percentage of fertilized ovules that became viable seeds in each fruit (brood size) depended mainly on the effect of ants (20.4% vs. 39.5%). *P. longiseta* is both the most abundant and the spatio-temporally most predictable flower visitor of *H. spinosa*. These characteristics, weighted by their flower visitation rate, make the worker ants the pollinators that maintain the strongest interaction with *H. spinosa*. The key factor is the great density of worker ants throughout the flowering period of the plant; thus the interaction basically depends on its high probability of occurrence (Gómez and Zamora 1992).

In order to determine plant and ant traits and the composition and abundance of the pollinator assemblage, Gómez et al. studied seven plant species in the Mediterranean highlands and arid lands of southeastern Spain. The flowers of these plant species were visited by insects from 29 families, including Formicidae, in five orders. In the Mediterranean highlands, *Alyssum purpureum* (Cruciferae), *Arenaria tetraquetra* (Caryophyllaceae), and *Sedum anglicum* (Crassulaceae) were visited by the ants *Proformica longiseta* and *Tapinoma nigerrimum*. In the arid lands, *Lepidium subulatum* (Cruciferae) and *Gypsophyla struthium* (Caryophyllaceae) were visited mainly by the ants *Crematogaster auberti, Plagiolepis schmitzii,* and *Camponotus foreli,* whereas the plants *Frankenia thymifolia* (Frankeniaceae) and *Retama sphaerocarpa* (Fabaceae) were visited by the ants *Leptothorax fuentei* and *Camponotus* sp. In all but two, *L. subulatum* and *G. struthium,* the ants accounted for 70% to 100% of the flower visits. The two *Camponotus* species and *L. fuentei* lack metapleural glands, and pollen was not found on the bodies of *C. auberti, P. schmitzii,* and *T. nigerrimum*. Results clearly showed that flying insects acted as pollinators for every plant species studied, although in the five species pollinated by ants, the pollination efficiency of winged insects was similar to the efficiency of ants. Moreover, the quantity of seeds produced without any pollinator was much lower than the quantity produced in the five species pollinated by ants. The role of the ants as pollinators seems to depend heavily on the relative abundance of the ants with respect to the other species on the pollinator assemblage, ant pollination becoming evident when ants outnumber other floral visitors. Gómez et al. (1996) suggest that this ant-pollination system is the result of prevailing ecological conditions more than an evolutionary result of a specialized interaction.

Another Cruciferae in Spain, *Lobularia maritima,* is also pollinated by ants (Gómez 2000). As in above examples, ants are not the exclusive flower visitors. *Camponotus micans* is one of more than 50 species of pollinators belonging to 30 families that visited the flowers of *L. maritima* during a two-year study. There is a significant seasonal variation in insect abundance, and ant visitation represented 81.2% of flower visitors during the summer. The overall effect of the ant *C. micans* is as high as that of winged insects and contributes both to seed production and to recruitment of new flowering offspring (fig. 5.3). A similar system is that of *Euphorbia cyparissias* (Euphorbiaceae) in Switzerland (Schürch, Pfunder, and Roy 2000), where the plant is visited by an array of insects including the ants *Formica cunicularia, F. pratensis,* and *Lasius alienus.* It was shown, using insect exclusions, that the ants were effective pollinators of *E. cyparissias* and that the seed set of plants exclusively visited by ants had a similar reproductive success compared to plants visited by all types of insects (Schürch, Pfunder, and Roy 2000).

Workers of *Formica neorufibarbis gelida* visit the flowers of three plant species inhabiting the alpine vegetation of Colorado, creating a very interesting system of mutualistic and antagonistic interactions. Puterbaugh indicates that the ants pollinate the gynodioecious *Paronychia pulvinata* (Caryophyllaceae), are herbivores of the gynodioecious *Eritrichum arentioides* (Boraginaceae), and appear to have little effect on the hermaphroditic *Oreoxis alpina* (Apiaceae). Given that floral nectar and lipids are important resources for alpine ants, the effect of all three plant species on the ants is positive. In the ant-pollinated *P. pulvinata,* ants effect pollination in females more than in hermaphrodites; however, ants are not the only flower visitors. Ants appeared to contact the stigma during nearly all flower visits to *P. pulvinata,* and stigmas of pistillate

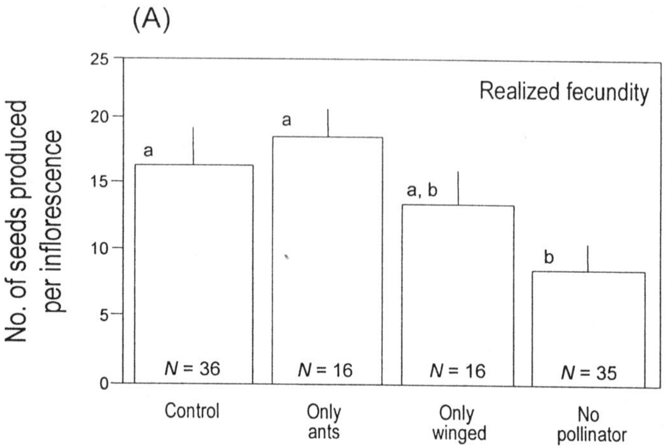

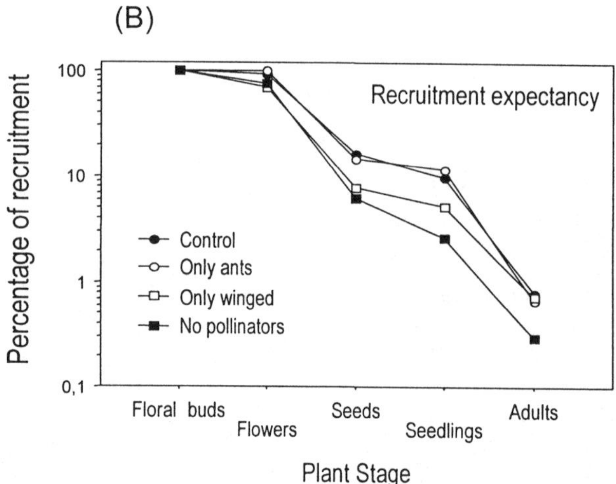

FIGURE 5.3. Effect of ant pollinators on seed production and seedling performance in *Lobularia maritima* in Spain. (A) Plant fecundity is significantly affected by the exclusion of pollinators, with an effect due to the presence or absence of ants. Pollinator-excluded flowers produced fewer seeds compared to controls, whereas those pollinated only by ants or only by winged insects produced as many seeds as did controls. Data are means ± SE; bars with different letters are statistically different (post hoc Tukey-Kramer HSD test). (B) Recruitment expectancies for *L. maritima* seeds depend on the pollination regimes. Ants and winged insects had similar overall effects on host-plant recruitment probability (0.6% and 0.7% of initial ovules produced flowering offspring, respectively), similar to that of open-pollinated flowers (1.0%; Dunnett's tests). Final recruitment probability was much lower in the pollinator-excluded, autogamous-pollinated flowers. Modified from Gómez 2000.

flowers that were visited once by an ant obtained significantly more pollen grains than unvisited flowers. Of those ant-visited flowers with pollen grains on the stigma, an average of 75% of the grains germinated, sending tubes down to the style. Ant-visited flowers also had more pollen tubes per style than control flowers. Only 40% of all ant-visited flowers received pollen that germinated and sent tubes down the style. As a consequence, the frequency of flowers having pollen tubes in the style after an ant's single visit was bimodal; the average numbers of pollen grains per stigma and pollen tubes per style for these ant-visited flowers were 2.48 ± 0.84 and 1.55 ± 0.42, respectively. Single ant visits to unpollinated pistillate flowers show that ants are inefficient pollinators, transferring no pollen (viable or inviable) to stigmas 55% of the time. Low pollen loads and infrequent movement between plants probably explain this result. In contrast, single ant visits to previously unpollinated perfect flowers on hermaphrodites produce seed-set similar to that in flowers open to pollination (Puterbaugh 1998).

Ants frequently visit flowers but rarely pollinate them (Peakall, Handel, and Beattie 1991; Rico-Gray 1989, 1993). Flower-visiting ants have received less attention than other insect flower visitors, despite the fact that the relationships between these ants and plants are rich in the diversity of fitness outcomes for plants (Puterbaugh 1998). Even though pollination by ants can be specialized (e.g., pollination of the orchid *Leporela fimbriata* by pseudocopulation by ants; see Peakall 1989), in most cases ant-pollinated flowers are also simultaneously visited by an array of other insects (e.g., Gómez and Zamora 1992; Gómez et al. 1996; Gómez 2000; Schürch, Pfunder, and Roy 2000). With regard to those cases, the traits conducive to pollination by ants (flower shape and size) may be an adaptation to small insect pollinators in general rather than ants specifically. One generalization that has arisen from these studies is that effective ant-pollinators often show high flower visitation frequency (e.g., Gómez and Zamora 1992; Gómez et al. 1996; Gómez 2000). However, flower-visiting ants have been mostly ignored; more thorough research is needed to demonstrate that ants are quite common pollinators (Beattie 1991). Hickman (1974) predicted that pollination by ants should be most common in hot and dry habitats because ants are usually very active in these environments. Later studies have documented that nectar can be an important liquid source to ants in deserts (Ruffner and Clark 1986) and dry tropical environments (Rico-Gray 1989, 1993), also suggesting that ant visitation and pollination might be more frequent in dry habitats.

Most published cases of pollination by ants involve plant species living in dry habitats such as deserts and arid montane, sandy, and rocky regions (Petersen 1977a, 1977b; Wyatt 1981; Svensson 1985; Gómez and Zamora 1992; García, Antor, and Espadaler 1995; Gómez et al. 1996), supporting Hickman's (1974) view. The few experimental studies that have been done in the North American alpine vegetation show that flower-visiting ants actually represent predators or parasites, rather than pollinators (Galen 1983). However, Puterbaugh (1998) has shown that ants do act as pollinators in the same type of environment, which supports the idea of a close association between mutualistic and antagonistic interactions (chapter 1).

Given the ability of ants to generate pollen flow, the reasons for the rarity of ant pollination appear to lie elsewhere (Peakall and Beattie 1991). Not all ant species have metapleural glands whose secretions disrupt pollen. It seems that plant pollination by ants is easier if plants have small, actinomorphic, hermaphroditic flowers, with easily accessible nectaries (Gómez and Zamora 1992) and are located in dry regions. Some ant-pollination systems may be the result of prevailing ecological conditions rather than an evolutionary result of a specialized interaction (Gómez et al. 1996).

Discouragement of Floral Visits by Ants

Janzen (1977, 1983) stated that in tropical lowlands there is a general failure of ants to forage nectar from floral nectaries. He hypothesized that the answer lies in nectar toxicity, difficulty in entering flowers, the "on/off" nature of nectar flow, nectar dilution, and the inadequacy of nectar as a pure diet. Nevertheless, several examples of floral-nectar foraging by ants have been described, and these authors tend to think that the lack of evidence is due to the limited amount of observation (Baker and Baker 1978; Feinsinger and Swarm 1978; Schubart and Anderson 1978; Rico-Gray 1980, 1989, 1993; Guerrant and Fiedler 1981; Haber et al. 1981; Rico-Gray, García-Franco, et al. 1998; Rico-Gray, Palacios-Rios, et al. 1998).

Ants are typically considered to be robbers of floral nectar that decrease plant fitness (e.g., McDade and Kinsman 1980; Fritz and Morse 1981; Wyatt 1981; Galen 1983; Herrera, Herrera, and Espadaler 1984; Norment 1988). Some evolutionary trends in floral morphology that have been tied to a decrease in the range of effective pollinators also have been thought to increase plant adaptedness by excluding nonpollinating nectarivores such as ants (Guerrant and Fiedler 1981; Stephenson 1981; Herrera, Herrera, and Espadaler 1984). For instance, as reported by Ghazoul, two acacia-ant mutualists are repelled by floral tissue chemicals from their own host plants as well as those from 13 other plant genera, whereas 13 of 25 ant species from several subfamilies are repelled by acacia floral chemicals. He suggests that floral ant repellents are widespread among plants, repel most ant species, and can prevent ants from parasitizing plant-pollinator mutualisms (Ghazoul 2001). Nevertheless, the high number (up to 40%) of plant species in certain habitats whose flowers are visited by ants foraging for nectar (Rico-Gray 1989, 1993; Rico-Gray, García-Franco, et al. 1998; Rico-Gray, Palacios-Rios, et al. 1998) should stimulate more focus on the role of ants as consumers of floral nectar. If ants were merely robbing nectar, there should be a considerable decrease in fitness of many individual plants. Alternatively, besides possible pollination, other subtle mutualistic interactions (e.g., related to herbivore deterrence) could be taking place. Furthermore, ants are also discouraged from visiting flowers if they drive away potential pollinators as they seek flower nectar. In *Acacia hindsi* in western Mexico, this possible conflict is prevented by a spatial segregation of resources (those exploited by the ants are predominantly concentrated on newly growing shoots, while inflorescences are present only on shoots from the previous year) and by an ant-repellent stimulus from young inflorescences (Raine, Willmer, and Stone 2002; see also Ghazoul 2001).

Although ants are present in most communities, there are few confirmed cases of pollination by ants. Published work suggests that natural selection has not favored ant pollination in many habitats, but has favored avoidance of nectar thievery or conflicts among pollinators and ants. However, ants do visit flowers in various environments, and more work is needed to clarify the nature of these interactions. Based on Hickman's (1974) summary of the characteristics of ant pollination systems, future studies should (1) search for more examples of ant-pollinated plants so that a broad generalization can be approached; (2) analyze the basis (chemical and physical) of the discouragement of floral visits by ants and its ecological consequences; and (3) examine in more detail the conflicts among pollinators and ant-guards and determine the effect of such conflicts on plant fitness.

APPENDIX 5.1 **Plant species in which ant pollination has been demonstrated**

Plant	Growth form	Location	Associated ant species	Source
Caryophyllaceae				
Arenaria tetraquetra	herb	Spain	*Proformica longiseta, Tapinoma nigerrimum*	Gómez et al. 1996
Gypsophyla struthium	herb	Spain	*Camponotus foreli, Crematogaster auberti, Plagiolepis schmitzii*	Gómez et al. 1996
Paronychia pulvinata	herb	USA	*Formica neorufibarbis*	Puterbaugh 1998
Crassulaceae				
Diamorpha smallii	herb	USA	*Formica schaufussi, F. subsericea*	Wyatt 1981; Wyatt and Stoneburner 1981
Sedum anglicum	herb	Spain	*Proformica longiseta, Tapinoma nigerrimum*	Gómez et al. 1996
Cruciferae				
Alyssum purpureum	herb	Spain	*Proformica longiseta, Tapinoma nigerrimum*	Gómez et al. 1996
Hormathophylla spinosa	shrub	Spain	*Proformica longiseta*	Gómez and Zamora 1992
Lepidium subulatum	herb	Spain	*Camponotus foreli, Crematogaster auberti, Plagiolepis schmitzii*	Gómez et al. 1996
Lobularia maritima	herb	Spain	*Camponotus micans*	Gómez 2000
Euphorbiaceae				
Euphorbia cyparissias	herb	Switzerland	*Formica cunicularia, F. pratensis, Lasius alienus*	Schürch, Pfunder, and Roy 2000
Fabaceae				
Retama sphaerocarpa	shrub	Spain	*Camponotus* sp., *Leptothorax fuentei*	Gómez et al. 1996
Frankeniaceae				
Frankenia thymifolia	shrub	Spain	*Camponotus* sp., *Leptothorax fuentei*	Gómez et al. 1996
Liliaceae				
Blanfordia grandiflora	herb	Australia	*Iridomyrmex* sp.	Ramsey 1995
Orchidaceae				
*Leporella fimbriata**	herb	Australia	*Myrmecia urens*	Peakall, Beattie, and James 1987; Peakall 1989; Peakall and James 1989; Peakall, Angus, and Beattie 1990
*Microtis parviflora**	herb	Australia	*Iridomyrmex gracilis*	Peakall and Beattie 1989, 1991
Polygonaceae				
Polygonum cascadense	herb	USA	*Formica argentea*	Hickman 1974

*Except for these two cases, the flowers of all plant species were also visited by organisms other than ants.

CHAPTER SIX

Antagonism and Mutualism

Direct Interactions

Although the origin of some mutualisms (e.g., pollination and seed dispersal) mostly involves a change in the outcome of the interaction (from antagonism to mutualism), other mutualisms are built on interactions involving at least an antagonistic pair of species and a mutualistic pair of species. Most studies, however, consider directly the interaction between two species, even though the evolutionary unit of many mutualisms involves at least three species in a way that emphasizes the evolutionary relationships between antagonism and mutualism (Thompson 1982). This is the case for two broad classes of ant-plant interactions: (1) direct ant-plant interactions, in which the plants provide an array of resources (e.g., food and domatia), and (2) indirect ant-plant interactions, which are mediated by honeydew-producing hemipterans (or lepidopteran caterpillars) (see chapter 7). Most direct ant-plant interactions are based on the array of resources provided as rewards by the plant. These range from food resources such as extrafloral nectar and a variety of food bodies, to domatia or nesting sites, in exchange for protection against herbivores (e.g., Koptur 1984; Rico-Gray and Thien 1989b) and/or encroaching vines (e.g., Davidson, Longino, and Snelling 1988; Suarez, de Moraes, and Ippolito 1998). In many associations, however, not one but a combination of these resources may be involved, and differences in specialization and specificity may be resource-based.

In this chapter we first describe the now-classic case of the *Pseudomyrmex-Acacia* association, in which plants offer inquiline ants a whole array of resources (i.e., food bodies, extrafloral nectar, and domatia) in exchange for defense against herbivores and competitors. We then review and discuss other interactions between ants and plants in which plants offer both food bodies and domatia, or only domatia, and follow with those interactions between ants and plants in which plants offer mainly extrafloral nectar (a few also provide domatia). Finally, we address plant defensive strategies, induced responses, and the nature of conditionality.

The *Pseudomyrmex-Acacia* Association

The best-known and most widely used example of ant-plant mutualism in which a plant offers all categories of resources (i.e., extrafloral nectar, Beltian food bodies, and domatia in hollow thorns) in exchange for defense from herbivores and encroaching vines, is the interaction, once thought to be species-specific, between the swollen-thorn *Acacia cornigera* (Fabaceae) and its ant inhabitant *Pseudomyrmex ferrugineus* (Pseudomyrmecinae) in eastern Mexico (Janzen 1966, 1967a, 1967b, 1969b, 1973b) (fig. 6.1). This ant species, however, may interact with 10 *Acacia* species (Keeler 1981c; Ward 1993). The *Acacia-Pseudomyrmex* association has been described in many text books and reviews on ant-plant interactions and has stimulated a number of studies on *Acacia* and its associated ants (e.g., Knox et al. 1986; Ward 1991; Willmer and Stone 1997; Young, Stubblefield, and Isbell 1997; Suarez, de Moraes, and Ippolito 1998), as well as on many other ant-plant systems (see also chapter 1).

The genus *Acacia* is comprised of ca. 1,500 species of shrubs and trees worldwide, but ant-inhabited acacias seem to have evolved separately in the tropics of America and Africa (Keeler 1989). Some acacias are adapted for feeding as well as housing acacia ants (Janzen 1966, 1967b). The ants feed on the solution of water, sugars, and amino acids secreted by the nectaries located at the petiole and on protein and lipids from the Beltian bodies (see below) produced by the plant at the leaf tips (fig. 6.1). The ant colony is distributed among the numerous, greatly enlarged thorns of a tree, which the ants hollow. Worker ants patrol the tree 24 hours per day, both guarding the colony against predators and searching for food. Since the tree contains their nest, obligate acacia ants

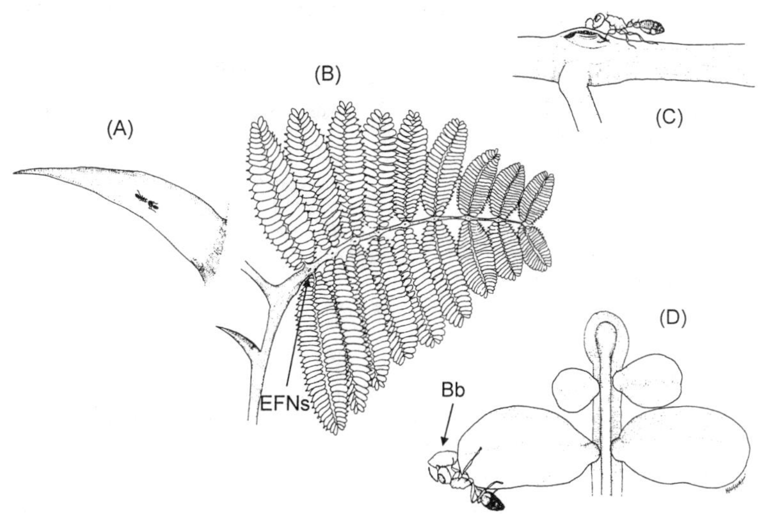

FIGURE 6.1. *Acacia cornigera* and *Pseudomyrmex ferrugineus.* (A) Ants on the swollen thorns. (B) *Acacia* branch showing extrafloral nectaries (EFNs) along its rachis. (C) Ant foraging for extrafloral nectar. (D) Ant foraging for a Beltian food body (Bb).

react strongly to any disturbance of the tree. If alarmed, foragers will recruit others from within the thorns to attack the intruder, pouring out of the thorns to attack, defending the acacia by both biting and inflicting a painful sting on any animal in contact with the tree (Janzen 1967b). The ants keep the plant free not only of insect and vertebrate herbivores, but of fungi and other plants as well. The latter is an unusual behavior in ants but an important adaptation of *Pseudomyrmex* ants that prevents small trees from being overwhelmed and killed by vines (Janzen 1966, 1967b). The colonies of *Pseudomyrmex* may contain more than 30,000 ants (up to 1.8–3.6 million workers in some polygynous species [see Janzen 1973b]) and may spread over several trees, which the ants of a single colony reach by crossing the ground.

Not all ant species living on ant acacias are equally aggressive defenders of the plant, and many arboreal ant species will inhabit swollen-thorn acacias if acacia ants do not exclude them (Janzen 1966). Ant inhabitants of acacias include obligate highly adapted *Pseudomyrmex* species that vigorously protect the plant, obligate or facultative *Pseudomyrmex* species that provide less effective or no defense at all, and species from other genera (e.g., *Camponotus* and *Crematogaster*) that inhabit the thorns but do not defend the plant (Ward 1991; Raine et al. 2004). For

TABLE 6.1 **Traits in swollen-thorn *Acacia* plants directly related to the ant-acacia coevolution**

A. General features	B. Specialized features
1. Woody shrub or tree life form	1. Woody but with very high growth rate
2. Reproduce from suckers	2. Rapid and year-round sucker production
3. Moderate seedling and sucker mortality	3. Very high unoccupied seedling and sucker mortality
4. Plants of dry areas	4. Plants of moister areas
5. Ecologically widely distributed	5. Ecologically very widely distributed
6. Leaves shed during dry season	6. *Year-round leaf production
7. Shade-intolerant, sometimes covered by vines	7. Shade-intolerant and free of vines
8. Stipules often persistent (thorns)	8. *Stipules longer persistent, woody with soft pith
9. Bitter-tasting foliage	9. Bland-tasting foliage
10. Each species with a group of relatively host-specific phytophagous insects, able to feed in the presence of the physical and chemical properties of the acacia	10. Each species with a few host-specific phytophagous insects, able to feed in the presence of the ants
11. Foliar nectaries	11. *Very enlarged foliar nectaries
12. Compound unmodified leaves	12. *Leaflets with tips modified into Beltian bodies
13. Flowers insect-pollinated, outcrossing	13. Same as A 13
14. Seeds dispersed by water, gravity, and rodents	14. Seeds dispersed by birds
15. Lengthy seed maturation period	15. Same as A 15
16. Not dependent upon another species for survival	16. Dependent upon another species for survival

Source: Modified from Janzen 1966.
*Traits regarded as essential for the interaction.

instance, the obligate resident *Pseudomyrmex nigropilosa* has been reported as a nonprotective parasite of the ant-acacia system in Mexico, and ant-occupied plants can be soon killed from severe herbivore attack (Janzen 1975). Also in Mexico (Los Tuxtlas, Veracruz), *Camponotus planatus* can parasitize the *Pseudomyrmex ferrugineus–Acacia mayana* association by inhabiting the swollen thorns and harvesting extrafloral nectar from the plant. However, as opposed to the host-tree effective defense provided by *P. ferrugineus, C. planatus* does not attack phytophagous insects and appears ineffective as an ant guard (Raine et al. 2004).

For several *Acacia* and *Pseudomyrmex* species, the interaction is an obligate mutualism, because acacia ants are found only on ant acacias, and the latter do not survive if deprived of defending ants (Janzen 1966). The origin of the mutualism is unknown, but it clearly has evolved over a long period of time, since a number of characteristics of both acacias (table 6.1) and ants (table 6.2) seem to have evolved in response to the interaction (Janzen 1966). For example, (1) acacias associated with *Pseu-*

TABLE 6.2 **Traits in *Pseudomyrmex* ants related to the ant-acacia coevolution**

A. General Features	B. Specialized Features
1. Fast and agile, not aggressive	1. Very fast and agile, aggressive
2. Good vision	2. Same as A 2
3. Independent foragers	3. Same as A 3
4. Smooth sting, barbed sting sheath not inserted	4. Smooth sting, barbed sting sheath often inserted
5. Lick substrate, form buccal pellet	5. Same as A 5
6. Prey items retrieved entire	6. Same as A 6
7. Ignore living vegetation	7. Maul living vegetation contacting the acacia
8. Workers without morphological castes	8. Same as A 8
9. Arboreal, highly mobile colony	9. Same as A 9
10. Larvae resistant to mortality by starvation	10. Same as A 10
11. One queen per colony	11. Sometimes more than one queen per colony
12. Colonies small	12. Colonies large
13. Diurnal activity outside nest	13. 24-hour activity outside nest
14. Few workers per unit plant surface	14. Many workers active on small plant surface area
15. Discontinuous food sources and unpredictable new nest site	15. Continuous food source and predictable nest sites
16. Founding queens forage far for food	16. Founding queens forage short distances for food
17. Not dependent on another species	17. Dependent on another species group

Source: Modified from Janzen 1966.
Note: Unless otherwise indicated, traits refer to worker ants.

domyrmex ants do not possess a significant chemical defense arsenal and depend heavily on ant activity (a similar pattern is observed in myrmecophytic and nonmyrmecophytic *Macaranga* [Euphorbiaceae] species [see Nomura, Itioka, and Itino 2000; Itino and Itioka 2001]); (2) most *Pseudomyrmex* species are arboreal ants that sting, but only obligate acacia-inhabiting ants show aggressive defense of the tree and insert the sheath as well as the stinger when stinging; and (3) most acacias have thorns and petiolar nectaries, but only ant acacias have swollen thorns and protein-containing Beltian bodies (fig. 6.1, tables 6.1, 6.2; Heil, Baumann, et al. 2004). Finally, even though leaf damage induces nectar secretion in various *Acacia* species, those species obligately inhabited by mutualistic ants nourish them by secreting extrafloral nectar constitutively at high rates that are not affected by leaf damage. The phylogeny of the genus *Acacia* indicates that the inducibility of extrafloral nectar is the pleisomorphic state, whereas the constitutive extrafloral nectar flow is derived. Thus a constitutive resistance trait evolved from an inducible one in response to particular functional demands (Heil, Greiner, et al. 2004). It has been recently shown by Heil, Rattke, and Boland that the control of the chemical composition of extrafloral nectar by ant-inhabited acacias plays a crucial role in excluding less desirable ant partners. Because myrmecophytic *Acacia* produces extrafloral nectar with high invertase (sucrose-cleaving) activity, the food reward contains no sucrose and is thus unattractive to nonsymbiotic ants. *Pseudomyrmex* ants that are specialized to live on *Acacia,* however, do not show invertase activity in their digestive tracts and prefer sucrose-free nectar. These findings demonstrate that postsecretory regulation of the carbohydrate composition of nectar contributes to specialization in the ant-acacia mutualism and illustrates how the absence of a simple and common compound such as sucrose can be used to exclude less profitable partners and to stabilize a specific symbiotic mutualism (Heil, Rattke, and Boland 2005).

Ants, Plants, and Food Bodies

Food bodies comprise a variety of small epidermal structures that have been interpreted as adaptations to attract ant foragers (Beattie 1985). These structures in higher plants are so diverse in form and function that food bodies constitute a nondiscrete and nonuniform category, and alternative interpretations are possible (O'Dowd 1982; Beattie 1985). For example:

1. The Beltian bodies of *Acacia* are vascularized and contain high amounts of protein (8%–14% dry mass), in addition to lipids (1%–9% dry mass) and carbohydrates (3%–11% dry mass) (Rickson 1969, 1975; Heil, Baumann, et al.2004).
2. The Müllerian bodies of *Cecropia* (Cecropiaceae) contain 39% glycogen (glucose residues normally found in animal tissues) and 8% lipids (Rickson 1971, 1973, 1976).
3. The Beccarian bodies of *Macaranga* (Euphorbiaceae) are especially rich in lipids (Rickson 1980), as well as the food bodies of *Piper* (Piperaceae), which contain 46% lipid by weight (Rickson and Risch 1984).
4. Pearl bodies in general are rich in lipids (Rickson 1980; O'Dowd 1982); however, the pearl bodies of *Ochroma* (Bombacaceae) are particularly rich in sterols (O'Dowd 1980). Water and carbohydrates (67% dry mass) are the main constituents of the pearl bodies of *Urera* (Urticaceae) (Dutra, Freitas, and Oliveira 2006).

Recent research has demonstrated that the chemical contents of food bodies may undergo evolutionary change as an adaptation to their role in ant attraction and may come to match special nutrition requirements of the consumers, which seems like a similar strategy to that of elaiosome production and chemical contents (chapter 3). For instance, the food bodies of myrmecophytic and nonmyrmecophytic *Macaranga* species are specifically adapted to their respective roles in ant attraction and nutrition (Heil et al. 1998). Myrmecophytic plants provide ants with high amounts of lipids (up to 40% in dry tissue) and proteins, and between 24% and 46% of the soluble sugar fraction is composed of a variety of sugars and sugar alcohols (arabinose, mannitol, and nine mono- or oligosaccharides). In nonmyrmecophytic *Macaranga* species, however, 85% of the soluble sugar fraction in their food bodies is mainly composed of common sugars (glucose, fructose, and maltose). In addition, *Acacia* food bodies contain all amino acids and all fatty acids that are considered essential for insects, and they contain more protein and lipids than the leaves from which they are derived, indicating an adaptive enrichment of nutritionally valuable compounds in structures functioning as ant food (Heil, Baumann, et al. 2004). Species in the genera *Macaranga, Piper,* and *Cecropia* are among the main representatives of plants producing food bodies.

Macaranga *(Euphorbiaceae)*

In the humid tropics of Southeast Asia, more than 23 species of the very diverse pioneer tree genus *Macaranga* are myrmecophytic and exhibit a variety of associations with ants, from facultative, nonspecific interactions (in species with solid stems) to protection by specialized plant-ants (in species with hollow stems), most in the genus *Crematogaster* (*Decacrema*) but also in a few other genera such as *Camponotus* (Smith 1903; Fiala et al. 1989, 1994, 1999; Fiala, Maschwitz, and Pong 1991; Fiala and Maschwitz 1992; Federle, Maschwitz, and Fiala 1998; Itioka et al. 2000; Feldhaar, Fiala, Hashim, and Maschwitz 2003). *Crematogaster borneensis* is highly specialized and is associated with several *Macaranga* species, but the current state of the taxonomy of this *Crematogaster* group does not allow for species-specific inferences (Fiala, Maschwitz, and Pong 1991; Feldhaar, Fiala, Gadau, et al.2003). By contrast, *Camponotus* (*Colobopsis*) sp. 1 is specifically associated with *Macaranga puncticulata* (Federle, Maschwitz, and Fiala 1998). In many Southeast Asian myrmecophytic *Macaranga* species, the epicuticular wax crystals that cover the stem surface of the plant act as an ecological isolation mechanism for the symbiotic ants. Because only the specific ant partners of glaucous *Macaranga* host plants (*Crematogaster* and *Camponotus* species) are able to walk on the slippery stems without difficulty, the epicuticular coatings of myrmecophytic species seem to have a selective role and protect the ant inhabitants against competitors (Federle et al. 1997).

The ants nest inside the hollow internodes and feed mainly on the food bodies provided by the plants. A series of experimental field studies have shown that the ants protect their host plants effectively from herbivores (sap-sucking insects, leaf miners, beetles, lepidopteran larvae, stem-borers, and gallers) and plant competitors (mainly vines) (Fiala et al. 1989, 1994, 1999). The degree to which ant inhabitants depend on the honeydew produced by associated hemipteran trophobionts for nutrition may vary widely among ant species (Itino et al. 2001). The patrolling behavior of the ants results in the removal of potential herbivores, and their pronounced aggressiveness and mass recruitment system enable the ants to defend the host plant against many types of herbivorous insects, with a significant decrease in leaf damage on ant-occupied compared to ant-free myrmecophytic and nonmyrmecophytic *Macaranga* species (Fiala et al. 1989). The intensity of ant defense, however, can vary among sympatric species of myrmecophytic *Macaranga* (Itioka et al. 2000). Moreover, the ants bite off any foreign plant part coming into contact with their host plant, and thus both ant-free myrmecophytic and nonmyrmecophytic *Macaranga* species had a significantly higher incidence of vine growth than individuals with active ant colonies (Fiala et al. 1989). Food-body production in myrmecophytic *Macaranga* species is usually concentrated on protected parts of the plants such as recurved stipules, whereas in nonmyrmecophytic species, food bodies are produced on leaves and stems and are collected by a variety of ants. The levels of food-body production differ not only between facultatively and obligatorily ant-associated species, but also among the various nonmyrmecophytic species. This may be related to the degree of interaction with ants, since food-body production is regulated and maintained at high rates only when ants are present; i.e., it is induced by ant presence (Heil et al. 1997).

Food-body production in the myrmecophytic species starts at a younger age than in the transitional or nonmyrmecophytic *Macaranga*. Food bodies of noninhabited *Macaranga* species are collected by a variety of ants, with no evidence of species-specific association (Fiala and

Maschwitz 1992); the degree of protection is quite variable, with only one out of three plant species showing significantly reduced herbivore loads and leaf-damage levels when ants were experimentally excluded from branches (Mackay and Whalen 1991). Extrafloral nectaries and food bodies are widespread in *Macaranga* and were important prerequisites for the evolution of the association with ants (Fiala and Maschwitz 1991, 1992). However, since both still occur simultaneously on nonmyrmecophytes, the presence of these two kinds of rewards alone does not explain myrmecophytism (but see Fiala, Maschwitz, and Linsenmair 1996). Although production of food bodies may have enhanced the evolution of myrmecophytism in *Macaranga*, the availability of protected nest sites on the host plant may have been the decisive additional factor leading to the obligatory associations (Fiala and Maschwitz 1992).

Cecropia *(Cecropiaceae)*

Species in the genus *Cecropia* are very successful early successional trees in tropical forests of Latin America, and most (ca. 100 species) host symbiotic ants (De Andrade and Carauta 1982; Davidson and Fisher 1991). All myrmecophytic *Cecropia* provide similar benefits to ants. Early in sapling growth, hollow stems expand and internodes develop small areas of unvascularized tissue (prostomata), where queens enter stems without rupturing phloem and flooding internodes with mucilage (Sagers, Ginger, and Evans 2000). Ant colonies inhabit these stems, and emerging workers collect the glycogen-rich Müllerian bodies developing at the bases of petioles on hairy platforms called trichilia (Rickson 1971, 1973, 1976). *Cecropia* plants also produce pearl bodies on the abaxial surface of new leaves (Sagers, Ginger, and Evans 2000).

Production of Müllerian bodies varies in *Cecropia*. For instance, Folgarait, Johnson, and Davidson have shown that *Cecropia* sp. B responds to Müllerian body removal (simulating ant presence) by increasing the production rate of new Müllerian bodies and that this response depends on the nutrient levels under which the plants are grown in greenhouse conditions. It is suggested that the plants may economize resources by controlling the production of Müllerian bodies and that benefits of this control may extend to prevention of fungal colonization on active, unattended trichilia (Folgarait, Johnson, and Davidson 1994). Also in greenhouse experiments, the carbon-based defenses (tannins and phenolics) of *Cecropia* reached higher concentrations at low nutrient levels, while the production of glycogen-rich Müllerian bodies increased with greater levels of both light and nutrients, according to Folgarait and Davidson. However, lipid- and amino acid–rich pearl bodies were produced in greater numbers under conditions of low light and high nutrient levels (Folgarait and Davidson 1994, 1995). Antiherbivore defenses in *Cecropia* vary with leaf age and habitat. Folgarait and Davidson found that mature leaves are mainly protected by chemical (tannins and phenolics) and physical (leaf toughness) defenses, whereas new leaves are protected by biotic defenses (ants). Plant investment in immobile defenses (tannins, phenolics, and leaf toughness) and in a defense with high initial construction costs (differentiation of trichilia to produce Müllerian bodies) were greater in slow-growing *Cecropia* typical of small openings in primary forest than in closely related and fast-growing pioneer *Cecropia* species of large-scale disturbances, where leaf-expansion rates and pearl-body production were greater (Folgarait and Davidson 1995).

Species in the ant genus *Azteca* (Dolichoderinae) predominate on fast-growing *Cecropia* in open, sunny environments (Janzen 1969a; De Andrade and Carauta 1982; Schupp 1986; Longino 1991). Trees growing slowly in shaded habitats are often occupied by specialized *Cecropia* ants in other genera, including *Pachycondyla* (Ponerinae), *Camponotus* (Formicinae), and *Crematogaster* (Myrmicinae) (Davidson, Snelling, and Longino 1989; Davidson et al. 1991; Davidson and Fisher 1991; Longino 1991). Some *Azteca* species are obligately associated with *Cecropia* and are considered competitively superior to nonobligate ants; however, there seems to be no host plant specificity (Longino 1989). Yu and Davidson (1997) analyzed the relative roles of host- and habitat-specificity in determining the match between *Cecropia* trees and their obligate ants in Peruvian rain forests and reported that pairings between *Cecropia* species and their ant inhabitants seem to be determined mainly by coincident habitat affiliations of ants and plants and by preadapted capacities of ants to distinguish among host-plant species.

Cecropia seedlings receive significant protection from at least some ant associates that repel insect herbivores (Downhower 1975; Schupp 1986) and/or remove encircling vines (Janzen 1969a; Davidson, Longino, and Snelling 1988). Plant protection varies depending on the size of the ant colony, and larger colonies have more individuals available to patrol a plant and recruit defenders against herbivores. Rocha and Bergallo (1992) demonstrated that larger colonies of *Azteca muelleri* inhabiting *Cecropia pachystachya* in Brazil recruit larger numbers of ants and are quicker to discover and dislodge the herbivore *Coelomera ruficornis* (Coleoptera) from plants; they also detected a negative and significant relationship between herbivore damage on leaves and ant colony size. It has been suggested that when species of *Cecropia* inhabit areas with relatively low herbivore pressure (e.g., islands and high elevations), the plants lose traits that serve as ant attractants, such as the production of trichilia that produce the glycogen-rich Müllerian bodies (Janzen 1973a; Rickson 1977; Wetterer 1997). Moreover, Putz and Holbrook (1988) have shown that ant-free *Cecropia peltata* trees in Malaysia grow faster and suffer rates of vine infestation and herbivory similar to or lower than indigenous secondary forest tree species, suggesting that other attributes of *Cecropia* trees may decrease their vulnerability to both vines and herbivores and make them successful pioneers (e.g., hollow stems, rapid growth rates, and abundant small seeds capable of prolonged dormancy in the soil).

The defense system of *Cecropia* needs further research to clarify the shifts between biotic (production of food bodies to attract ants) and immobile or inducible (chemical and physical) herbivore defenses, both under habitat change and in the presence or absence of specialized and nonspecialized ant partners. Additionally, the relationship between ant inhabitants and honeydew-producing coccids inside the domatia needs a cost-benefit analysis (see Gaume, McKey, and Terrin 1998).

Finally, it has recently been shown that the ant-*Cecropia* mutualism could be highly asymmetric. Using stable-isotope analysis, Sagers, Ginger, and Evans calculated that only about 18% of worker ant carbon is derived from *Cecropia,* whereas 93% of the nitrogen in ant-occupied host plants is derived from debris deposited by ants. However, if all the resources exchanged in this system are taken into account, *Cecropia* plants fully exploit the presence of *Azteca* ants, but at no cost to the ant colony; and in turn *Azteca* fully exploits the hollow stems for nest cavities and larval food, but at little cost to the host (Sagers, Ginger, and Evans 2000) (see also chapter 8). It seems that the real evolutionary challenge for mutualism may not be the exploitation of one party by the other, but rather the ability to survive in intimate association in the first place (Doebeli and Knowlton 1998; Sagers, Ginger, and Evans 2000).

Piper *(Piperaceae)*

Species in the genus *Piper* inhabiting light gaps or the interior of the tropical rain forest of Costa Rica are associated with ants of the genus

Pheidole (Risch et al. 1977). The ants live in cavities formed by the appressed, curled margins of the petiole and in the hollowed-out stems, and they feed on the lipid-rich food bodies produced by the plant inside the petiolar cavities (Rickson and Risch 1984). The number of food bodies that are differentiated depend directly on the presence of ants. When ants are removed, production of food bodies nearly ceases, and when ants reinvade, production begins; thus food tissue of the plant is induced by the insect (Risch and Rickson 1981; Rickson and Risch 1984).

Pheidole ants function as antiherbivore defense on young *Piper* leaves by disrupting the activity of herbivores during their egg or early instar stages. More ants are found on young leaves (on average 2.0 ants/leaf) than on mature leaves (0.51 ants/leaf); unoccupied plants have significantly greater leaf damage than occupied plants; and damage to newer leaves tends to decrease with increased ant activity (Letourneau 1983, 1998). Moreover, ants attack herbivores at early stages of development. Over 75% of all egg baits on young and mature leaves are taken up by ants in observation periods of 30 or 60 min, and eggs are found significantly faster on young than on mature leaves. In more than half of the trials, the eggs were taken to the edge of the leaf and dropped to the ground rather than retrieved as a nutrient source (Letourneau 1983). In summary, *Piper* plants are defended by ants, but production of rewards by the plants depends mainly on ant presence; i.e., it is induced by the ants.

By conducting predator-addition experiments in a Costa Rican rain forest, Dyer and Letourneau (1999b) have recently demonstrated top-down cascading effects through three trophic levels within the community associated with *Pheidole*-inhabited *Piper* shrubs. The interaction system involves *Piper cenocladum* and three consumer groups: (1) the second trophic level includes many species of arthropod herbivores (Marquis 1991); (2) the third trophic level comprises scavenging, predatory, and parasitic arthropods, especially *Pheidole bicornis,* which lives inside the plant's stem and petioles; and (3) the fourth trophic level includes the beetle *Tarsobaenus letourneauae* (Cleridae), a top predator whose larvae feeds on *P. bicornis* brood and *Piper* food bodies (Letourneau 1990). Periodic additions of the specialist top predator to *Piper* shrubs caused a reduction in the predatory beetle's prey (*P. bicornis* ants), increased herbivory, and reduced leaf area of the plant (Dyer and Letourneau 1999b). Cascading effects of top predators were evident both in experiments with transplanted *Piper* fragments and in naturally occurring, established shrubs. Direct effects of predatory beetles on *Pheidole* caused 2.2 times lower ant densities in beetle-treated *Piper* fragments as compared to control fragments treated only with insecticides, resulting in increased herbivory levels in the former experimental group. A similar pattern was observed when predators were added to naturally occurring shrubs. Thus, reduction in the density of ant mutualists by top predatory beetles can affect the fitness of *Pheidole*-inhabited *Piper* shrubs by making the plant more vulnerable to herbivore damage as compared to plants with undisturbed ant colonies (Dyer and Letourneau 1999b; see also Letourneau and Dyer 1998a, 1998b). The parasitic effect of predatory beetles on the *Piper-Pheidole* mutualism is further enhanced by the fact that beetles can stimulate the plant to produce food bodies as if ants were present (Risch and Rickson 1981)—the beetles thus deprive *Piper* ant-plants of both food resources and protective ant mutualists (Letourneau 1990).

The following patterns can be summarized from the above examples of direct interactions between ants and plants offering domatia and food bodies. Even though the number of ant species is quite low per association and they are quite specialized plant-ants, these interactions are rarely species specific. Plant defense by ants has been demonstrated for several species in the genera *Macaranga, Cecropia,* and *Piper.* Although there are more plant species within these genera that have been described as offering domatia and food bodies, they have not yet been studied (see also chapter 8). There may be high levels of variation in resources offered by plant species within a genus, e.g., the continuum formed by the *Macaranga* species. Environmental conditions (light and soil) may influence the production of food bodies and alter the balance between immobile defenses and defenses with high initial construction costs (e.g., food bodies in *Cecropia*). Plants respond to the absence of ants by decreasing or halting the production of costly ant rewards, but specialized code breakers can parasitize the ant-plant mutualistic system and disrupt plant defense by ant inhabitants. Benefits have been assessed only for the plants; benefits have only been assumed for the ants (i.e., if they receive domatia and food, then their fitness increases, but see Sagers, Ginger, and Evans 2000).

Ant-Inhabited Plants Offering No Direct Food to Ants

Myrmecophytes can produce a variety of food resources that are consumed by their ant inhabitants, and these may include food bodies and pearl bodies (above), extrafloral nectar (Janzen 1966; McKey 1984, 1991), glandular trichomes (Vasconcelos 1991; Svoma and Morawetz 1992), floral nectar, and nutrient-rich plant wounds (Davidson and McKey 1993 and citations therein). In addition, numerous ant-plants offering direct food resources to ants can harbor inside their domatia hemipteran trophobionts that play a key role in ant nutrition (e.g., Wheeler 1942; Benson 1985; Oliveira, Oliveira-Filho, and Cintra 1987; Davidson, Longino, and Snelling 1988; Davidson and McKey 1993; Heil and McKey 2003).

Many myrmecophytic plant species, however, do not offer any direct food reward for their ant inhabitants and provide only the domatia space where the colony develops (Davidson and Mckey 1993). Nevertheless, occurrence of honeydew-producing coccoids inside the domatia may constitute the main energy source for ant inhabitants of many species of myrmecophytes (Wheeler 1921; Davidson and Mckey 1993), and it has been suggested that such trophobionts might have played a role in the evolution of secondary domatia (leaf pouches) of ant-plants (Benson 1985).

Tachigali *(Caesalpinaceae)*

The Amazonian canopy tree *Tachigali myrmecophila* harbors the stinging ant *Pseudomyrmex concolor* in its hollow leaf rachis and petioles (Fonseca 1994). The ants do not receive any food resource directly from the host plant and do not consume insect prey on the plant surface; apparently all the nutrients and energy requirements of the ants are obtained from honeydew-producing coccoids kept inside the domatia. By experimentally removing *P. concolor* colonies from host plants, Fonseca (1994) has demonstrated that incidence of herbivores was 4.3 times higher on ant-excluded plants than on ant-inhabited control plants and that rates of herbivory were 10 times higher on plants lacking ants compared to controls, about 3 times higher on immature than mature leaves, and about 2.5 times higher in the wet than in the dry season. After 18 months, plants from which ants were removed presented an accumulated level of leaf herbivory about twice as high as did the plants housing ants; and leaves from ant-inhabited plants lived much longer than those of ant-excluded or naturally unoccupied plants (fig. 6.2). The size and structure of the colonies of *P. concolor* inhabiting *Tachigali* plants is primarily controlled by the total domatia space offered by the host plant, explaining why plant-ants demonstrate host-limited foraging territories and suggesting that domatia size can be selected in a way that increases the plant's net benefit from the association (Fonseca 1993, 1999).

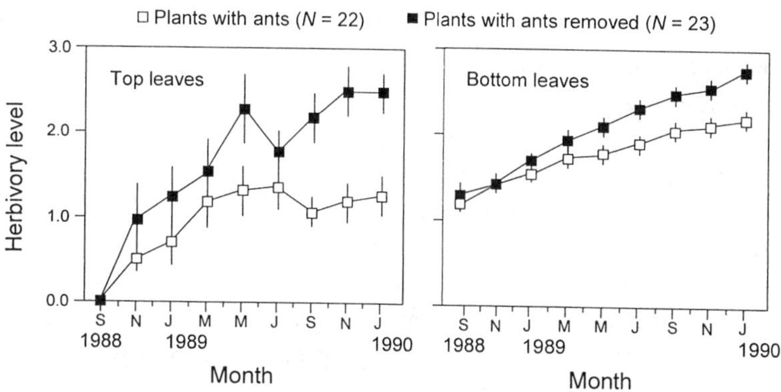

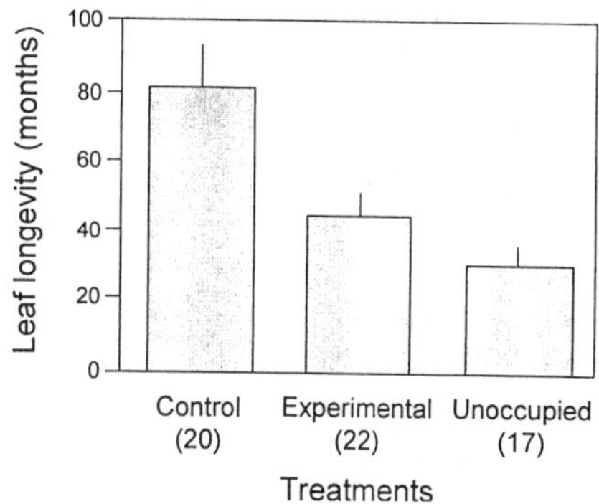

FIGURE 6.2. (A) Intensity of the leaf herbivory levels on plants of *Tachigali myrmecophila* naturally occupied by the ant *Pseudomyrmex concolor* and on plants from which the ants had been experimentally removed. Bottom leaves are those present at the beginning of the experiment, and top leaves are those emerging subsequently. In either case ant presence significantly affected herbivore levels. (B) Leaf longevity of ant-occupied plants (control), of plants from which ants were experimentally removed, and of naturally unoccupied plants differed significantly among plant groups during the 18-month period. Data are means ± 1 SE. Modified from Fonseca 1994.

Hirtella *(Chrysobalanaceae)*

Although myrmecophytes can receive a range of benefits from ant inhabitants, including protection against herbivores and encroaching vines, and nutrients essential for plant growth (Davidson and McKey 1993;

chapter 8), some associated ant species act as parasites by utilizing domatia and food resources without providing benefits (e.g., Janzen 1975), while others can negatively affect plant growth and reproduction by pruning reproductive or vegetative structures of their host-plants (Yu and Pierce 1998; Stanton et al. 1999; Yu 2001). Many plant-ant species forage and nest entirely on their hosts, and in such cases colony growth and reproductive success can depend strongly on host plant growth (Fonseca 1993, 1999). However, myrmecophytes can theoretically control the production of domatia and keep the available space for the colony at an optimum size for the interest of the plant in terms of defense allocation (Fonseca 1993). Within such ant-plant systems, cheating (i.e., the use of mutualistic resources or services without providing any benefits in return) can evolve if the allocation to domatia by the plant is not the optimal solution for the ants, thus creating a conflict of interests between the two partners (Fonseca 1999). Izzo and Vasconcelos (2002) detected such a conflict in the relationship between the Amazon plant *Hirtella myrmecophila,* which produces leaf-pouches as domatia, and its obligate ant partner, *Allomerus octoarticulatus* (Myrmicinae). The plant offers no food resources, and the ants do not tend scale insects or forage away from their host plant; their major food source appears to be insects found on foliage. Because *H. myrmecophila* drops domatia from older leaves, a unique trait among myrmecophytes, branches with and without ants can be found within the same plant individual. Older branches generally bear only old leaves with no domatia and therefore have no ants, whereas younger branches have leaves of various ages. Ants forage mainly on new leaves, and experimental removal of ants demonstrated that they effectively defend the leaves against herbivores. The ants, however, also behave as castration parasites and severely damage the host plant's inflorescences. Only ant-free older branches produced mature flowers and fruits, and flower production was eight times as high on plants from which ants were experimentally removed than on control plants. Due to the reproductive costs caused by ant-inhabitants, Izzo and Vasconcelos (2002) suggested that domatia abortion is a strategy developed by *Hirtella* to minimize the effects of cheating by *Allomerus* ants, thus supporting the view that evolutionary conflicts of interest between mutualistic species often impose selection for cheating on the partner, as well as for mechanisms to retaliate or to prevent super-exploitation. The stability of the *Hirtella-Allomerus* relationship is presumably maintained by the effect of the above opposing selection pressures operating independently on the two partners.

Numerous other cases of symbiotic ant-plant relationships have been described and investigated experimentally, involving a diversity of plant and ant taxa. Lists and detailed accounts can be found in numerous review articles and books (Bequaert 1922; Wheeler and Bequaert 1929; Wheeler 1942; Buckley 1982a; Beattie 1985; Benson 1985; Huxley 1986; Jolivet 1986; McKey 1989; Hölldobler and Wilson 1990; Davidson and McKey 1993; Heil and McKey 2003).

Ants, Plants, and Extrafloral Nectaries

Extrafloral nectaries are sugar-producing glands found outside flowers (including those found on reproductive structures, but not including pollinator-rewarding flower nectaries [see Rico-Gray 1993]). These structures have been present in flowering plants at least since the Oligocene (Pemberton 1992). Ants interact with the foliar nectaries of some fern species (fig. 6.3) and with the multiple types of nectaries found in flowering plants (figs. 6.4, 6.5). The study of ant-plant interactions mediated by extrafloral nectar can be divided into four categories or approaches that are not clear-cut:

1. The study of nectar contents and anatomy of extrafloral nectaries (reviewed by Koptur 1992a; see also Jeffrey, Arditti, and Koopowitz 1970; Elias, Rozich, and Newcombe 1975; Elias and Gelband 1976; Keeler 1977, 1980b; Elias and Prance 1978; Rhoades and Bergdahl 1981; Elias 1983; Anderson and Symon 1985; Morellato and Oliveira 1994; Koptur and Truong 1998; Díaz-Castelazo et al. 2005).
2. The taxonomic and geographic distribution of plants with extrafloral nectaries (Keeler 1979a, 1979b, 1980a, 1981a, 1985; Pemberton 1988, 1990, 1998; Oliveira and Leitão-Filho 1987; Morellato and Oliveira 1991; Oliveira and Oliveira-Filho 1991; Koptur 1992a, 1996; Blüthgen and Reifenrath 2003; Oliveira and Freitas 2004; Díaz-Castelazo et al. 2004, 2005).
3. Surveys of the ant fauna visiting the extrafloral nectaries of particular plant species or patterns of ant visitation in the general foliage (e.g., Schemske 1982; Oliveira and Brandão 1991; Rico-Gray 1993; Oliveira, Klitzke, and Vieira 1995; Oliveira and Pie 1998; Rico-Gray, García-Franco, et al. 1998; Blüthgen et al. 2000; Cogni, Raimundo, and Freitas 2000; Dejean et al. 2000; Hossaert-McKey et al. 2001; Cogni and Freitas 2002; Blüthgen, Gebauer, and Fiedler 2003; Díaz-Castelazo et al. 2004).

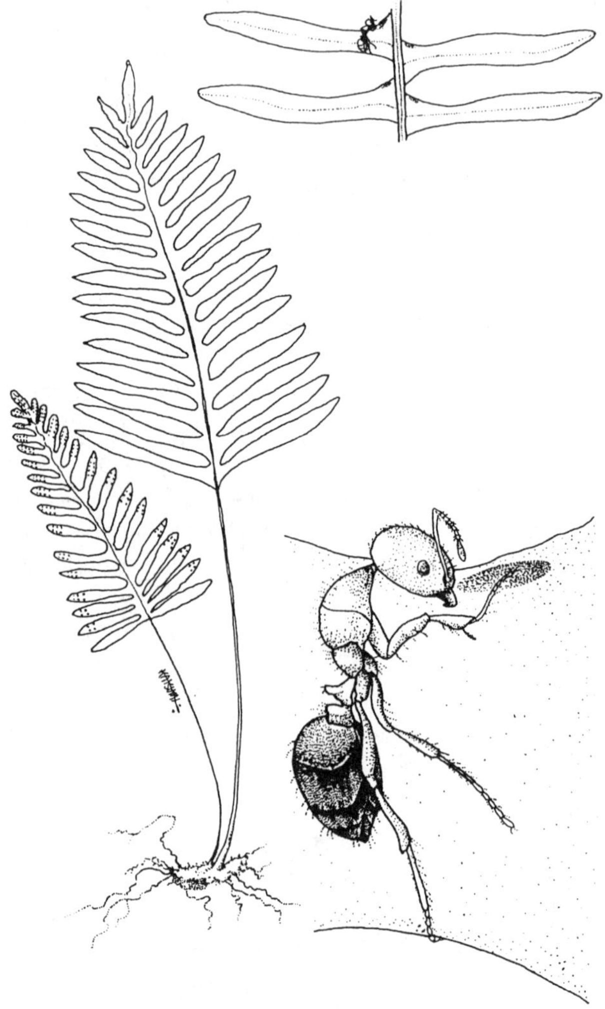

FIGURE 6.3. Ants foraging for nectar on the foliar nectaries of the fern *Polypodium plebeium* (Polypodiaceae). The presence of nectaries is sometimes suggested by the marks of a black sooty mold (*Ceratocystis* sp.; Ophiostomales, Ascomycotina) on the acroscopic lobes of the simple frond.

4. Studies based on observation and experimentation to evaluate defense of plants by ants (see below).

Ants forage for foliar (fig. 6.3), circumfloral (fig. 6.4), and extrafloral (fig. 6.5) nectar. The nectar produced by these structures represents the

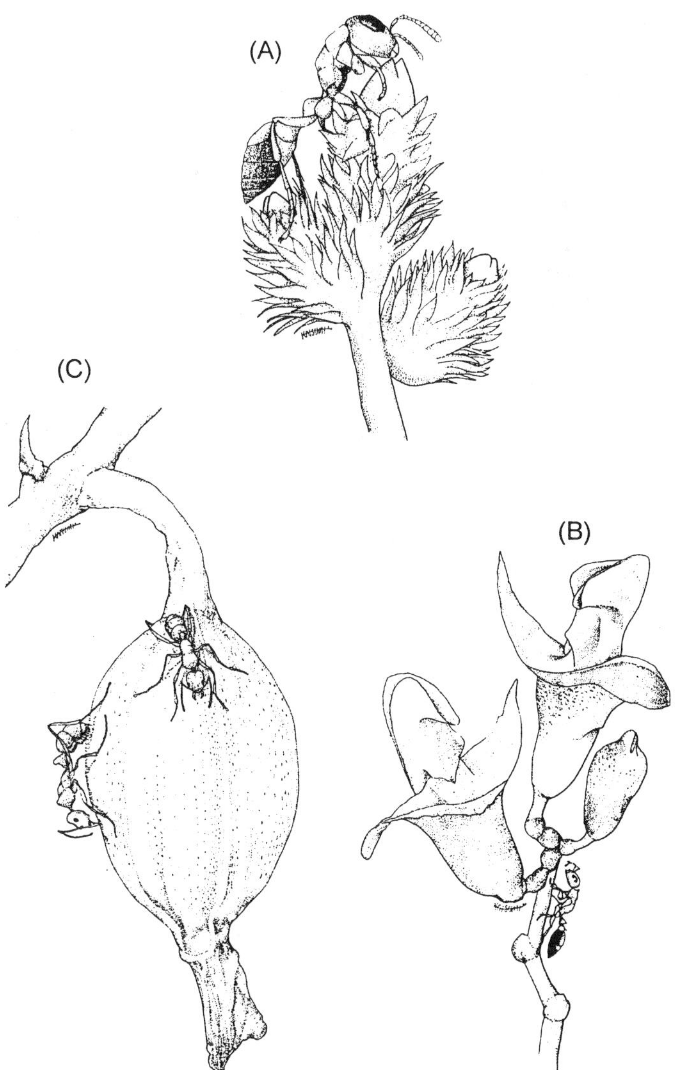

FIGURE 6.4. Examples of ants foraging for circumfloral nectar. (A) *Pseudomyrmex* sp. foraging on the calix and petals of *Iresine celosia* (Amaranthaceae). (B) *Pseudomyrmex* sp. foraging on the calix and floral peduncle of *Canavalia rosea* (Fabaceae). (C) *Ectatomma tuberculatum* foraging on the fruit of *Myrmecophila christinae* (Orchidaceae). Modified from Rico-Gray 2001; Rico-Gray et al. 2004.

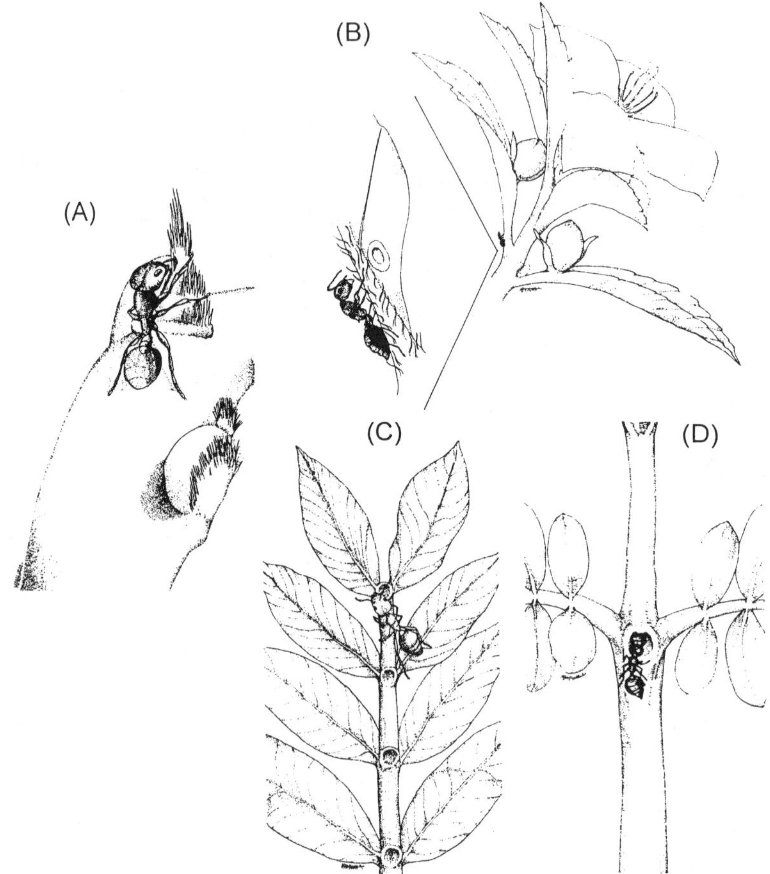

FIGURE 6.5. Ants foraging for extrafloral nectar. (A) *Crematogaster brevispinosa* foraging for the extrafloral nectar produced at the base of the areoles of cladodes of *Opuntia stricta* (Cactaceae). (B) *C. brevispinosa* at the base of the petioles of *Turnera ulmifolia* (Turneraceae). (C) *Camponotus abdominalis* collecting extrafloral nectar on the rachis of an *Inga* species (Fabaceae). (D) *C. brevispinosa* at a nectary located on the rachis of the foliole of a legume shrub. Modified from Rico-Gray 2001.

most important food resource for ants in many areas (e.g., Rico-Gray 1989, 1993; Rico-Gray, García-Franco, et al. 1998; Rico-Gray, Palacios-Rios, et al. 1998). The division between floral nectar, nectar from other reproductive structures (circumfloral), and extrafloral nectar is here modified from Elias (1983). Floral nectar is here classified as the nectar involved in pollination; extrafloral nectar is produced by nectaries associated with vegetative structures (leaves, stem, and stipules). The latter are called foliar nectaries for the particular case of the ferns. Although some authors consider the nectar produced on the surface of plant reproductive structures (spike, pedicel, bud, calyx, and fruit) to be extrafloral (see discussion in Elias 1983), we place it in a separate category because of differences in position and in time of production, relative to nectaries associated with leaves (Rico-Gray 1993). We will first address the interaction between ants and ferns (foliar nectaries) and then the interaction between ants and flowering plants (extrafloral nectaries).

Foliar nectaries in ferns have been observed in various genera (Koptur, Smith, and Baker 1982; Paterson 1982; Borger and Hoenecke 1984), and their role in interspecific interactions has been studied mostly in the bracken fern (*Pteridium aquilinum*) in various parts of the world (Tempel 1981; Lawton and Heads 1984; Rashbrook, Compton, and Lawton 1991, 1992). The result of most ant-exclusion experiments has been no evidence for ant-derived protection of the ferns (Tempel 1983; Rashbrook, Compton, and Lawton 1991, 1992). For instance, finding no evidence for ant protection to bracken, whose chief herbivores seem adapted against ant predation, Lawton and Heads (1984) suggested that the nectaries of these ferns prevent colonization by new herbivores. Only in one case has a field experiment clearly demonstrated the protection of a fern by ants. Koptur, Rico-Gray, and Palacios-Rios found that fronds of the Mexican epiphytic fern *Polypodium plebeium* (Polypodiaceae) (fig. 6.3), from which ants were excluded during development, suffered significantly greater damage (over 6 times greater) from sawfly larvae (Diprionidae) than fronds with ants present. Moreover, fronds of *P. plebeium* with ants excluded lost on average 4.0% of their leaf area, whereas fronds with ants present lost only 0.4%. The seasonal production of nectar in *P. plebeium,* which is synchronous with the production of new leaves and visits by the sawfly larvae, provides an effective defense against leaf-feeding larvae during the critical period of frond unfurling and maturation (Koptur, Rico-Gray, and Palacios-Rios 1998).

Certain authors have given the long evolutionary history between ferns and herbivores as the reason for an elaborate chemical defense (Cooper-Driver, Finch, and Swain 1977; Hendrix 1977; Jones and Firm 1979) and also a perceived paucity of herbivores (but see Hendrix and Marquis 1983). However, ferns (especially polypod ferns, which make up >80% of living fern species) diversified in the shadow of angiosperms (Schneider et al. 2004), suggesting perhaps a history similar to that of angiosperms and herbivores. Finally, ferns often receive substantial damage from certain herbivores adapted to a fern diet (Lawton 1976; Balick, Furth, and Cooper-Driver 1978; Cooper-Driver 1978; Gerson 1979; Hendrix 1980; MacGarvin, Lawton, and Heads 1986; Lawton, MacGarvin, and Heads 1987; Shuter and Westoby 1992). Moreover, Hendrix and Marquis (1983) have suggested that the hypothesis of underutilization of ferns by herbivores is largely an artifact of inadequate sampling.

* * *

Evidence for a protective role of ants visiting the extrafloral nectaries (EFNs) of flowering plants has increased markedly since Janzen's (1966, 1967a, 1967b) pioneer experimental work on the interaction between *Acacia* plants and *Pseudomyrmex* ants (see references in Bentley 1977a; Buckley 1982a; Beattie 1985; Jolivet 1986; Koptur 1992a; Whitman 1994; Heil and McKey 2003). Protection, however, is not universal, and some studies failed to demonstrate a deterring effect of visiting ants on herbivores (appendix 6.1). Some studies have addressed the protection of foliage from herbivores (Bentley 1976; Tilman 1978; Koptur 1979; Stephenson 1982), while others have evaluated the protection of the reproductive structures of plants (Bentley 1977b; Deuth 1977; Keeler 1981b; Oliveira 1997). The experimental design employed by most ant-guard studies to test whether ant presence increases plant reproductive fitness includes a plant group from which ants have been removed and their further access blocked off (usually by applying a sticky resin and by pruning plant bridges) and a plant group to which foraging ants have free access. Plant fitness is estimated as the difference in herbivory levels (percentage of leaf tissue removed) and/or in the number of fruits and/or seeds produced by each plant group (e.g., Bentley 1977b; Inouye and Taylor 1979; Koptur 1979; Rico-Gray and Thien 1989b; Vasconcelos 1991; Oliveira et al. 1999). For new methodological approaches not involving ant-exclusion from plants, see recent studies by Rudgers (2004) and Kost and Heil (2005).

Protection from ants varies among habitats (Bentley 1976; Barton 1986; Alonso 1998; Cogni, Freitas, and Oliveira 2003) and is more effective at lower altitudes (Koptur 1985). Moreover, ant-plant interactions are more common in dry-deciduous tropical lowlands than in tropical mountains (Stein 1992; Rico-Gray, García-Franco, et al.1998). An increase in plant fitness may depend on the different components of an ant assemblage (Ness 2003a, 2003b; Mody and Linsenmair 2004). Usually larger ants offer better protection against insect herbivores than smaller

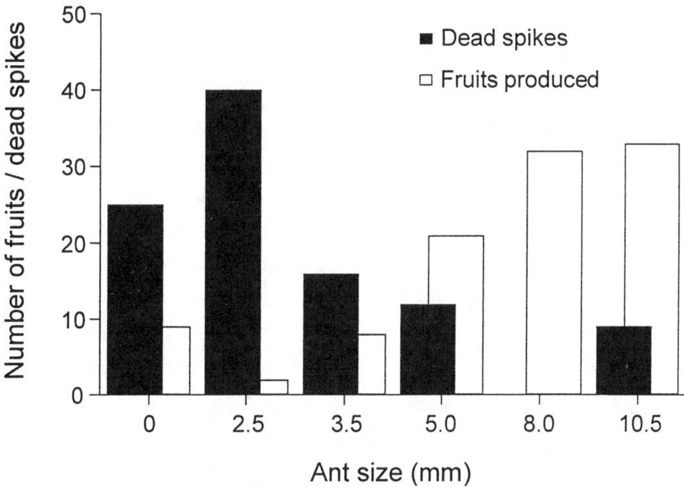

FIGURE 6.6. Number of dead inflorescence spikes and of fruits produced by individuals of *Myrmecophila christinae* (Orchidaceae) visited by ant species with different worker body length on the coast of Yucatán, Mexico. 0 = no ants; 2.5 mm = *Crematogaster brevispinosa;* 3.5 mm = *Camponotus planatus;* 5.0 mm = *C. rectangularis;* 8.0 mm = *C. abdominalis;* 10.5 mm = *Ectatomma tuberculatum.* Modified from Rico-Gray and Thien 1989a, 1989b.

ants (fig. 6.6) (Schemske 1982; Horvitz and Schemske 1984; Oliveira, Silva, and Martins 1987; Koptur and Lawton 1988; Rico-Gray and Thien 1989b). Furthermore, the effect of ants on herbivory may not be consistent. Mody and Linsenmair (2004) demonstrated that plants dominated by different ant species differed significantly in leaf damage caused by herbivorous insects and that ants significantly reduced the abundance of different arthropod groups (Araneae, Blattodae, Coleoptera, Hemiptera, and non-ant Hymenoptera), whereas other groups, including important herbivores, seemed not to be affected (Lepidoptera, Orthoptera, Thysanoptera, and Heteroptera). Moreover, the presence of ants benefits plants only when specific ant species are attracted and protection by these ants is not counterbalanced by their negative effect on other beneficial arthropods (Mody and Linsenmair 2004; see chapter 10). Finally, a number of specialized herbivores utilizing ant-visited plants possess morphological and behavioral mechanisms that circumvent ant predation or ant-induced injury on host plants (Janzen 1966; Koptur 1984, 1992a; Heads and Lawton 1985; Freitas and Oliveira 1992, 1996; Salazar and Whitman 2001; Machado and Freitas 2001; Oliveira, Freitas, and Del-Claro 2002; Oliveira and Freitas 2004).

Plant protection by ants has been demonstrated in both temperate and tropical environments (e.g., Inouye and Taylor 1979; Stephenson 1982; Ibarra-Manríquez and Dirzo 1990; Mackay and Whalen 1991; Maschwitz and Fiala 1995; Dejean, Olmsted, and Snelling 1995). However, most examples come from tropical habitats, especially the neotropics, where species of myrmecophytes and plants with extrafloral nectaries have been reported to be more common (Bentley 1977a; Rico-Gray, García-Franco, et al. 1998; McKey and Davidson 1993; Fonseca and Ganade 1996; Oliveira and Freitas 2004; and references therein). Plant protection by ants visiting EFNs has been studied across a variety of plant growth-forms such as herbs, shrubs, trees, vines, and cacti (appendix 6.1). In the dry-deciduous forest of Costa Rica, Bentley (1981) demonstrated that the interaction between ants and plants bearing EFNs could be influenced by plant growth-form because extrafloral nectaries, ant foragers, and removal of potential herbivores were more common on vines than on nonvine species (see also Bentley and Benson 1988; Schupp and Feener 1991). In a lowland tropical dry-deciduous forest in eastern Mexico, however, ants visited shrubs more frequently, followed by herbs and trees (Rico-Gray 1993; Rico-Gray, García-Franco, et al. 1998).

* * *

The functions usually ascribed to floral nectar (attraction of pollinators) and extrafloral nectar (attraction of ants or other insects that protect the host plant against herbivores) cross paths in some examples. Extrafloral nectar in *Acacia pycnantha* has been reported specifically to attract pollinators (Vanstone and Paton 1988). In other cases extrafloral nectar has been reported to attract not ants but a variety of parasites and parasitoids (Hespenheide 1985; Whitman 1994; and references therein) that defend plants against insect herbivores (Pemberton and Lee 1996). The floral nectar of *Rorippa indica* (Cruciferae) attracts ants that defend the plant from herbivorous insects (Yano 1994). The "postfloral" nectaries (i.e., floral nectaries secreting after flowering) of *Mentzelia nuda* (Loasaceae) attract ants whose presence deters ovipositing or feeding herbivores and significantly enhances seed set (Keeler 1981b). Finally, the floral nectaries of the wind-pollinated *Croton suberosus* (Euphorbiaceae) attract wasps that defend the plant from herbivores (Domínguez, Dirzo, and Bullock 1989).

The following patterns can be summarized regarding direct interactions between ants and plants offering extrafloral nectar. The number of ant species can sometimes be relatively high per association (e.g., Schemske 1982; Oliveira and Brandão 1991; Blüthgen et al. 2000). Therefore it seems that this type of ant-plant interaction is diffuse (no extreme specialization), largely fortuitous, and facultative; specialization between particular ant and plant species is rare, suggesting that only occasionally does selection favor obligate mutualisms (Schemske 1983; Beattie 1985). Selective benefits should accrue to plants that attract a broad array of ants; the greater the diversity of ants, the greater the variety of plant enemies that they are likely to remove or disrupt the activities of, and the greater the probability that in any given habitat, season, or time of day, some ant species will forage on the plant (Beattie 1985). The ubiquity and abundance of ants in certain areas (e.g., Rico-Gray 1993; Rico-Gray, García-Franco, et al. 1998; Rico-Gray, Palacios-Rios, et al. 1998) may indicate that plants offering food rewards to ants will receive benefits (i.e., protection) from some array of ant species (Rico-Gray 1993). Most studies have not demonstrated that both the ant and the plant participants benefit significantly and therefore cannot conclude that a given interaction is mutualistic (Cushman and Beattie 1991). Although the vast majority of the studies were not conceived for the purpose of testing mutual benefits between ants and plants, the existence of mutualistic interactions should be subject to testing because authors assume mutualism from their results.

Plant defense by ants has been demonstrated for ferns and for flowering plants, and in only a few cases the plants did not receive benefits from the ants (see appendix 6.1). However, defense may be associated with only a certain ant or ant species in an ant assemblage, or with a certain time or place (chapter 10). Moreover, there are many more plant species that produce extrafloral nectar in habitats where a positive effect of ants on plant fitness has not been demonstrated. Most studies have not focused on the dynamic nature of defense in ant-plants, and many cues and signals controlled by the plant and the environment can enhance defense by influencing recruitment, patrolling, and persistence of ants (Agrawal 1998b; Agrawal and Rutter 1998; Cronin 1998). In addition, few studies have addressed defensive systems that include both EFNs and plant-released volatile chemicals; some of the chemicals deter herbivores and others attract parasitoids to herbivores (Kawano et al. 1999). And although in the vast majority of the studies, the ants are shown to defend plants against insect herbivores, there are cases in which the ants defend plants against mammals as well (e.g., monkeys [McKey 1974], giraffes [Madden and Young 1992], and goats [Stapley 1998]).

Plant Defensive Strategies and Induced Responses

A variety of plant defenses have evolved which enable plants to avoid or reduce herbivory and increase their fitness (Ågren and Schemske 1993): (1) morphological defenses (e.g., thick cuticle, trichomes, thorns, and spines); (2) chemical defenses, both "qualitative" (toxins, effective against specialists) and "quantitative" (digestibility-reducing compounds, effective against generalists) (e.g., Heil, Baumann, et al.2002); and (3) the less-studied biotic defenses, such as ants and wasps (e.g., Koptur 1991; Powell and Stradling 1991; Cuautle and Rico-Gray 2003; Mody and Linsenmair 2004).

Depending on the attacking herbivore, most plants use an array of defenses that can vary in intensity and effectiveness and can operate on different temporal and spatial scales (Ågren and Schemske 1993; Futuyma and Mitter 1997; Agrawal and Rutter 1998; Cronin 1998). The efficacy of defensive characters will then depend on the identity of the herbivore species, and the composition of a component community (i.e., plant, herbivores, and natural enemies of herbivores [Root 1973]) should affect the course of plant evolution (Futuyma and Mitter 1997). Many studies have proposed trade-offs between plant defenses, that is, a lack of redundancy of defenses that act on the same temporal, spatial, and/or herbivore scales (e.g., McKey 1979; Davidson and Fisher 1991; Ågren and Schemske 1993; Nomura, Itioka, and Itino 2000; Itino and Itioka 2001). However, an absolute trade-off between defense systems is not always found (e.g., Steward and Keeler 1988).

The production and maintenance of defenses when plants have limited resources are associated with a cost, which, apart from being a function of the concentration of defense substances in leaf tissues, will also depend on the turnover rate of these substances and on the relative overlap between resources used for production of defenses and resources used for growth (Skogsmyr and Fagerström 1992; Heil, Hilpert, et al.2002). For instance, production of food rewards by the obligate myrmecophyte *Macaranga triloba* (Euphorbiaceae) is limited by nutrient supply, and thus nutrient availability can directly influence the plant's investment in its ants (Heil, Hilpert, et al.2001). A general model for allocation to different defenses in ant-plant systems should take into account the relative effectiveness of each defense, since no single defense is likely to eliminate damage by all herbivores (Agrawal and Rutter 1998). For instance, in the association between *Cecropia* trees and *Azteca* ants, the plant may use a variety of defensive strategies and adjust them as needed to the environmental conditions (Coley 1986; Folgarait and Davidson 1995; Agrawal 1998b). Induced responses to herbivory, i.e., modifications to plants' defensive strategies triggered by changes in environmental cues, have been characterized extensively and demonstrate that plant defenses are very dynamic and create an unpredictable environment for herbivores (Milewski, Young, and Madden 1991; Karban and Baldwin 1997; Agrawal 1998a; Agrawal and Rutter 1998; Heil 1999, 2004; Heil, Greiner, et al.2004).

Ants can select and prefer sugar solutions containing a complex mixture of amino acids over sugar-only solutions (Lanza 1988; Koptur and Truong 1998; Völkl et al. 1999; Blüthgen and Fiedler 2004a, 2004b; Blüthgen, Gottsberger, and Fiedler 2004). In damaged plants, the amount and quality (sugar and amino acid concentration) of extrafloral nectar increase induced by the damage (Smith, Lanza, and Smith 1990; Swift and Lanza 1993), thus increasing patrolling by defending ants as well (Bentley 1977a; Tilman 1978; Stephenson 1982; Swift, Bryant, and Lanza 1994; Ness 2003a; Heil, Greiner, et al.2004). In broad bean, *Vicia faba,* Mondor and Addicott (2003) have demonstrated experimentally that the overall number of EFNs on a plant increases dramatically (59% to 106%) over one week depending on the damage treatment.

According to Kawano et al., the perennial herbs *Fallopia japonica* and *F. sachalinesis* (Polygonaceae) have EFNs at the base of the leaf petiole that secrete nectar that attracts nine ant species, wasps, flies, and beetles. When leaves are damaged by herbivores (Coleoptera and Hemiptera), the plants release two contrasting types of volatile substances. Within the first two hours after damage, the leaves release insect deterrents, and two hours after damage they start releasing a chemical signal attracting parasitoids to herbivores, creating a complicated defensive strategy that presents a complex and unpredictable environment for herbivores (Kawano et al. 1999). Volatile compounds (e.g., monoterpenes and fatty acid derivatives) commonly occur both in leaves when damaged and in flowers, suggesting that floral volatiles may have originated from general leaf volatiles in the meshing of the life cycles of insects and plants (Pellmyr and Thien 1986; Azuma et al. 1997). The emission by plants of different-purpose volatiles after leaf damage suggests that the interaction between flower and pollinator and the chemical communication between the first and the third trophic levels (e.g., plant signals to predators and parasitoids as a defensive mechanism) may be interrelated (Price et al. 1980; Azuma et al. 1997; Adler 2000).

Induced responses have also been experimentally tested in the Amazonian ant-plants *Hirtella myrmecophila* (Chrysobalanaceae) (Romero and Izzo 2004) and *Maieta poeppiggi* (Melastomataceae) (Christianini and Machado 2004). Results show a significant increase in the number of patrolling ants after experimental damage to leaves, with a stronger response in young than in mature leaves; no increment in patrolling was observed in undamaged adjacent leaves. Aqueous leaf extracts of *H. myrmecophila* placed on undamaged leaves also induced increased ant patrolling compared to water controls, with ants responding more strongly to extracts from young leaves (fig. 6.7). Apparently visual and chemical cues associated with herbivory are involved in the induction of ant recruitment; the continuous patrolling activity by ants and their rapid response to foliar damage results in the detection and capture (or repellence) of most insect herbivores before they can inflict significant damage to the leaves (Christianini and Machado 2004; Romero and Izzo 2004). In *Maieta guianensis,* however, the two most common ant associates respond in different ways to experimental cues associated with herbivory. While *Pheidole minutula* is induced by both physical damage and extracts of leaf tissue, *Crematogaster laevis* is induced by leaf damage only, suggesting that this interspecific variation in induced responses could influence the quality of defense provided by ants (Lapola, Bruna, and Vasconcelos 2003; see also Bruna, Lapola, and Vasconcelos 2004).

Using an elegant experimental design, Heil tested whether adult *Phaseolus lunatus* (Fabaceae) plants could be induced to produce herbivore-induced volatiles or secretion of extrafloral nectar after the application of the phytohormone jasmonic acid. His results show that the treatment reduced both the number of dead shoot tips and leaf damage by herbivores. Also, treated tendrils grew faster and produced more leaves than controls. At the end of the experiment, treated tendrils bore two times as many inflorescences and three times as many fruits as controls (Heil 2004; see also Heil 1999; Heil, Koch, et al. 2001; Kost and Heil 2006). Although defensive ant responses are stimulated by cues associated with herbivory, the local and regional variation in the composition of potential partner taxa could influence the ecology and evolution of defensive mutualisms in ways that have hardly been explored (Bruna, Lapola, and Vasconcelos 2004). Finally, Agrawal and Rutter (1998) propose that dynamic ant responses to damage should be controlled by the plant in facultative interactions (mediated by extrafloral nectar) but should be predominantly controlled by the ants in obligate interactions (mediated by food bodies and domatia); overall, they suggest that

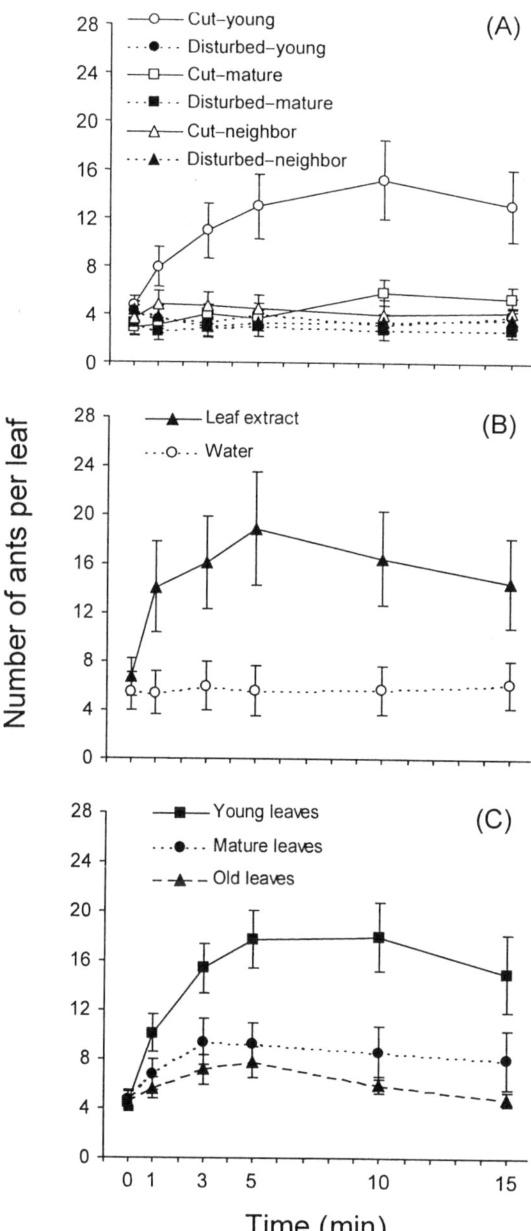

FIGURE 6.7. Leaf damage induces ant recruitment in the Amazon ant-plant *Hirtella myrmecophila*. (A) More workers of *Allomerus octoarticulatus* are recruited to young and mature damaged (cut) leaves than to disturbed leaves (control) or to leaves neighboring those cut and disturbed. (B) Ants are recruited to leaves with extracts of young leaves (experimental), but not to leaves receiving only water (control). (C) Extracts of young leaves induce 2 to 3.5 times as many ants to recruit relative to extracts of mature and old leaves. Data are means ± 1 SE. Modified from Romero and Izzo 2004.

induced responses to herbivory should be common in ant-plants. Indeed, Blüthgen and Wesenberg have shown that ants can induce twig structures that resemble classical ant domatia in the Amazonian rain forest tree *Vochysia vismiaefolia* (Vochysiaceae). Because myrmecophytes can develop domatia regardless of ant activity, domatia induction by ants in V. *vismiaefolia* may have represented an important initial step in the evolution of ant-plants (Blüthgen and Wesenberg 2001).

The Nature of the Associations and the Importance of Conditionality

Extrafloral nectar-based ant-plant associations tend to be more generalized, while the associations involving domatia and food bodies, sometimes in addition to nectar, tend to be more specialized and specific (Schemske 1983; Beattie 1985). Plants do not usually offer the whole array of resources to ants, and thus most ant-plant interactions are not specific and tend to be more generalized and opportunistic (Beattie 1985; Rico-Gray 1993). When the value of the reward is high, ant colonies will gain more by acting as mutualists. However, this value will be conditioned by various factors, including the composition and quality of rewards, the distance of the rewards from an ant nest, the nutritional status of the ant colonies, and the availability of alternative resources (Cushman and Addicott 1991). Furthermore, the evolutionary implications of this conditionality are twofold. Given the variation in mutualistic systems, it seems probable that selection will favor the evolution of flexible, nonobligate associations between species, such that the cost of having an unpredictable mutualist are minimized (Beattie 1985; Cushman and Addicott 1991). The preponderance of conditionality in nature suggests that species-specific coevolution in mutualistic systems will be the exception rather than the rule, due to the disruptive effects of spatial and temporal variation in the strength and direction of selection pressures (Horvitz and Schemske 1984; Cushman and Addicott 1991). Blüthgen and Fiedler have shown that both interspecific variability in ant gustatory preferences (i.e., type and concentration of different sugars and amino acids) and conditional effects such as competition and colony requirements affect resource selection in multispecies ant communities, which in turn may be crucial in niche partitioning of species-rich nectarivore assemblages. For instance, ant selectivity between different nectar solutions was significantly reduced when different ant species co-occurred on the same bait, and preferences for single amino acids were also reduced when colonies fed extensively on the same compound for two days prior to the experiment (Blüthgen and Fiedler 2004b).

Another factor that should condition the outcome of ant-plant interactions is the hardly-considered fact that not all plant individuals in a given population may be associated with ants. Most studies have considered ant-plant interactions as if all individuals in the associated species were, in fact, interacting. The question is, Will selective pressures differ if only a fraction of individual plants of a given species are involved in an association? And the answer is yes, they will. Initial research in Mexican coastal forests of the state of Veracruz suggests that in a given year only a fraction of the individuals in many plant populations are involved in associations with ants, and not all are long-lasting. Most associations between ants and nectaries located in the reproductive structures of plants were short-lived; and, interestingly, 66% of the nectary-bearing plant species were visited by ants, whereas 84% of the recorded ant species visited nectaries (Rico-Gray 1993; Díaz-Castelazo et al. 2004). Exceptions were the few cases of obligate associations (e.g., *Acacia cornigera* and *Myrmecophila tibicinis*) and cases in which the resource offered by the plants was nectar associated with leaves (e.g., *Ipomoea alba, Passiflora* sp., and *Inga* sp.) (Rico-Gray 1989, 1993; Rico-Gray and Thien 1989b; Torres-Hernández et al. 2000). The latter is an ephemeral resource, but when nectaries are active, ants visit most individual plants. It seems that one of the benefits of these "obligate" associations is the higher plant-occupation frequency. Moreover, associations between plants, ants, and hemipterans behave the same way (table 6.3). Finally, Heil, Fiala, et al. (2001) have demonstrated that the efficacy of defense in obligate associations (*Macaranga triloba* and *M. hosei*) is higher than in facultative ones (*M. tanarius*) and that ant-derived protection detected in obligate associations through a long-term study is orders of magnitude higher than previously observed in a short-term study.

Topics for Future Consideration

Based on his study of the ant-plant *Leonardoxa africana* (Caesalpinaceae), Mckey (1984) proposed the hypothesis that in ant-plants whose leaves are very long-lived, the cost of providing leaves with permanent

TABLE 6.3 **Percentage of plant individuals or inflorescences from different species hosting ant-Hemiptera associations in the tropical dry-deciduous coastal forest at La Mancha, Veracruz, Mexico**

Plant species	Type of association	No. of plants (P) or inflorescences (I) sampled	Percentage with ant-Hemiptera association	Duration of association
Croton punctatus (Euphorbiaceae)	Ant-mealybug	150 (P)	3.0	Year-round
Bursera graveolens (Burseraceae)	Ant-mealybug	20 (P)	0.5	Year-round
Ambrosia texana (Asteraceae)	Ant-mealybug	250 (P)	0.4–1.0	3–4 months
Tecoma stans (Bignoniaceae)	Ant-membracid	1,385 (I)	18.0	2–3 months
Paullinia fuscescens (Sapindaceae)	Ant-aphid	4,312 (I)	45.0	2–3 months

Source: Modified from Rico-Gray 1993.

chemical or mechanical protection decreases relative to the cost of maintaining a large worker force of ants throughout the life of the leaf; therefore, a small worker force is maintained that patrols only the young leaves. Future studies should explicitly be designed to address and discuss the evolution of indirect defenses involving ants (and other EFN-feeding organisms, like wasps) versus direct defenses involving secondary compounds and plant physical traits (McKey 1984; Ågren and Schemske 1993; Heil 2004). Are there clear trade-offs between defenses? Which are the selective pressures that result in the presence, in a plant genus, of species with EFNs, species with secondary compounds, and species with both (e.g., *Cnidoscolus;* see chapter 12)? A phylogeny for such groups is clearly needed. Do ants become more effective in protecting plants over evolutionary time? What is the role of induced defenses in ant-plants? Moreover, the association with ants may also cause a reproductive trade-off for the plant in relation to costs and benefits to male and female function (Wagner 1997, 2000).

Another topic needing further research is the role of plant ontogeny and the evolution of caulinary domatia in myrmecophytes. Brouat and McKey, studying both myrmecophytic and nonmyrmecophytic species, report that over ontogeny the primary cross-sectional area of a terminal internode and the surface area of the leaf borne by it increase in isometric proportion when the functions of the twig are limited to vascular supply and biomechanical support of the leaf. However, when a new selection pressure acts on twigs, for instance to provide shelter to ants, the leaf-stem size relationship is modified and becomes allometric. These authors suggest that the allometry of the leaf-stem size relationship could be general for all hollow-stemmed plants, whether myrmecophytic or not. They also suggest the need for comparative histological and biomechanical studies on hollow- and solid-stemmed plants to better understand the allometry and its evolutionary implications (Brouat and McKey 2001; see also Blüthgen and Wesenberg 2001).

APPENDIX 6.1 **Flowering plants that maintain a nonsymbiotic relationship with visiting ants mediated by the offer of extrafloral nectar, and where the ant-guard theory has been studied**

Plant	Growth form	Geographic location	Location of nectaries	Associated ant genera (number of species)	Protection by ants	Source
Acanthaceae						
Aphelandra deppeana	Shrub	Costa Rica	RE	Genera not provided (8)	Yes	Deuth 1977
Anacardiaceae						
Anacardium occidentale	Tree	Sri Lanka, India, and Malaysia	LE, RE	*Anoplolepis* (1), *Camponotus* (7), *Cardiocondyla* (2), *Cataulacus* (2), *Crematogaster* (6), *Diacamma* (1), *Dolichoderus* (1), *Lepisiota* (1), *Monomorium* (4), *Oecophylla* (1), *Paratrechina* (2), *Pheidole* (1), *Plagiolepis* (1), *Polyrhachis* (3), *Tapinoma* (2), *Technomyrmex* (1), *Tetraponera* (7)	Yes	Rickson and Rickson 1998
Asteraceae						
Helianthella quinquenervis	Herb	Colorado, USA	RE	*Formica* (3), *Myrmica* (1), *Tapinoma* (1)	Yes	Inouye and Taylor 1979
Helichrysum bracteatum	Herb	Southeast Australia	LE, RE	*Iridomyrmex* (> 1)	No	O'Dowd and Catchpole 1983
Helichrysum viscosum	Herb	Southeast Australia	LE, RE	*Camponouts, Crematogaster, Iridomyrmex* (> 1), *Monomorium, Notoncus* (no. of species not provided)	No	O'Dowd and Catchpole 1983
Bignoniaceae						
Campsis radicans	Vine	Illinois, USA	LE, RE	*Crematogaster* (1), *Formica* (1), *Lasius* (1)	Yes	Elias and Gelband 1975

Catalpa bignonioides	Tree	Georgia, USA	LE	*Solenopsis* (1), *Forelius* (1), plus native species (*Camponotus, Prenolepis, Crematogaster, Formica*)	Yes	Ness 2003a, 2003b
Catalpa speciosa	Tree	Michigan, USA	LE	*Camponotus* (2), *Crematogaster* (1), *Formica* (1), *Prenolepis* (1)	Yes	Stephenson 1982
Bixaceae						
Bixa orellana	Shrub	Costa Rica	ST, RE	*Azteca* (1), *Camponotus* (3), *Cephalotes* (1, = *Paracryptocerus*), *Crematogaster* (1), *Dolichoderus* (1, = *Monacis*), *Ectatomma* (1), *Pachycondyla* (1, = *Neoponera*), *Pseudomyrmex* (3)	Yes	Bentley 1977b
Bombacaceae						
Ochroma pyramidale	Tree	Costa Rica	LE	*Azteca* (1), *Dolichoderus* (1), *Crematogaster* (1), *Solenopsis* (1)	Yes	O'Dowd 1979
Bromeliaceae						
Dyckia floribunda	Herb	Argentina	LE	*Camponotus* (2), *Cephalotes* (1, = *Zacryptocerus*), *Crematogaster* (1), *Pseudomyrmex* (1)	Yes	Vesprini, Galetto, and Bernardello 2003
Cactaceae						
Ferocactus gracilis	Cactus	Baja California, Mexico	LE, RE	*Brachymyrmex* (1), *Camponotus* (1), *Crematogaster* (1), *Dorymyrmex* (1, = *Conomyrma*), *Forelius* (1, = *Iridomyrmex*), *Monomorium* (1), *Myrmecocystus* (1), *Pheidole* (2), *Solenopsis* (1), *Tetramorium* (1, = *Xiphomyrmex*)	?	Blom and Clark 1980
Opuntia acanthocarpa	Cactus	Arizona, USA	LE, RE	*Crematogaster* (1), *Forelius* (1, = *Iridomyrmex*)	Yes	Pickett and Clark 1979

(*continued*)

APPENDIX 6.1 **(continued)**

Plant	Growth form	Geographic location	Location of nectaries	Associated ant genera (number of species)	Protection by ants	Source
Opuntia stricta	Cactus	Veracruz, Mexico	LE, RE	*Camponotus* (3), *Crematogaster* (1), *Forelius* (1), *Monomorium* (1), *Paratrechina* (1), *Pseudomyrmex* (2)	Yes	Oliveira et al. 1999
Caryocaraceae						
Caryocar brasiliense	Shrub	São Paulo, Southeast Brazil	LE, RE	*Acanthoponera* (1), *Azteca* (4), *Brachymyrmex* (1), *Camponotus* (10), *Cephalotes* (1, = *Zacryptocerus*), *Crematogaster* (2), *Dorymyrmex* (1, = *Conomyrma*), *Ectatomma* (1), *Leptothorax* (1), *Ochetomyrmex* (1), *Pachycondyla* (1), *Paratrechina* (2), *Pheidole* (3), *Pseudomyrmex* (2), *Solenopsis* (1), *Tapinoma* (1), *Wasmannia* sp.	Yes	Oliveira 1997; Oliveira and Brandão 1991
Convolvulaceae						
Ipomoea carnea	Vine	Costa Rica	LE, RE	*Camponotus* (4), *Crematogaster* (1), *Forelius* (1, = *Iridomyrmex*), *Monomorium* (1), *Pseudomyrmex* (1), *Solenopsis* (2)	?	Keeler 1977
Ipomoea leptophylla	Vine	Nebraska, USA	LE, RE	*Crematogaster* (1), *Dorymyrmex* (1), *Formica* (4), *Lasius* (1), *Myrmica* (1), *Pheidole* (1)	Yes	Keeler 1980b
Ipomoea pandurata	Vine	North and South Carolina, USA	RE	*Camponotus* (2), *Crematogaster* (2), *Forelius* (1, = *Iridomyrmex*), *Formica* (2), *Monomorium* (1), *Pheidole* (1), *Solenopsis* (1)	Yes	Beckmann and Stucky 1981

Dioscoreaceae						
Dioscorea praehensilis	Vine	Cameroon, Africa	ST	*Crematogaster* (1), *Pheidole* (1), plus 6 other genera (25)	Yes	Di Giusto et al. 2001
Euphorbiaceae						
Adriana tomentosa	Shrub	Southeast Australia	LE, RE	*Camponotus* (2), *Crematogaster* (1), *Iridomyrmex* (2), *Ochetellus* (1), *Myrmecia* (1), *Polyrhachis* (2) *Rhytidoponera* (2),	No	Mackay and Whalen 1998
Croton sarcopetalus	Shrub	Argentina	LE, RE	*Brachymyrmex* (1), *Camponotus* (2), *Cephalotes* (1, = *Zacryptocerus*), *Crematogaster* (2), *Linepithema* (1), *Paratrechina* (1), *Pseudomyrmex* (2)	No	Freitas et al. 2000
Homolanthus novo-guianeensis	Tree	Papua New Guinea	LE	*Crematogaster* (1), *Paratrechina* (3), *Pheidole* (3), *Rhoptromyrmex* (1)	No	Whalen and Mackay 1988
Macaranga aleuritoides	Tree	Papua New Guinea	LE	*Crematogaster* (2), *Leptomyrmex* (1), *Meranoplus* (1), *Paratrechina* (3), *Pheidole* (3), *Polyrachis* (1), *Tapinoma* (1)	Yes	Whalen and Mackay 1988
Macaranga punctata	Tree	Papua New Guinea	LE	*Crematogaster* (2), *Meranoplus* (1), *Paratrechina* (1), *Pheidole* (2), *Polyrachis* (1)	No	Whalen and Mackay 1988
Macaranga quadriglandulosa	Tree	Papua New Guinea	LE	*Leptomyrmex* (1), *Meranoplus* (1), *Paratrechina* (2), *Pheidole* (1), *Rhoptromyrmex* (1)	Yes	Whalen and Mackay 1988
Mallotus philippensis	Tree	Papua New Guinea	LE	*Crematogaster* (1), *Paratrechina* (2), *Pheidole* (2)	Yes	Whalen and Mackay 1988
Fabaceae						
Acacia constricta	Shrubs	Arizona, USA	LE	*Formica* (1)	No[a]	Wagner 1997, 2000
Cassia fasciculata	Herb	Florida and Iowa, USA	LE	Florida: *Camponotus* (1), *Dorymyrmex* (1), *Crematogaster* (2), *Forelius* (1, = *Iridomyrmex*),	Yes	Barton 1986; Kelly 1986

(continued)

APPENDIX 6.1 **(continued)**

Plant	Growth form	Geographic location	Location of nectaries	Associated ant genera (number of species)	Protection by ants	Source
				Formica (1), *Paratrechina* (1), *Monomorium* (1), *Pheidole* (2), *Pseudomyrmex* (1), *Solenopsis* (1) Iowa: *Aphaenogaster* (1), *Crematogaster* (1), *Formica* (1), *Lasius* (1), *Leptothorax* (1), *Monomorium* (1), *Myrmica* (1), *Paratrechina* (1)		
Erythrina flabelliformis	Shrub	Arizona, USA	LE, RE	*Camponotus* (2), *Crematogaster* (1), *Forelius* (1), *Monomorium* (1), *Paratrechina* (1), *Pseudomyrmex* (1)	Yes	Sherbrooke and Scheerens 1979
Inga densiflora	Tree	Costa Rica	LE	*Camponotus* (1), *Ectatomma* (1), *Mymelachista* (1), *Paratrechina* (1), *Pheidole* (1), *Solenopsis* (1)	Yes	Koptur 1984
Inga punctata	Tree	Costa Rica	LE	*Camponotus* (3), *Crematogaster* (3), *Dolichoderus* (1, = *Monacis*), *Mymelachista* (1), *Pheidole* (2), *Procryptocerus* (1)	Yes	Koptur 1984
Pentaclethra macroloba	Tree	Costa Rica	LE	*Paraponera* (1)	?	Bennett and Breed 1985
Vicia angustifolia	Herb	California, USA	LE, RE	*Myrmica* (1)	Yes	Koptur 1979
Vicia faba	Herb	Japan	LE	*Tetramorium* (1)	Yes	Katayama and Suzuki 2004
Vicia sativa	Herb	California, USA; and England	LE	*Iridomyrmex* (1), plus native species	Yes/No	Koptur 1979
Loasaceae						
Mentzelia nuda	Herb	Nebraska, USA	RE	*Dorymyrmex* (1, = *Conomyrma*), *Formica* (2), *Lasius* (1), *Pheidole* (1)	Yes	Keeler 1981b

Malvaceae						
Gossypium thurberi	Shrub	Arizona, USA	LE, RE	*Forelius* (1), plus 11 species	Yes	Rudgers 2004; Rudgers and Strauss 2004
Hibiscus pernambucensis	Tree	São Paulo, Brazil	LE	*Azteca* (1), *Brachymyrmex* (1), *Camponotus* (5), *Crematogaster* (2), *Leptothorax* (1), *Paratrechina* (1), *Procryptocerus* (1), *Pseudomyrmex* (3), *Solenopsis* (1),	Yes?[b]	Cogni and Freitas 2002; Cogni, Freitas, and Oliveira 2003
Marantaceae						
Calathea ovandensis	Herb	Veracruz, Mexico	RE	*Brachymyrmex* (1), *Crematogaster* (1), *Dolichoderus* (1, = *Monacis*), *Pachycondyla* (1), *Paratrechina* (> 1), *Pheidole* (1), *Solenopsis* (1), *Wasmannia* (1)	Yes	Horvitz and Schemske 1984
Meliaceae						
Pseudocedrela kotschyi	Tree	Ivory Coast, Africa	LE	*Camponotus* (3)	No	Mody and Linsenmair 2004
Mimosaceae						
Stryphnodendron microstachyum	Tree	Costa Rica	LE	*Camponotus* (2), *Crematogaster* (2), *Ectatomma* (1), *Monomorium* (1), *Octostruma* (1), *Pachycondyla* (1), *Paraponera* (1), *Pseudomyrmex* (3), *Solenopsis* (2), *Wasmannia* (2)	Yes	De la Fuente and Marquis 1999
Ochnaceae						
Ouratea spectabilis	Tree	São Paulo, Southeast Brazil	SL, RE	*Acanthoponera* (1), *Brachymyrmex* (1), *Camponotus* (10), *Cephalotes* (1, = *Zacryptocerus*), *Crematogaster* (1), *Ectatomma* (1), *Pachycondyla* (1), *Pheidole* (2),	Yes	Ferreira 1994; Oliveira and Freitas 2004

(*continued*)

APPENDIX 6.1 **(continued)**

Plant	Growth form	Geographic location	Location of nectaries	Associated ant genera (number of species)	Protection by ants	Source
				Pseudomyrmex (4), *Solenopsis* (1), *Tapinoma* (1), *Wasmannia* (2)		
Ouratea hexasperma	Tree	Central Brazil	SL, RE	*Azteca* (1), *Camponotus* (11), *Cephalotes* (4, = *Zacryptocerus*), *Crematogaster* (2), *Pachycondyla* (1), *Pheidole* (2), *Pseudomyrmex* (2), *Solenopsis* (1)	?	Oliveira, Klitzke, and Vieira 1995
Orchidaceae						
Aspasia principissa	Epiphyte	Panama	RE	*Azteca* (2), *Camponotus* (2), *Crematogaster* (2), *Dolichoderus* (1, = *Hypoclinea*), *Ectatomma* (2), *Pachycondyla* (1), *Paratrechina* (1), *Pheidole* (1), *Tapinoma* (1)	?	Fisher and Zimmerman 1988
Spathoglotis plicata	Epiphyte	Guadelupe, French Antilles	RE	*Brachymyrmex* (1), *Camponotus* (1), *Ectatomma* (1), *Wasmannia* (1)	?	Jaffe et al. 1989
Passifloraceae						
Passiflora auriculata	Vine	Costa Rica	LE	*Brachymyrmex* (> 1), *Camponotus* (> 1), *Crematogaster* (> 1), *Ectatomma* (2), *Pheidole* (> 1), *Pseudomyrmex* (> 1), *Solenopsis* (> 1), *Wasmannia* (1), Unidentified (2)	Yes?[b]	Apple and Feener 2001
Passiflora biflora	Vine	Costa Rica	LE	*Brachymyrmex* (> 1), *Crematogaster* (> 1), *Pseudomyrmex* (> 1), *Solenopsis* (> 1), *Wasmannia* (1), Unidentified (2)	Yes?[b]	Apple and Feener 2001

Passiflora coccinea	Vine	French Guiana	LE, RE	*Camponotus* (2), *Crematogaster* (1), plus 6 other species	Yes?[b]	Wirth and Leal 2001
Passiflora glandulosa	Vine	French Guiana	LE	*Camponotus* (4), *Crematogaster* (1)	Yes	Labeyrie et al. 2001
Passiflora incarnata	Vine	Georgia, USA	LE	*Camponotus* (1), *Crematogaster* (1), *Forelius* (1, = *Iridomyrmex*), *Formica* (1), *Solenopsis* (1)	Yes	McLain 1983
Passiflora oerstedii	Vine	Costa Rica	LE	*Brachymyrmex* (> 1), *Camponotus* (> 1), *Crematogaster* (> 1), *Solenopsis* (> 1), *Wasmannia* (1), Unidentified (2)	Yes?[b]	Apple and Feener 2001
Passiflora quadrangularis	Vine	Costa Rica	LE	*Camponotus* (2), *Crematogaster* (3), *Dorymyrmex* (1, = *Conomyrma*), *Pheidole* (1), *Pseudomyrmex* (5), *Solenopsis* (2)	Yes	Smiley 1986
Passiflora vitifolia	Vine	Costa Rica	LE	*Azteca* (1), *Camponotus* (3), *Crematogaster* (7), *Ectatomma* (1), *Leptothorax* (1), *Pachycondyla* (1), *Paratrechina* (1), *Pheidole* (1), *Pseudomyrmex* (2), *Wasmannia* (1), unidentified Ponerinae (1)	Yes	Smiley 1986
Polygonaceae						
Fallopia japonica	Herb	Japan	LE	*Camponotus* (2), *Formica* (2), *Lasius* (1), *Leptothorax* (1), *Pheidole* (1), *Pristomyrmex* (1)	No	Kawano et al. 1999
F. sachalinensis	Herb	Japan	LE	*Camponotus* (2), *Formica* (2), *Lasius* (1), *Leptothorax* (1), *Pheidole* (1), *Pristomyrmex* (1)	No	Kawano et al. 1999
Rosaceae						
Prunus serotina	Tree	Michigan, USA	LE	*Formica* (1)	Yes	Tilman 1978
Rubiaceae						
Psychotria limonensis	Shrub	Panama	LE, RE	*Ectatomma* (2)	Yes	Altshuler 1999

(continued)

APPENDIX 6.1 **(continued)**

Plant	Growth form	Geographic location	Location of nectaries	Associated ant genera (number of species)	Protection by ants	Source
Tocoyena formosa	Tree	Minas Gerais, Southeast Brazil	RE	*Brachymyrmex* (2), *Camponotus* (8), *Cephalotes* (1, = *Zacryptocerus*), *Ectatomma* (3), *Pachycondyla* (1), *Pseudomyrmex* (3), *Wasmannia* (1), Unidentified Dolichoderinae (2), Unidentified Myrmicinae (1)	No	Santos and Del-Claro 2001
Tiliaceae						
Triumfetta semitriloba	Shrub	Minas Gerais, Southeast Brazil	LE	*Brachymyrmex* (1), *Camponotus* (1), *Cephalotes* (1), *Crematogaster* (1), *Linepithema* (1)	Yes	Sobrinho et al. 2002
Turneraceae						
Turnera ulmifolia	Shrub	Veracruz, Mexico	LE	*Camponotus* (2), *Crematogaster* (1), *Dorymyrmex* (1), *Forelius* (1), *Pseudomyrmex* (1)	Yes	Torres-Hernández et al. 2000; Cuautle and Rico-Gray 2003
Vochysiaceae						
Qualea grandiflora	Tree	São Paulo, Southeast Brazil	ST, RE	*Camponotus* (4), *Cephalotes* (2, = *Zacryptocerus*), *Ectatomma* (1), *Pseudomyrmex* (2), *Wasmannia* (1), Unidentified Dolichoderinae (1), Unidentified Formicinae (1)	Yes	Oliveira, Silva, and Martins 1987; Costa, Oliveira-Filho, and Oliveira 1992
Qualea multiflora	Tree	Minas Gerais, Southeast Brazil	ST, RE	*Camponotus* (3), *Cephalotes* (1, = *Zacryptocerus*), *Crematogaster* (1), *Ectatomma* (1), *Pseudomyrmex* (1), *Solenopsis* (1)	Yes	Del-Claro, Berto, and Réu 1996

Zingiberaceae						
Costus allenii	Herb	Panama	RE	*Azteca* (1), *Camponotus* (4), *Crematogaster* (1), *Ectatomma* (2), *Odontomachus* (1), *Pachycondyla* (1), *Paraponera* (1), *Pheidole* (3), *Solenopsis* (1), *Tapinoma* (1), *Wasmannia* (1)	?	Schemske 1982
C. laevis	Herb	Panama	RE	*Azteca* (1), *Camponotus* (5, = *Dendromyrmex*), *Crematogaster* (5), *Dolichoderus* (2, = *Monacis*), *Ectatomma* (2) *Odontomachus* (2), *Pachycondyla* (2), *Paraponera* (1), *Pheidole* (2), *Tapinoma* (1), *Wasmannia* (1)	?	Schemske 1982
C. pulverulentus	Herb	Panama	RE	*Azteca* (1), *Crematogaster* (3), *Ectatomma* (2), *Odontomachus* (1), *Pachycondyla* (1), *Paraponera* (1), *Paratrechina* (1), *Pheidole* (1), *Pseudomyrmex* (1)	?	Schemske 1982
C. scaber	Herb	Panama	RE	*Azteca* (2), *Camponotus* (3), *Dolichoderus* (1, = *Monacis*), *Ectatomma* (1), *Odontomachus* (1), *Pachycondyla* (1), *Paraponera* (1), *Pheidole* (2), *Wasmannia* (1)	?	Schemske 1982
C. woodsonii	Herb	Panama	RE	*Camponotus* (4), *Crematogaster* (1), *Ectatomma* (1), *Wasmannia* (1)	Yes	Schemske 1980, 1982

Note: Location of nectaries: LE = leaf (blade and/or petiole); SL = stipules; ST = stem; RE = reproductive structures (floral bracts, calix, spike, peduncle, fruit).

[a] Ants do not affect herbivory levels, but proximity to ant nest enhances plant nutrition and seed production; contact with ants decreases pollen viability.

[b] Visiting ants attack and remove simulated herbivores (live termite baits) from the plants.

CHAPTER SEVEN

Antagonism and Mutualism

Indirect Interactions

Some plants attract natural enemies of herbivores (e.g., ants and predatory and parasitoid flies and wasps) by bearing honeydew-producing insects such as hemipterans (fig. 7.1), lepidopteran larvae, and gallmakers (Cuautle et al. 1999; Pierce et al. 2002; Inouye and Agrawal 2004). Ant-tended Hemiptera (aphids, scales, coccids, white-flies, leafhoppers, and treehoppers) are sap-sucking herbivores that excrete the excess liquid and sugars as energy-rich honeydew (Auclair 1963; Way 1963; Buckley 1987a, 1987b). Most Hemiptera are herbivores and their deleterious effect on plants is not only due to sap-sucking, which decreases plant fitness; they are also important vectors of plant pathogens (Buckley 1987a, 1987b; Delabie 2001). The presence of Hemiptera in low-diversity systems (e.g., greenhouses, crop fields, and orchards) has been associated with high plant damage (Beattie 1985). It has been suggested, however, that under natural conditions hemipterans do not reach high densities, and therefore their presence can even be beneficial to some plants rather than harmful (Janzen 1972, 1973a). In this chapter we address and review the general characteristics of ant-hemipteran-plant interactions, including their conditional nature and possible outcomes, and their effects on the fitness of the various participants. In addition, using fossil and current evidence, we analyze the association between ants and hemipterans as it relates to the evolution of extrafloral nectaries.

FIGURE 7.1. The ant *Camponotus sericeiventris* foraging for the honeydew produced by *Notogonioides* treehoppers (Membracidae) on the plant *Nectandra coriaceae* (Lauraceae) in La Mancha, Veracruz, Mexico. From Rico-Gray 2001.

Ants, Plants, and Hemipterans

By avidly consuming honeydew, foliage-dwelling ants often increase the survival of honeydew-producing hemipterans and, consequently, may augment their deleterious effect on plants (Way 1963; Compton and Robertson 1988, 1991; Jordano and Thomas 1992; Jordano et al. 1992; Jahn and Beardsley 1996). There is, however, a high degree of conditionality in the outcome of such interactions (Cushman and Whitham 1989; Bach 1991; Cushman and Addicott 1991; Andersen 1991; Del-Claro and Oliveira 2000; Billick and Tonkel 2003). For instance, the effect of tending ants on honeydew-producing Membracidae not only increases treehopper survival but also confers a direct reproductive benefit to treehopper females, which may abandon the first brood to ants and lay an additional clutch next to the original brood (McEvoy 1979; Bristow 1983). Ant-derived benefits related to both protection and fecundity may vary yearly, though, and the ant species may differ in their effect on treehopper survival (Del-Claro and Oliveira 2000). Moreover, ants and other insects that are attracted to hemipteran honeydew may benefit plants when they attack and/or disrupt the activity of other, nonhemipteran herbivores (Messina 1981; Beattie 1985; Buckley 1987b; Oliveira and Del-Claro 2005). If protection to the plant outweighs the damage caused by the honeydew-producer, then an indirect mutualism is established, with benefits for ants, hemipterans, and host plant (Carroll and Janzen 1973; Compton and Robertson 1988, 1991).

Theoretically, in order for ant-derived plant protection to arise from the interaction between ants and honeydew-producers, three criteria should be met: (1) the ant-tended partner should not be the main herbivore (i.e., other herbivores should be present as well); (2) the ants should not allow excessive feeding rates or high densities in the hemipteran populations; and (3) the ants should be very effective in removing non-hemipteran herbivores and seed predators (Messina 1981; Horvitz and Schemske 1984). The benefits of ant protection, however, may be both unevenly and unpredictably distributed over time, especially if the main threat to the plant is due to maximum rather than to average herbivory (Rosengren and Sundström 1991). In summary, a positive outcome for the plant in an ant-Hemiptera-plant interaction will usually depend on the development and outcome of the ant-Hemiptera portion of this tritrophic interaction (see below).

Hemiptera-mediated mutualisms may offer some advantages over nectary-mediated mutualisms (Whitman 1994; but see Becerra and Venable 1989, 1991). On the one hand, with honeydew-producing Hemiptera, the plant does not need to evolve or maintain specialized secretory glands, and the hemipterans can provide both carbohydrates and protein to ant bodyguards (ants sometimes eat Hemiptera). On the other hand, sap-feeding hemipterans have so many natural enemies that they rarely proliferate in natural conditions unless ants are present (for reports on agricultural systems, see Gerling 1990; Ben-Dov and Hodgson 1997; Delabie 2001). Moreover, the dispersal capabilities of some honeydew-producing Hemiptera are so low that they probably would not appear on plants unless placed there by attendant ants (Way 1963). The presence of ants may increase the fitness of hemipterans by altering life-table parameters or influencing where they feed on the host plant (Addicott 1978; Völkl 1994; Stiefel and Margolies 1998). However, protection from predators and parasitoids is probably the most important benefit obtained by exudate-producing insects from their association with ants (Bristow 1984; Buckley 1987a, 1987b, 1990; Buckley and Gullan 1991; Itioka and Inoue 1996; Müller and Godfray 1999; Pierce et al. 2002).

Variable Outcomes in Ant-Hemipteran Systems

Although mutualism is defined as an interaction between two species that is beneficial to both (Boucher, James, and Keeler 1982), some mutualisms can be understood only in the context of the community and by assessing the influence of other species and other trophic levels on the pairwise relationship (reviewed by Bronstein and Barbosa 2002). A number of factors such as time, identity, abundance and behavior of participant species, and degree of herbivore damage may influence the final results of ant-hemipteran associations, and only by considering the variation of associated costs and benefits inherent to these systems can we understand their dynamics and the range of possible outcomes (Cushman 1991; Bronstein 1998; Gaume, McKey, and Terrin 1998; Stadler et al. 2001).

Publilia membracids can receive a range of benefits from their honeydew-gathering ant attendants, including protection from natural enemies and enhanced fecundity (McEvoy 1979; Bristow 1983, 1984). A series of experimental studies conducted by Cushman and co-workers in Arizona with *Publilia modesta* has shown that the strength of this ant-hemipteran mutualism can depend on a number of factors (Cushman and Whitham 1989, 1991; Cushman and Addicott 1991). Although the presence of tending ants reduces the occurrence of natural enemies (salticid spiders) on membracid-infested plants, the impact on *P. modesta* survival is age-specific and only nymphs benefit directly from ant tending (fig. 7.2A). Additionally, ant-derived benefits on treehopper survival are density-dependent, and nymphs in large aggregations benefit more from ant-tending than those in small aggregations (fig. 7.2B). In their three-year study with *P. modesta* treehoppers, Cushman and Whitham (1989) further demonstrated that protective services by tending ants vary between years and that the ant-membracid mutualism apparently becomes weak or absent when salticid spiders are scarce and/or are not deterred by tending ants. The protective services of ants can also turn out to be a limited resource for *P. modesta,* and intraspecific competition for ant attendants between *P. modesta* aggregations can significantly reduce ant-derived protection to nymphs and dramatically decrease the production of newly eclosed adult treehoppers (fig. 7.2C).

Ants differ in the benefits they provide to different honeydew-producers, and density-dependent effects may also work in the opposite way, as detected for ant-tended *P. modesta* membracids. For instance, while ant tending significantly increases the growth of low-density aphid

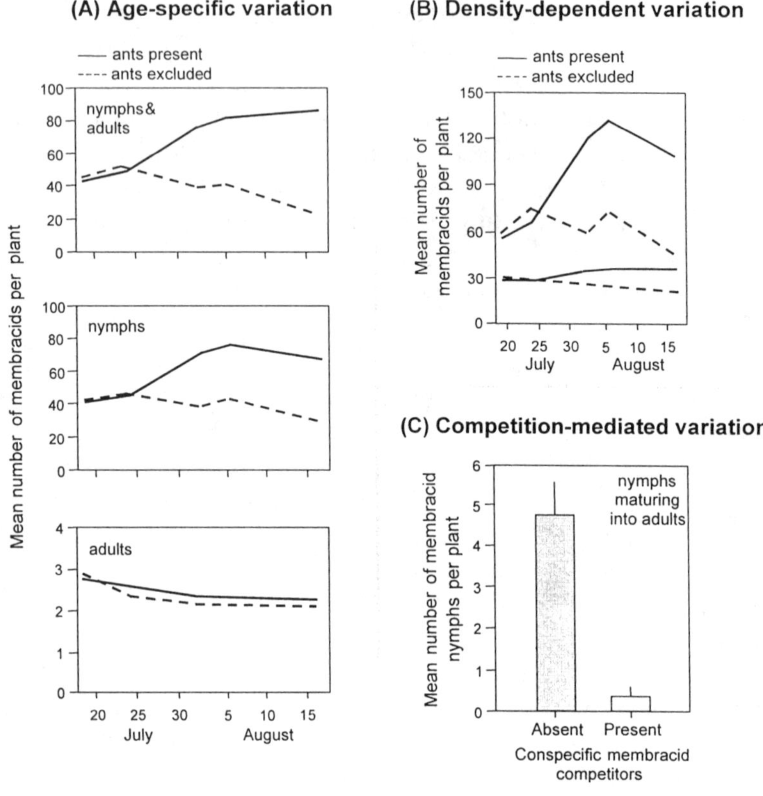

FIGURE 7.2. Sources of variation in the impact of tending ants on honeydew-producing *Publilia modesta* membracids, in ant-exclusion experiments carried out in Arizona (USA). (A) Number of *P. modesta* per plant through time in July–August 1985. Although ants increased overall membracid survival, only nymphs benefit from ant tending; no positive effect was detected for adult membracids. (B) Number of *P. modesta* nymphs per plant through time in July–August 1985, in large and small aggregations. Nymphs in large aggregations benefit more from ant tending than nymphs in small aggregations. (C) Number of *P. modesta* nymphs maturing into adults after 3 weeks in the presence or absence of experimentally added membracid competitors. The presence of competitors decreases production of newly eclosed adults by 92%. (A) and (B) modified from Cushman and Whitham 1989; (C) modified from Cushman and Whitham 1991.

populations, this positive effect from ants decreases as aphid density increases. Density dependence of the ant-aphid interaction is associated with decreased per capita tending levels as aphid density increases (Breton and Addicott 1992; Stadler and Dixon 1998). The decline may occur because the ants are unable to respond to the rapid increase in aphid density or because they have a limited requirement for honeydew (Addicott 1979). In contrast, large membracid aggregations are tended by more ants than smaller ones because large aggregations produce more honeydew, which in turn attracts a larger number of tending ants, resulting in increased benefit (Wood 1982; Cushman and Whitham 1989). Ant-derived protection has been shown to affect density-dependent parasitism rates in the ant-tended scale insect *Ceroplastes rubens* (Itioka and Inoue 1996). While the parasitism rate significantly decreases as scale density increases, irrespective of ant presence, the difference in parasitism rate between density levels is strikingly less without ant attendance. Thus the density-dependent decrease of the parasitism rate was more pronounced with ant attendance (Itioka and Inoue 1996; see also McNeil, Deslisle, and Finnegan 1977). Consequently, for any given population of hemipterans, the interaction with ants can be mutualistic for only a limited period of time (Andersen 1991a; Cushman and Whitham 1989, 1991; Breton and Addicott 1992).

Mutualistic associations between ants and hemipterans are usually diverse multispecies systems, which become very complex when temporal and spatial heterogeneity are considered (Addicott 1978, 1979; Andersen 1991a; Bristow 1991; Gullan, Buckley, and Ward 1993; Heckroth et al. 1998; Del-Claro and Oliveira 1999). However, even though ant-Hemiptera associations are usually multispecies systems, certain species-specific ant-hemipteran pair associations may be favored (Addicott 1979; Bristow 1984; Heckroth et al. 1998). For example, the aphid *Aphis vernoniae* and the membracid *Publilia reticulata* are tended by the ants *Tapinoma sessile, Myrmica lobicornis fracticornis,* and *M. americana* on their host plant *Vernonia noveboracensis* (Compositae). The benefits accruing to the hemipterans are unequal and asymmetric; aphids benefit more when tended by *Tapinoma,* while membracids benefit more when tended by *Myrmica.* In conclusion, the survival of hemipteran colonies was greatest when attended by a certain ant species and lowest when ants were excluded (Bristow 1984).

The honeydew-producing treehopper *Guayaquila xiphias* and its ant attendants form one of the best-studied neotropical ant-Hemiptera associations. In the cerrado savanna of Brazil, aggregations of *G. xiphias* commonly infest shrubs of *Didymopanax vinosum* (Araliaceae) and are tended by a diverse assemblage of honeydew-gathering ants. Honeydew flicking by brood-guarding females and incipient aggregations effectively attracts prospective tending ants onto the host plant (Del-Claro and Oliveira 1993, 1996). Across diurnal and nocturnal censuses, a total of 21 ant species have been recorded collecting honeydew from *G. xiphias* aggregations, the most frequent ones being *Camponotus rufipes, C. crassus, C. renggeri,* and *Ectatomma edentatum* (Del-Claro and Oliveira 1999). The relevance of *Guayaquila* honeydew as an energy supply for the ants is such that some species (*C. rufipes* and *E. edentatum*) tend the treehoppers continuously for 24 hours, and *C. rufipes* may even build satellite nests of dry grass to house the membracids (see Oliveira, Freitas, and Del-Claro 2002). The treehoppers are attacked by three main types of natural enemies: salticid spiders prey on nymphs and adults, predatory larvae of *Ocyptamus arx* (Diptera: Syrphidae) suck empty the entire body contents of the treehoppers, and *Gonatocerus* wasps (Myrmaridae) parasitize the egg masses (Del-Claro and Oliveira 2000). Controlled ant-exclusion experiments demonstrated that tending ants have a positive impact on treehopper survival and decrease the abundance of the natural enemies of *Guayaquila* on the host plant. Two years of experimental manipulations, however, revealed that ant-derived effects on hemipteran survival can vary both with time and with the species of tending ant (fig. 7.3A, B). Whereas in 1992 *Camponotus* and *Ectatomma* species were equally beneficial to *Guayaquila,* in 1993 only *C. rufipes* had a positive effect on treehopper survival (fig. 7.3B). At an initially high abundance of natural enemies (as in the experiment of 1993) the ant species differ in their impacts on treehopper survival, and apparently only the most aggressive ant attendant (*C. rufipes*) can confer protection to the membracids. The experiments of 1992 also revealed that ant-tending can positively affect *Guayaquila* fecundity, because brood-guarding females transfer parental care to ants and lay an additional clutch more often than untended females do (91% versus 54% of the cases, $N = 22$ females in each group; on *Publilia* membracids, see also McEvoy 1979; Bristow 1983). However, ant-derived impact on *Guayaquila* fecundity also varied between years, no significant positive effect being detected in 1993. It is possible that membracid females refrain from deserting their first brood in periods of initially high abundance of natural enemies.

The study with ant-tended *G. xiphias* is unique because it simultaneously demonstrates temporal conditionality related to both protection and fecundity in an ant-hemipteran system, and because it shows that species-specific effects from tending ants may present temporal variation associated with shifts in the abundance of natural enemies (Del-Claro and Oliveira 2000; see also Cushman and Whitham 1989).

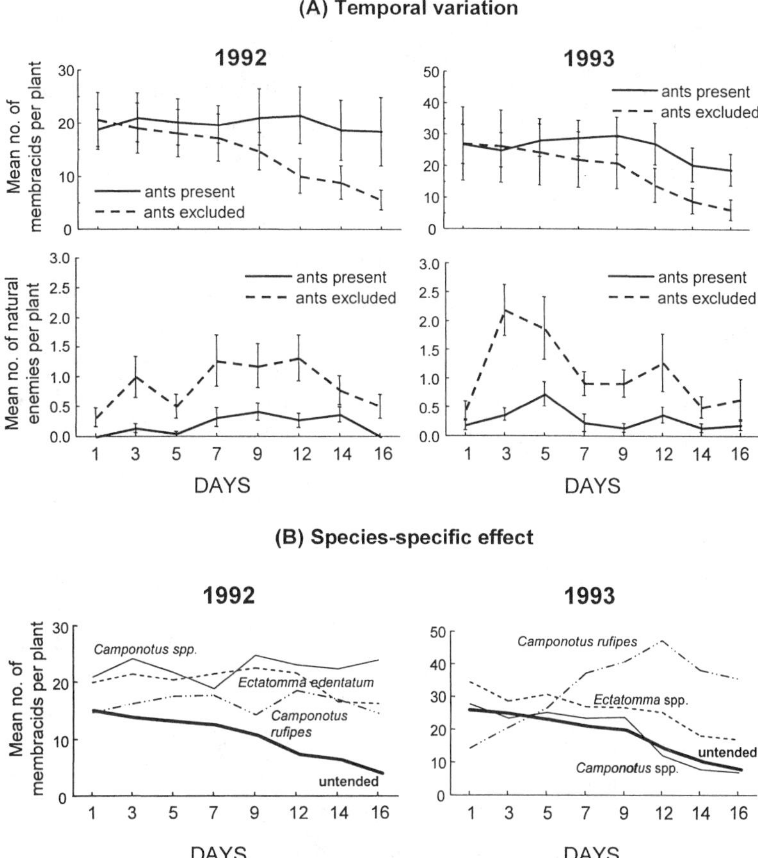

FIGURE 7.3. Number of *Guayaquila xiphias* membracids per plant (*Didymopanax vinosum*) through time, in ant-exclusion experiments ($N = 22$ in each group) carried out in the Brazilian cerrado in 1992 and 1993. (A) Ants had a positive effect on membracid survival in 1992 but not in 1993 and decreased significantly the numbers of natural enemies of membracids on plants (note the increased abundance of natural enemies in the first week of the 1993 experiment). (B) Membracid density per plant through time as a function of the species identity of ant partners. While in 1992 all ants had a positive impact on membracid survival, in 1993 only *Camponotus rufipes* benefited the hemipterans. Modified from Del-Claro and Oliveira 2000.

Another factor influencing the formation and maintenance of ant-Hemiptera interactions is the seasonal patterns of host plant use by hemipterans, particularly the selection of feeding sites (Bristow 1991; Dansa and Rocha 1992; Del-Claro and Oliveira 1999). For example, aphid colonies on the floral tips of *Nerium oleander* (Apocynaceae) attracted three to four times as many ants (*Iridomyrmex humile*) as colonies on leaf tips, even though the latter frequently contained more aphids (Bristow 1991). Host plants may modify the potential for ant-tended insects to attract ants and thus influence the formation and outcome of interactions between ants and exudate-producing insects (Price et al. 1980; Strong, Lawton, and Southwood 1984; Pierce 1985; Cushman 1991). Indeed, host plant characteristics or species identity can determine sugar composition of hemipteran honeydew (Campbell 1986; Hendrix, Wei, and Leggerr 1992; Fischer and Shingleton 2001), and this, in turn, can affect the level of ant tending (Völkl et al. 1999). More recently, however, Quental, Trigo, and Oliveira found no evidence of host plant mediation in the ant–*Guayaquila xiphias* system. Because the treehoppers prefer to feed near the inflorescences, these authors tested whether membracid aggregations located on plants (*Didymopanax vinosum*) with flowers would consume a phloem fluid of increased quality (higher sugar concentration), produce a higher-quality honeydew, and as a result receive increased ant attendance and better protection than treehoppers located on plants without flowers. Results showed that host-plant flowering status did affect the food quality gathered by the membracids; i.e., phloem sugar concentration was higher in plants with flowers compared to plants without flowers. This difference, however, did not translate into richer or more copious honeydew for tending ants on plants with flowers. Although treehoppers survived better when tended by ants, plant flowering status had no effect on ant tending levels or on hemipteran survival. Plants with flowers accumulated more natural enemies through time than plants without flowers, though, suggesting that the mode through which the host plant could possibly affect ant-*Guayaquila* interactions is complex and likely involves a range of indirect effects from other participant species (Quental, Trigo, and Oliveira 2005). Moreover, plant quality and the relative importance of different nutrients, as related to metabolic constraints faced by sap-feeding hemipterans (Douglas 2003), also need to be taken into account in the analyses of plant-hemipteran-ant interactions. We should therefore expand our view of tritrophic interactions so as to include the potential contribution of the host plants to the attraction of ant attendants toward exudate-producing insects (Cushman 1991). Due to the great variation in outcome exhibited by ant-Hemiptera interactions, it should be no surprise that ant-Hemiptera-plant interactions are unstable, rendering an uncertain outcome for the host plant.

The Effect of Ant-Hemiptera Interactions on Host Plants

Ant foraging on vegetation can be promoted by the occurrence of predictable and immediately renewable food sources, including extrafloral nectar, honeydew from sap-feeding hemipterans, and secretions from lepidopteran larvae (see Way 1963; Bentley 1977a; Buckley 1987a, 1987b; Koptur 1992a; Pierce et al. 2002). Intense ant foraging on leaves has resulted in a multitude of ant-plant-herbivore interactions, ranging from facultative to obligate associations (Davidson and McKey 1993). From the plant's standpoint, the outcomes of many of these interactions are largely mediated by how ant behavioral patterns can affect herbivore performance on a given host plant (Oliveira, Freitas, and Del-Claro 2002). Liquid food is typically supplied on foliage in the form of extrafloral nectar and insect-derived exudates, and aggressive behavior exhibited by ants at these food sources ("ownership behavior" [see Way 1963]) can produce variable consequences, both for the herbivore and for the plant (Beattie and Hughes 2002; Oliveira and Freitas 2004; and references therein). Ant behavior required to deter herbivores on plants is similar to that needed to protect ant-tended insects from their natural enemies, and in either case ownership behavior by the ants is sufficient to expel intruders from the ants' immediate patrolled area (Way 1963; Bentley 1977a). Indeed, several ant species associated with extrafloral nectaries (EFNs) or honeydew-producing insects probably confer protection to visited plants and tended insects through similar behaviors (DeVries 1991; Koptur 1992a). For instance, the aggressive *Camponotus rufipes,* an abundant ant on cerrado foliage (Oliveira and Brandão 1991), is very effective at both deterring herbivores on plants with EFNs (Oliveira, Silva, and Martins 1987) and protecting *Guayaquila* treehoppers from their natural enemies (fig. 7.3B; see Del-Claro and Oliveira 2000). Overt aggression by ants, however, may not be essential for herbivore deterrence on foliage—"timid," minute *Petalomyrmex* ants can effectively protect *Leonardoxa* trees from chewing and sucking herbivores (Gaume, McKey, and Anstett 1997), whereas the "passive" *Pheidole* confers protection to ant-occupied *Piper* saplings by removing eggs of insect folivores (Letourneau 1983).

Although few studies have demonstrated full benefits to plants harboring ant-tended hemipterans (see below), some studies support the hypothesis that honeydew-producing hemipterans can protect plants by attracting tending ants that deter other herbivores, as first proposed by

Carrol and Janzen (1973). For instance, Room (1972) demonstrated that *Crematogaster* ants tending hemipterans on a mistletoe species protect the plant from other herbivores and allow increased shoot growth of the mistletoe. By attacking leaf-chewing beetles on goldenrod (*Solidago*), *Formica* ants associated with *Publilia* membracids decrease defoliation to the plant, which results in increased seed production and growth by stems with membracids (Messina 1981). In a study involving different types of herbivores, Oliveira and Del-Claro (2005) demonstrated that ants tending honeydew-producing *Guayaquila xiphias* treehoppers on *Didymopanax vinosum* shrubs significantly reduce damage by chewing and mining insects to leaves, and by thrips to the apical meristem of host plants (fig. 7.4). Another type of benefit conferred by tending ants to a scale's host plant was demonstrated by Bach (1991), who showed that sanitation by honeydew-gathering ants reduces leaf death and abscission caused by fungal infection on accumulated honeydew (see also Queiroz and Oliveira 2001).

Beneficial effects from ants tending hemipterans have also been demonstrated for plants regularly housing ant colonies (i.e., myrmecophytes). For instance, *Myrmelachista* ants tending mealybugs inside *Ocotea* trees remove insect eggs from young stems and leaves (Stout 1979). *Crematogaster* ants inhabiting *Macaranga* trees tend scales inside the stems and not only remove herbivores from leaves but also prune foreign plants that come in contact with their host plant (Fiala et al. 1989). The type of ant-tended insect can also mediate ant-derived benefits to an ant-occupied plant: Gaume, McKey, and Terrin (1998) demonstrated that net benefits against herbivory conferred by *Aphomomyrmex* ants to myrmecophytic *Leonardoxa* depends on the type of phloem-sucking hemipteran (coccids or pseudococcids) tended by resident ants inside the tree's hollow stems.

Negative effects to plants from ants tending honeydew-producing insects have also been detected in temperate and tropical habitats. For example, Buckley reports that *Iridomyrmex* ants tending *Sextius* membracids in Australia also gather extrafloral nectar from the host plant (*Acacia*). Because treehoppers were more successful than EFNs in attracting the ants, protection by ants against other herbivores was disrupted and produced an overall negative effect of the ant-membracid interaction on plant growth and seed set (Buckley 1983; see also DeVries and Baker 1989). In Maryland, United States, Fritz demonstrated that although *Formica* ants attending *Vanduzea* membracids can decrease

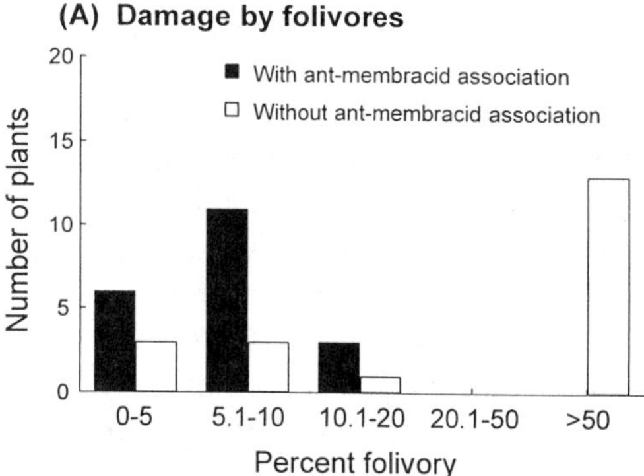

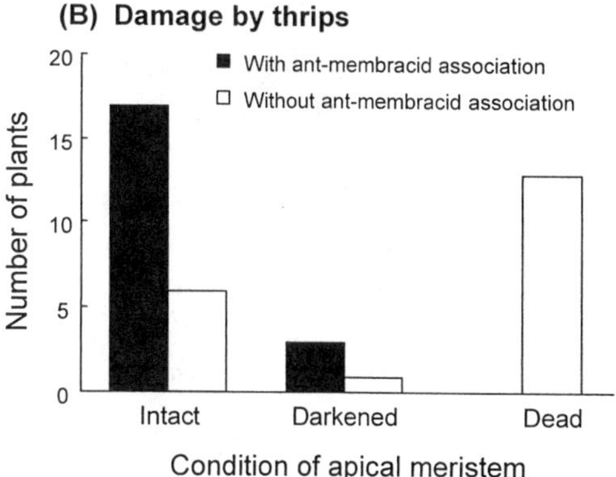

FIGURE 7.4. Effect of the association between ants and *Guayaquila xiphias* membracids on the host plant, *Didymopanax vinosum*, in the Brazilian cerrado. Damage to plants comprises 1 year of herbivore activity after the establishment of experimental plant groups. (A) Levels of pooled folivory by *Caralauca olive* beetles (Chrysomelidae) and leaf-mining lepidopteran larvae are much lower on plants hosting ant-tended membracids than on plants without the association. (B) Damage by *Liothrips didymopanicis* (Thysanoptera: Phlaeothripidae) to the apical meristem is considerably less severe on plants with ant-tended membracids compared to plants without the hemipterans. Modified from Oliveira and Del-Claro 2005.

adult density and oviposition by leaf-mining beetles on black locust (*Robinia pseudoacacia*), they also indirectly protect beetle larvae by removing their main hemipteran predator. Due to the opposite impacts of ants on adult and immature beetles, Fritz (1983) detected no apparent benefit or harm to black locust in hosting ants tending *Vanduzea*. In Mexican coastal sand dunes, Rico-Gray and Thien (1989a) report that honeydew-producing mealybugs shift ant attention away from the EFNs of *Myrmecophila* orchids, which results in augmented damage to plant reproductive organs and decreased fruit set. In the same habitat, Rico-Gray and Castro (1996) showed that the impact of ant-aphid interactions on *Paullinia* seed set varies yearly from negative to neutral. Temporal variation in the detrimental effects of aphid infestation to flower and fruit production has also been demonstrated for wild radish (*Raphanus sativus*) in California (Snow and Stanton 1988).

Because of the complicated nature of ant-Hemiptera-plant interactions and the many factors that render the outcome of such associations uncertain for the plants (Messina 1981; Beattie 1985; Gove and Rico-Gray 2006), these are not always stable interactions (Thompson 1982, 1994; Oliveira and Del-Claro 2005). Although in general ant-Hemiptera-plant interactions have been considered to generate a positive outcome for the plant, few studies have demonstrated full benefits to plants harboring ant-tended Hemiptera (Buckley 1987b; Zachariades 1994; Rico-Gray and Castro 1996). For instance, although shrubs of *D. vinosum* hosting ant-*Guayaquila* interactions present decreased damage levels to leaves and growing shoot tips (fig. 7.4), they become susceptible to infestation by the lycaenid butterfly *Parrhasius polibetes* (= *Panthiades polibetes*), whose ant-tended larvae destroy floral buds. As also shown for other ant-tended lycaenid butterflies (Atsatt 1981a, 1981b; Pierce and Elgar 1985), females of *P. polibetes* preferentially infest plants with ant-*Guayaquila* interactions (fig. 7.5), and the liquid-rewarding caterpillars shift ant attention partly away from the treehoppers. Bud-consumption by *P. polibetes* on plants with ant-*Guayaquila* interactions decreases the host plant's reproductive output by 84% (Oliveira and Del-Claro 2005). Clearly the multispecies system around ant-*Guayaquila* associations is similar to those studied by other authors, in which the analyses of pairwise interactions cannot predict the overall impact on the plant from all species involved (see Price et al. 1980; Messina 1981; Thompson 1988; Bronstein 1994a, 1994b; Cushman 1991; and references therein).

Furthermore, ants are not the only organisms interacting with honeydew-producing Hemiptera. Hemipteran honeydew has been suggested as a key resource in the evolution of Diptera (Downes and Dahlem 1987), but a suite of other organisms are also attracted to it, including bees and wasps (Krombein 1951; Zoebelein 1956a, 1956b; Evans and Eberhard 1970; Jirón and Salas 1975; Vinson 1976; Downes and Dahlem 1987; Moller and Tilley 1989; Godfray 1994; Cuautle et al. 1999; Camargo and Pedro 2002), fungi (Jirón and Salas 1975; Hughes 1976;

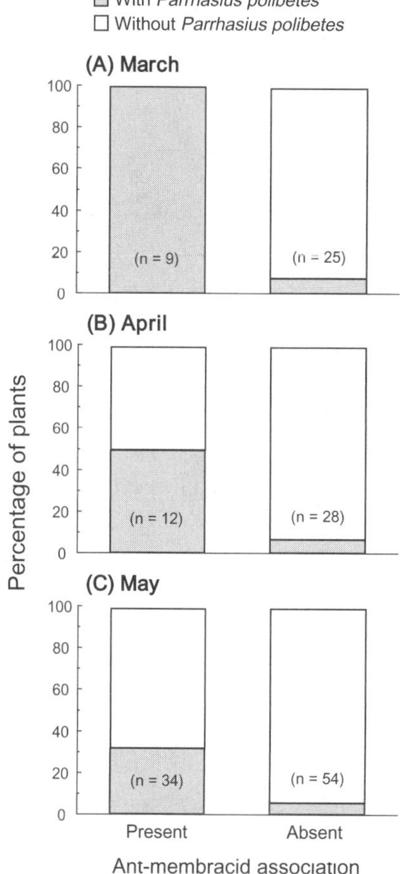

FIGURE 7.5. Monthly infestation patterns by larvae of the ant-tended *Parrhasius polibetes* butterfly on *Didymopanax vinosum* shrubs in the Brazilian cerrado, on individual plants with or without ant-tended *Guayaquila xiphias* membracids. In all months the butterflies preferably infested plants hosting ant-tended membracids. Modified from Oliveira and Del-Claro 2005.

Borror, De Long, and Triplehorn 1981; Greenberg, Macias-Caballero, and Bichier 1993; Cuautle, García-Franco, and Rico-Gray 1998; Queiroz and Oliveira 2001), and birds (Paton 1980; Edward 1982; Gaze and Clout 1983; Greenberg, Macias-Caballero, and Bichier 1993).

Ant-Hemiptera associations are among the most facultative, opportunistic, and variable mutualistic interactions, and since the seasonal nature of the environment and of food availability forces insects to move among food sources, decreasing the chance to form specific interactions, it is difficult to predict the effect of these interactions on the host plants and throughout the community (e.g., Gove and Rico-Gray 2006). Likewise, the evolutionary potential for future establishment of these mutualisms, or their potential for coevolution, is highly uncertain.

Ant-Hemiptera Associations and the Evolution of Extrafloral Nectaries

Ant-Hemiptera-plant interactions increase in complexity and uncertainty when plants have EFNs. The presence of hemipterans may disrupt the ant-extrafloral-nectary interaction and amplify the negative effects of the hemipterans on the plant (Buckley 1983; Rico-Gray and Thien 1989a). These effects, it is argued, are due to the increased attractiveness of insect exudates to ants, since honeydew is generally richer than extrafloral nectar (Buckley 1983; Beattie 1985; Horvitz and Schemske 1984). It is thus clear that the possible evolutionary relationship between Hemiptera and the origin and evolution of EFNs should be carefully analyzed using both fossil and current evidence.

Fossil Evidence

Despite the prevalence and importance of ant-plant interactions, not much is known about their evolutionary history (Beattie 1985; Pemberton 1992; see also chapter 1). Even though ant-plant interactions may have developed as early as the mid-Cretaceous, the fossil evidence is extremely scarce (C. C. Labandeira, pers. com.). The only published account dates ant-plant interactions back to the Oligocene and reports the presence of EFNs in 35-million-year-old fossil leaf impressions of the extinct *Populus crassa* (Pemberton 1992). The leaves come from the Florissant Formation of Colorado, where ants also lived. Thirty-two ant species are among the most abundant of the fossil insects at Florissant. Ten of these extinct ant species belong to five extant genera (*Camponotus, Formica, Iridomyrmex, Lasius,* and *Pheidole*), which have modern species acting as ant-guard mutualists of modern plants that have EFNs (Pemberton 1992 and references therein). In addition to ants, other insects lived at Florissant, including predatory insects (coccinellid beetles and syrphid flies) and parasitic insects (Braconidae and Ichneumonidae wasps), whose modern relatives visit EFNs (Pemberton 1992).

The presence of fossil ants and EFNs could be a clear evidence for the ant-guard antiherbivore defense and ant-plant interactions in general. However, because ants were not the only insect group associated with fossil extrafloral-nectaried *Populus,* and since the other groups present also forage for extrafloral nectar, the origin and evolution of EFNs should not be treated as exclusively related to ants, as it usually is treated (see chapter 6), or, for that matter, as a solution by the plant to decrease the effect of ant-Hemiptera associations (Becerra and Venable 1989, 1991). Even though extrafloral and circumfloral nectaries are visited mainly by ants, a variety of other arthropod groups (e.g., wasps, bees, flies, beetles, spiders, and mites) also regularly visit these liquid food sources (Koptur 1992a; Pemberton 1992; Pollard, Beck, and Dodson 1995; Pemberton and Lee 1996; Cuautle et al. 1999; Kawano et al. 1999; Ruhren and Handel 1999; Torres-Hernández et al. 2000), and the relative effects of these groups on the origin and evolution of nectaries and on plant fitness have rarely been comparatively analyzed (but see Cuautle and Rico-Gray 2003 and references therein).

Current Evidence

It is generally accepted that EFNs evolved as a generalized plant defense, attracting ants that in turn would repel or remove herbivores. Nevertheless, results of field studies have been contradictory, and plant defense by ants has not been demonstrated in all cases (see Koptur 1991, 1992a). An alternative hypothesis is that EFNs actually may have evolved as a specific defense against ant-Hemiptera mutualisms (Becerra and Venable 1989, 1991). In short, plants would shift ant attention away from honeydew-producing Hemiptera by offering nectar as an alternate food source. However, several studies do not support this hypothesis (Buckley 1983; Rico-Gray and Thien 1989a; Fiala 1990; Del-Claro and Oliveira 1993), demonstrating that at least some ant species actually prefer hemipteran honeydew (in general of better quality than extrafloral nectar) to the food offered by the plant (e.g., Katayama and Suzuki 2003) or forage for both in an opportunistic manner.

In the cerrado vegetation near Brasília, Brazil, more caterpillar species were found foraging on leaves of plants with EFNs than on plants without these glands (H. Morais, unpubl. data). Behavioral observations of myrmecophilous butterfly larvae that possess secreting "ant-organs" suggest that caterpillars are more attractive to ants than EFNs are (DeVries and Baker 1989). It would seem that caterpillars, or females when ovipositing, are seeking the protection of ants to deter predators and parasitoids (Atsatt 1981b; see also fig. 7.5).

At Los Tuxtlas, Mexico, Horvitz and Schemske carried out perhaps the most detailed study to date of the interactions involving a plant with extrafloral nectaries (*Calathea*), visiting ants, and an ant-tended insect (larvae of *Eurybia;* Lepidoptera, Riodinidae). They demonstrated that in the absence of tending ants, damage to reproductive tissues by *Eurybia* caterpillars reduced *Calathea* seed production by 66%, far more than the 33% reduction in the presence of tending ants. Because *Eurybia* is a specialist herbivore and frequently finds *Calathea* host plants, the ant-*Eurybia* association seems beneficial for the plant even though seed set is greatest without the caterpillars. Species-specific variation in the impact of ants on seed production and interhabitat differences in ant communities further complicate the unpredictable and variable character of this complex interaction system (Horvitz and Schemske 1984).

Ants tending Hemiptera may not abandon their food source when offered an alternative and rich food option; on the contrary, EFN-bearing plants may even indirectly benefit hemipteran aggregations by attracting more would-be tending ants than plants lacking these glands (Buckley 1987a, 1987b; Compton and Robertson 1988; Cushman and Addicott 1991). Del-Claro and Oliveira (1993) demonstrated that the discovery by ants (*Camponotus crassus* and *C. rufipes*) of an alternate sugar/liquid food source experimentally available on the host plant (*Didymopanax vinosum*) did not provoke desertion by the ants. The ants continued tending the aggregations of honeydew-producing membracids (*Guayaquila xiphias*) at nearly the same rate as ant visitation to the honey solution increased steadily within the same period.

Similarly, in order to assess whether ants would abandon honeydew-producing Hemiptera when offered an additional, considerably rich and abundant, food source, the association between the EFN-lacking shrub *Solanum lycocarpum* (Solanaceae), the predominantly diurnal ant *Camponotus crassus,* and two membracid species (*Enchenopa* spp.) was studied in the cerrado near Brasília, Brazil (Rico-Gray and Morais 2006). The results show that the mean number of ants visiting the simulated nectaries was significantly higher than those tending the membracids. However, the membracids were never left untended, suggesting that even when ants exploit a considerably rich and abundant liquid food source, they do not abandon other sources. Katayama and Suzuki (2003) demonstrated that ants shift their collection pattern from extrafloral nectar to honeydew at increased density of aphids per plant. It is reasonable to assume that honeydew is more attractive to ants than EFN at high aphid density, because ants react sensitively to differences in quality and/or quantity of both extrafloral nectar and honeydew (see below). Furthermore, ants (*Lasius japonicus* and *Tretamorium tsushimae*) are more efficient in excluding herbivores (*Hypera postica,* Coleoptera) on individuals of the EFN-bearing *Vicia angustifolia* (Fabaceae) parasitized by aphids (*Aphis craccivora*) than on plants without aphids. Whether ants switch food sources based on quality and/or quantity, and not merely based on the presence or absence of food sources, awaits further investigation.

The above results are not surprising since ants can select and prefer sugar solutions containing a complex mixture of amino acids (Lanza 1988; Smith, Lanza, and Smith 1990; Swift and Lanza 1993; Swift, Bryant, and Lanza 1994; Blüthgen and Fiedler 2004a, 2004b; but see also Woodring et al. 2004) or can discriminate between poor and rich hemipteran honeydew (Völkl et al. 1999; Blüthgen, Gottsberger, and Fiedler 2004). Moreover, herbivore damage can induce increased amounts and better quality (i.e., concentration of sugar and amino acids) of extrafloral nectar (Smith, Lanza, and Smith 1990; Swift and Lanza 1993), which in turn increases plant attractiveness to ants and, consequently, patrolling activity by defending ants (Bentley 1977a; Tilman 1978; Stephenson 1982; Swift, Bryant, and Lanza 1994). Rico-Gray (1993) has shown that 38% of the plant species in Mexican coastal dry-deciduous vegetation host ant-Hemiptera associations and that in over one-third of those cases, the ants also foraged for the nectar produced by the plant (whether floral, circumfloral, or associated with leaves). Furthermore, nectar was the most important food resource for ants in certain habitats in Mexico (Rico-Gray 1989, 1993; Rico-Gray, García-Franco, et al. 1998; Rico-Gray, Palacios-Rios, et al.1998), but seasonal changes in food availability (or the temporary presence of a richer food source) apparently allow ants to use a wide variety of food resources. Such seasonal shifts in turn may change the level of defense (if any) by ants of particular plant species.

The main liquid food resources used by ants (floral and extrafloral nectar, hemipteran honeydew, and secretion from lepidopteran larvae) are treated as separate items in most studies. Future studies should consider that some ant species include several or all such food resources as part of their diet at different times of the year and that not all ant species use the entire array of available resources within a community

(Rico-Gray 1993; Rico-Gray, García-Franco, et al. 1998). Moreover, there are few species-specific ant-plant associations, and in many instances the associations do not involve all the individuals of a given plant species (see examples in Rico-Gray 1993). Relatively recent evidence and newly emerging theory on the quantitative and qualitative allocation of resources to antiherbivore defense suggest that the distributions of particular defense strategies are functions of resource availability (Schupp and Feener 1991 and references therein). The possibility that fluctuations in the level of defense of plants by ants could reflect fluctuations in resource availability should be addressed. Furthermore, in most cases only a portion of the array of interactions between a plant and its visitors is studied at one time; we suggest taking a multispecies/multitrophic approach instead (e.g., Cuautle and Rico-Gray 2003; Cuautle, Rico-Gray, and Díaz-Castelazo 2005).

The evolution of EFNs deserves more attention, and results must be interpreted with caution because ant-plant associations tend to be facultative rather than obligate and specialized. For example, ant species may readily switch from hemipterans to nectaries depending on seasonal changes in food availability (Rico-Gray and Sternberg 1991; Rico-Gray 1993). Mutualisms often do not result in a high degree of specificity. Several unrelated ant species commonly share an easily exploitable resource, and only occasionally does selection favor the evolution of a single ant species and a single plant species specialized to exploit the mutualism (Thompson 1982). Furthermore, floral and extrafloral nectaries and honeydew-producing hemipterans are also visited by an array of organisms other than ants (Koptur 1992a; Pemberton 1992; Pollard, Beck, and Dodson 1995; Pemberton and Lee 1996; Cuautle et al. 1999; Kawano et al. 1999; Ruhren and Handel 1999). The ants tending EFNs may limit the presence of predatory and/or parasitoid wasps and flies (Koptur and Lawton 1988; Pemberton and Lee 1996) (fig. 7.6), and the interaction of these organisms with ants could change the level of defense and the overall outcome for the plant. For example, several ant species (e.g., *Camponotus* spp., *Crematogaster brevispinosa, Pheidole* spp.) and wasps (*Polistes instabilis, Polybia occidentalis*) visit the EFNs of the shrub *Turnera ulmifolia* (Turneraceae). The presence of both wasps and ants, when acting separately, exerts a positive effect in decreasing herbivory levels and increasing the number of unripe fruits. When acting together, however, their effect is not additive and suggests interference between groups (Cuautle and Rico-Gray 2003). Clearly ad-

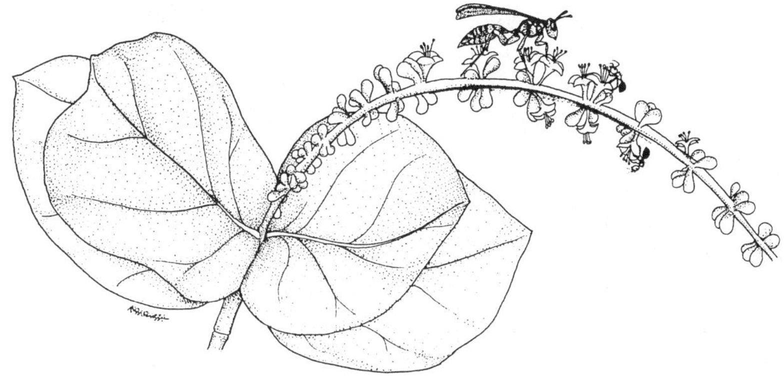

FIGURE 7.6. Inflorescences of *Coccoloba uvifera* (Polygonaceae) visited by ants (e.g., *Camponotus planatus*) and predatory/parasitoid wasps (e.g., *Polistes*) and beetles, on the coast of Yucatán, Mexico.

ditional studies are needed to compare the joint effect of ants, wasps, bees, and flies visiting EFNs (Beckmann and Stucky 1981) and the influence of ant-tended insects on plant fitness (Messina 1981; Horvitz and Schemske 1984; Oliveira and Del-Claro 2005).

The hypothesis that EFNs may have evolved as a specific defense against ant-Hemiptera mutualisms (Becerra and Venable 1989) has to be carefully reviewed, because ants are not the only organisms collecting exudates on foliage and potentially affecting plant fitness. Moreover, present-day interactions may overshadow the original selective pressures that created the relationship. Based on the scarce fossil record of ant-plant interactions in general, and of EFNs in particular, and on the fact that the first fossil record of an extrafloral nectary is associated with several insects and spiders, it is not clear which of these organisms created the selective pressures associated with the origin of EFNs.

* * *

Future work should deal in greater depth with the real nature of the ant-plant-Hemiptera associations, that is, as multispecies interacting systems (Bronstein and Barbosa 2002) and not as associated pairs of species (e.g., ant-hemipterans, ant-plant, plant-hemipterans). In this way we should hopefully clarify whether the origin and evolution of EFNs can be associated with ant-plant-hemipteran interactions. Because of the many factors that render the outcome of ant-Hemiptera-plant systems uncertain for the plants, these are not always stable interactions. Furthermore the seasonal nature of the environment, including variation in food availability, forces insects to use multiple resources (Suzuki, Ogura, and Katayama 2004) and decreases the chance to form specific interactions, making it difficult to establish and predict the effect of these interactions on host plants and communities, as well as to discern their evolutionary potential. Although in general ant-Hemiptera-plant interactions have been considered to generate a positive outcome for the plant (Oliveira and Del-Claro 2005 and included references), sap-feeding hemipterans are herbivores whose damage can indeed severely decrease plant fitness (Delabie 2001). The extent of this damage is counteracted only if deterrence of nonhemipteran herbivores by honeydew-gathering ants is enough to increase plant fitness above the level of damage caused by the hemipterans. Finally, plant-hemipteran-ant associations are of interest not so much because of the direct interactions between different pairs of organisms (e.g., plants-Hemiptera or Hemiptera-ants), but rather because of the diverse indirect effects that these complex multitrophic interactions can have on the plants. The variable character of such complex systems requires sampling at different sites and times for a more realistic evaluation of their impact on plants (Horvitz and Schemske 1984; Thompson 1988).

CHAPTER EIGHT

Nutrition of Plants by Ant Mutualists

Life History of Ant-Fed Plants and Ant-Garden Systems

Not all mutualisms have evolved from, or are related to, an antagonistic interaction, and it has been suggested that some components of life histories may predispose organisms to mutualistic interactions (Thompson 1982). For instance, organisms living in environments characterized by intermediate levels of disturbance and faced with intermediate survival abilities are expected to have a higher probability of evolving mutualisms with other species than organisms with very high or very low survival abilities in similar environments rich in biotic interactions (Roughgarden 1975). Under intermediate conditions, the small positive effects of a mutualist can increase survival or growth rates significantly (Roughgarden 1975; Thompson 1982). On a broad scale, some of the most obvious mutualisms in communities, including ant-plant mutualisms, are associated with intermediate levels of disturbance.

Light gaps in forests induce a high richness of biotic interactions through intense competition among plants and through animal-plant interactions. For instance, extrafloral nectaries (hence ant-plant associations) are more common along forest edges and in light gaps than in other forest sites in Costa Rica (Bentley 1976), and the density of plants with extrafloral nectaries (EFNs) is higher in relatively open vegetation types such as the cerrado savanna and Amazonian secondary-growth areas in Brazil (Oliveira and Oliveira-Filho 1991; Morellato and Oliveira 1991). EFNs are also more common in lowland open coastal sites than in mountain humid forests in Jamaica (Keeler 1979a) and Mexico (Rico-Gray 1989, 1993; Díaz-Castelazo and Rico-Gray 1998; Rico-Gray, García-Franco, et al. 1998; Rico-Gray, Palacios-Rios, et al. 1998; Rico-Gray et al. 2004). And there are more ant-plant interactions in dry open habitats than in close, more humid environments in Mexico (Rico-Gray, García-Franco, et al. 1998; see chapter 10). Somewhat similar patterns were detected in ant-hemipteran systems. For instance, ant-membracid associations are more frequent on plants growing in open and sunny areas than in shady sites in the Brazilian cerrado (Del-Claro and Oliveira 1999), and the proportion of membracid species that are dependent upon ants for defense declines with increasing altitude in Colombia (Wood 1984; Olmstead and Wood 1990).

Mutualism can also be favored when organisms with a high probability of encounter and very low premutualism growth rates live in environments that impose a high level of physical stress (e.g., nutrient-poor habitats) but lack the richness of antagonistic interactions that is the basis for selection in many other mutualisms (Thompson 1982). The mutualisms associated with this set of ecological conditions often involve nutrition of a host by a symbiont in nutrient-poor environments. In nutrient-poor environments, such as those characterized by intermediate disturbance regimes, small inputs by a mutualist can potentially have major effects on fitness of species with intermediate survival ability.

Probably the clearest examples of how stressful environments can favor novel forms of interaction, including mutualism, involve the reversal of the usual trophic order of life. Insectivorous plants and ant-fed plants are the two major examples of plants evolving to gain nutrients directly and actively from animals; the first in an antagonistic manner and the latter through mutualism with ants (Thompson 1981b, 1982). The mutualistic subset of these reversed trophic interactions involves ants that live within plant parts that appear specialized for harboring ants and for absorbing nutrients from the ants' debris piles (Huxley 1986; Beattie 1989; Benzing 1991). We will first review the published literature and examples of ant-fed plants and then describe the ant-plant associations commonly referred to as ant gardens.

Ant-Fed Plants

Myrmecotrophy, or the potential ability of plants to absorb nutrients from debris piles of ant nests (Beattie 1989; Benzing 1991), was originally demonstrated experimentally in a few species of flowering plants and ferns. The general view was that ant-fed plants were largely tropical epiphytes living in nutrient-poor soils in the families Rubiaceae (*Myrmecodia, Hydnophytum, Squamellaria, Anthorrhiza,* and *Myrmedoma*), Orchidaceae (*Myrmecophila* = *Schomburgkia*), Polypodiaceae (*Lecanopteris* and *Solanopteris*), Asclepiadaceae (*Dischidia*), and Bromeliaceae (*Tillandsia*) (appendix 8.1). However, the recent discovery that woody plants (shrubs and trees) such as *Tococa guianensis, Maieta guianensis* (Melastomataceae), *Cordia nodosa* (Boraginaceae), and *Hirtella physophora* (Chrysobalanaceae) in South America; *Piper fimbriulatum* and *P. obliquum* (Piperaceae) in Costa Rica; and *Cecropia peltata* (Cecropiaceae) in Trinidad can obtain nutrients from debris deposited by ant inhabitants is changing the general view of ant-fed plants (appendix 8.1). We will first review the epiphytic myrmecophytes and then the geophytes or nonepiphytic myrmecophytes.

Janzen was the first to report that species of two notable genera of the Rubiaceae (*Myrmecodia* and *Hydnophytum*) in Southeast Asia harbored ants (usually *Iridomyrmex cordatus*) in their chambered tubers. He also hypothesized that the animal debris placed by the ants in the rough-walled chambers breaks down, probably facilitated by fungi living in the debris piles, is subsequently absorbed, and contributes as a nutritive source for the plant (Janzen 1974). In these plants a tuber is developed and becomes chambered even in the absence of ants (Huxley 1978, 1980, 1986). The ants live and raise their brood in the smooth chambers of *Myrmecodia,* and part of the colony's catch composed of insect and small arachnid species is placed in the plant's rough-walled chambers (Huxley 1978; Rickson 1979). Isotope tracer studies (^{35}S sulphate and ^{35}S methionine [Huxley 1978]; ^{14}C amino acids and glucose [Rickson 1979]) have shown that phosphate and probably amino acids experimentally fed to the ants were deposited preferentially in the rough-walled chambers and were absorbed by the plant (Huxley 1978; Rickson 1979). This was the first demonstration of direct nutrient uptake from organic material (ant debris piles) deposited in hollow chambers of plants.

Uptake was also demonstrated for another Southeast Asian epiphyte (*Dischidia major,* Asclepiadaceae) living in nutrient-poor sandstone soils and inhabited by *Philidris* spp. (Dolichoderinae) ants. Treseder, Davidson, and Ehleringer found that the ants live in saclike "ant-leaves," where they raise their young and deposit debris (feces, dead ants, and scavenged insect parts). Adventitious roots from the plant grow through the cavity opening and proliferate wherever debris has accumulated. To test the hypothesis that the plant uses ant debris as a nitrogen source and that the stomata on the internal surfaces of leaf cavities may also absorb ant-respired carbon dioxide and thereby reduce transpiration water loss, Treseder and co-workers measured stable-isotope ratios ($\delta^{13}C$, $\delta^{15}N$) of ants, hosts, and substrates and capitalized on differences in isotope composition of possible nutrient sources. They concluded that 39% of the carbon in occupied host-plant leaves derives from ant-related respiration and that 29% of the host nitrogen is derived from debris deposited into the cavities by ants. Thus, in exchange for domatia, the ants provide significant amounts of two limiting resources (carbon dioxide and nitrogen) in a nutrient-poor environment (Treseder, Davidson, and Ehleringer 1995).

The first demonstration of nutrient uptake from ant debris piles for a nonflowering plant was studied by Gay in the epiphytic fern genus *Lecanopteris* (Polypodiaceae) in Southeast Asia. These ferns are regularly inhabited by five ant species (*Iridomyrmex* and *Crematogaster*) that nest and deposit debris in the hollow rhizomes that form domatia. The plants gain nutrients from the ants in two ways: by root absorption from carton runways that surround plants and by uptake of solutes from ant feces and debris through the inner rhizome walls. The rhizome cavity surface is black, minutely pitted, and bears no specialized absorptive structures. Radioactively labeled nutrients (14-glucose, 86-rubidium, and 32-phosphorus) that were experimentally injected into the rhizome cavity were translocated throughout the plant. Ants inhabiting *Lecanopteris* were fed radioactively labeled nutrients (15-nitrogen-labeled glycine and urea), and these also were incorporated into the fern tissues (Gay 1993).

Nutrient uptake for a neotropical species was first demonstrated for the large epiphytic orchid *Myrmecophila christinae* (= *Schomburgkia tibicinis*) (Orchidaceae), which inhabits coastal environments (sand dune scrub, mangroves, and tropical dry forest) in the state of Yucatán, Mexico (fig. 8.1). The large pseudobulbs of this orchid serve as permanent domatia for several ant species (*Camponotus* spp. and *Crematogaster brevispinosa*) and as seasonal domatia for *Ectatomma tuberculatum* (Thien et al. 1987; Rico-Gray et al. 1989; Rico-Gray and Sternberg 1991). A pseudobulb is developed and becomes chambered even in the absence of ants, and each pseudobulb has an opening at the base. Ants select certain chambers where they live and raise their brood, and they pack other pseudobulbs (ca. 5% per orchid plant) with debris that include dead ants, feces, a variety of insects, pieces of plant material,

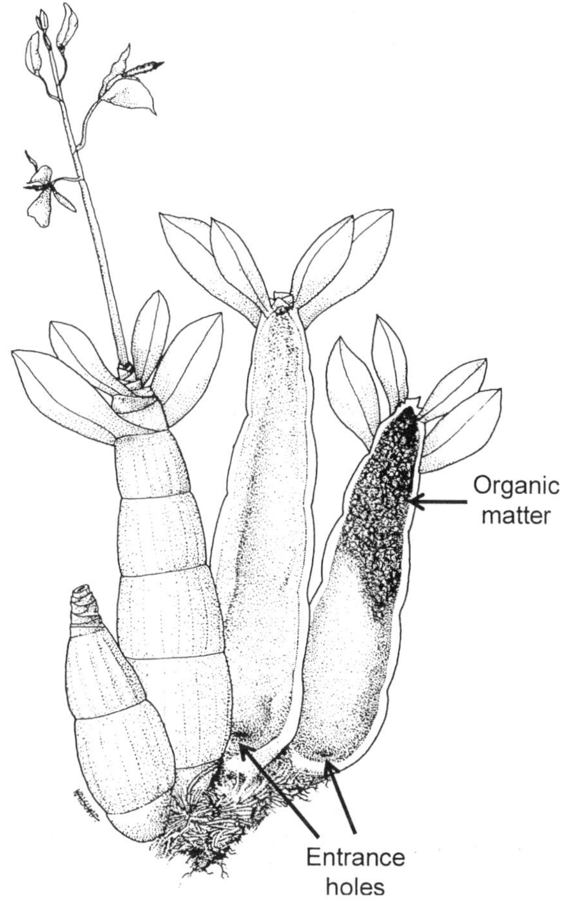

FIGURE 8.1. Pseudobulbs of *Myrmecophila christinae*. From left to right: external view of pseudobulb, pseudobulb with inflorescence, inside of pseudobulb without organic matter, inside of pseudobulb with organic matter. Entrance holes for the ants are indicated for two pseudobulbs. Modified from Rico-Gray et al. 2004.

seeds, and sand (Rico-Gray et al. 1989). Sometimes roots from the pseudobulb grow through the cavity opening and proliferate wherever debris has accumulated. As in the above example, it was hypothesized that material in the debris piles breaks down (fungi and bacteria are present) and is subsequently absorbed and constitutes a nutritive source for the plant. A uniformized set of radioactively labeled (^{14}C-labeled D-glucose) dead *Solenopsis invicta* ants was prepared and introduced into a set of pseudobulbs (control pseudobulbs were also marked). After two weeks

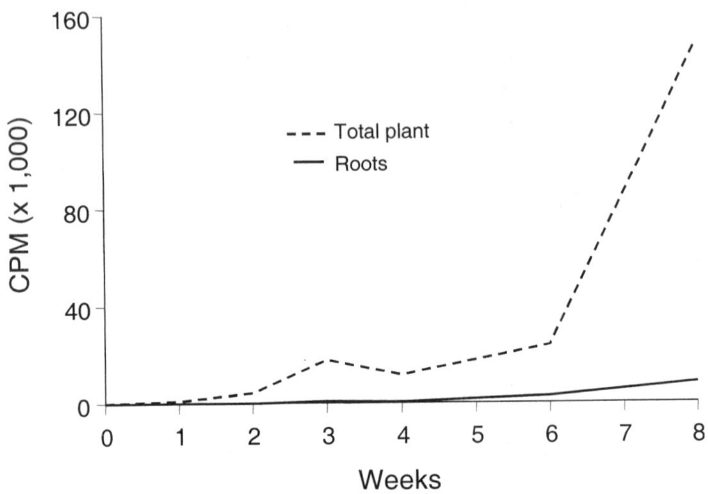

FIGURE 8.2. Amount of radioactivity incorporated into tissues of *Myrmecophila christinae* (= *Schomburgkia tibicinis*) following placement of ^{14}C-labeled ants into pseudobulbs. Total amount of radioactivity in counts per minute (CPM) per 5 gm fresh weight in combined supernatant and pellet fractions for total plant (roots and stems combined) and for roots only. Modified from Rico-Gray et al. 1989.

of exposure, the labeled organic material moved from the dead *S. invicta* ants and was recorded in the various plant parts (leaves, roots, and pseudobulbs) (fig. 8.2), thus demonstrating nutrient uptake from ant debris piles in this orchid species.

With the exception of the ant-*Dischidia* case, the ant-plant associations reviewed above are especially complex. Besides offering the plants organic matter from debris piles, ants also patrol and protect the plants from herbivorous insects. Ants move along the stems of the individuals of *Myrmecodia* and *Hydnophytum,* where they obtain nectar and are protected by outgrowths of the plant surface (stipules, stem tissue, or spines [Huxley 1986]); patrol and protect the individuals of *Lecanopteris* ferns from herbivorous insects; nurture the juveniles or disperse spores (Gay 1993); and move along the inflorescence spikes of the orchid *Myrmecophila christinae,* where they obtain nectar (Rico-Gray and Sternberg 1991) and protect the plant from herbivorous insects (Rico-Gray and Thien 1989a, 1989b).

As mentioned above, nutrient uptake is not restricted to epiphytes. The tropical tree *Cecropia peltata* (Cecropiaceae) in Trinidad obtains 98% of its nitrogen from debris deposited by ants, and *Azteca* inquilines derive 18% of workers' carbon from their host plant (Sagers, Ginger, and Evans 2000). Fischer et al. have shown in the shrubs *Piper fimbriulatum* and *P. obliquum* (Piperaceae) that there is a nitrogen flux from ants (*Pheidole bicornis*) to plants and that nutrient transfer occurred within six days after the ants had been fed with ^{15}N-labeled food. They also demonstrated that 25% of the labeled nitrogen ingested by the ants was incorporated by the plants (Fischer et al. 2003). Other geophyte examples where nutrient uptake has been demonstrated, all from South America, are presented in appendix 8.1.

Neotropical ant-fed geophytes (as well as the epiphyte *M. christinae*) may also receive the benefit of protection by the ants. For instance, as Alvarez et al. demonstrated, *Tococa* myrmecophytes (*T. spadaciflora, T. guianensis,* Melastomataceae) in the Colombian Chocó are protected from herbivory by ants (*Pheidole* sp., *Azteca* sp., *Brachymyrmex heeri, Crematogaster* sp., and *Wasmannia auropunctata*); each plant, however, was occupied by a single ant species. It was shown that *T. spadaciflora* was more vulnerable to herbivores in the absence of its associated ants, as demonstrated by a higher average herbivory in ant-excluded plants (91%) than in control plants (27%). Similarly, ant-free *T. guianensis* exhibited significantly higher herbivory levels than ant-occupied plants (55% vs. 22%, respectively) (Alvarez et al. 2001). In the Brazilian cerrado, *Tococa formicaria* (Melastomataceae) has also been reported as receiving antiherbivore protection from its ant inhabitants, especially *Azteca* sp. 1, which was present within the domatia of 47% of the sampled individuals (Bizerril and Vieira 2002).

In *Hirtella myrmecophila* (Chrysobalanaceae), its obligate ant partner (*Allomerus octoarticulatus*) inhabits younger branches and forages mainly on new leaves. By experimentally removing the ants, Izzo and Vasconcelos showed that *A. ortoarticulatus* is crucial for the defense of young leaves against herbivores. Furthermore, mature flowers and fruits were found only on older branches with no ants, and flower production was eight times greater on plants whose ants were experimentally removed, which suggests that the abortion of domatia in older leaves is a strategy developed by *H. myrmecophila* to minimize the effects of cheating by *A. octoarticulatus* (Izzo and Vasconcelos 2002; chapter 6).

Furthermore, other studies have reported that species in the genera *Maieta, Clidemia, Tococa* (Melastomataceae) are myrmecophytic and associated with ants in the genera *Azteca, Allomerus, Crematogaster, Dolichoderus, Gnamptogenys, Myrmelachista, Pheidole, Pseudomyrmex,* and *Solenopsis* (Whiffin 1972; Benson 1985; Vasconcelos 1991, 1993; Cabrera and Jaffe 1994; Clausing 1998; Vasconcelos and Davidson 2000; Belin-Depoux and Bastien 2002). *Maieta guianensis* is a small shrub that produces both leaf pouches as ant domatia and small food bodies (pearl bodies [O'Dowd 1982]). These resources are used by the ants *Pheidole minutula* and *Crematogaster* sp., which are obligate plant-ants but are not restricted to *M. guianensis* (Vasconcelos 1991, 1993; Vasconcelos and Davidson 2000). The ants patrol both young and mature leaves, and when the ants were experimentally removed from the plants, leaf damage increased significantly (about 25%), compared to less than 1% in plants with ants. Plant fitness was also reduced, as fruit production was 45 times as great in plants with ants as in plants without ants one year after ant removal (Vasconcelos 1991). It has also been shown that *M. guianensis* and *Tococa bullifera* exhibit a significant difference in size depending on the associated ant species. Those plants with *Crematogaster laevis* were on average smaller than those with *Pheidole minutula* (*M. guianensis*) and *Azteca* sp. (*T. bullifera*) (Vasconcelos and Davidson 2000). Interestingly, the ants (genus *Myrmelachista*) associated with *Tococa guianensis* and *Clidemia heterophylla* are extremely efficient at keeping their host plants free of encroaching vegetation by first biting it and then spraying into its leaves a poison with herbicidal effect (Renner and Ricklefs 1998).

Ant-fed plants are a varied and complex group in terms of their evolution. The specialized structures to utilize nutrients from ant debris piles have evolved in different plant families and at different times of their evolutionary history. These associations become more complex because in many cases the ants also protect the plant against herbivores, tend honeydew-producing insects, or feed directly on food rewards offered by the plant. Further generalizations on ant-fed plants await more natural history and experimental studies in different parts of the world.

Ant Gardens

Ants are probably the most common arboreal insects in tropical forests (Hölldobler and Wilson 1990; Majer 1990; Longino and Colwell 1997; see chapter 9). Consequently, it is quite possible that ants are the main animal contact for epiphytes, with both groups participating in a number of different types of interspecific interactions that have become prominent in epiphyte biology, such as ant-fed plants and ant gardens (Benzing 1991; Davidson and Epstein 1989).

Ant gardens (*Ameisengarten* [Ule 1902, 1905, 1906]) are one of the least studied, and probably least understood, ant-plant interactions. The term *ant garden* refers to the inferred mutualism between ants and the epiphytic plants that grow in association with the arboreal carton nests of ants in both New and Old World tropical forests (fig. 8.3; see Kleinfeldt 1978; Madison 1979; Davidson 1988; Davidson and Epstein 1989; Benzing 1991). The ants belong to a variety of advanced taxa (e.g., *Azteca, Camponotus, Solenopsis,* and *Crematogaster*), sharing high-calorie diets and the ability to produce nests of carton (Benzing 1991; Yu 1994). The

FIGURE 8.3. General view of the rain-forest epiphyte *Anthurium gracile* attached to a tree branch. Ants (e.g., *Odontomachus* and *Pachycondyla*) construct carton nests within the epiphyte's spherical root mass and create an ant garden by harvesting fleshy epiphyte seeds that are incorporated into the nest walls where they germinate.

ants are usually involved in many aspects of the life of the plant—they may disperse seeds, place seeds in a suitable substrate (carton nests), forage for floral and extrafloral nectar, and occasionally pollinate or defend the plant. Ant-garden plants are similarly diverse, including ferns and angiosperms (Benzing 1991). Numerous reports exist for species of *Anthurium* and *Philodendron* (Araceae), *Codonanthe* and *Codonanthopsis* (Gesneriaceae), *Epiphyllum* (Cactaceae), *Ficus* (Moraceae), *Peperomia macrostachya* (Piperaceae), and *Streptocalyx longifolia* (Bromeliaceae) (Kleinfeldt 1978, 1986; Madison 1979; Benson 1985; Davidson 1988; Orivel and Dejean 1999; Yu 1994).

The establishment of several common epiphytes appears to be mainly or exclusively restricted to ant gardens (Davidson 1988). In general, wind, birds, or ants disperse the seeds from epiphytes, and juvenile plants appear on tree trunks or branches with their roots usually exposed to the air. Ant-garden ant species construct nests around these roots, bringing soil particles and plant debris to the young root cluster, which is soon embedded in an organic matrix and surrounded by carton (Kleinfeldt 1978, 1986; Beattie 1985; Davidson 1988; Davidson and Epstein 1989; Orivel, Dejean, and Errard 1998). The carton is considered quite nutritive for plants, particularly if it contains vertebrate feces or decayed foliage (Davidson 1988; Benzing 1991). The plants supply a variety of resources to ants, ranging from seed attachments and pearl bodies to sugary exudates of extrafloral nectaries and/or hemipteran honeydew. Moreover, the fibrous roots of some ant-garden epiphytes branch through the ant carton, offering greater structural integrity (Kleinfeldt 1978, 1986; Davidson 1988; Davidson and Epstein 1989; Orivel, Dejean, and Errard 1998; Yu 1994).

Several examples of associations between epiphytes and ants are mentioned in the literature, but many seem to be more casual than those referred to as ant gardens (Beattie 1985; review in Davidson and Epstein 1989). The first detailed study of this process concerns the arboreal, carton-building ant *Crematogaster longispina* (Myrmicinae) on the roots of the epiphytic vine *Codonanthe crassifolia* (Gesneriaceae) in a Costa Rican rain forest (Kleinfeldt 1978). It is assumed that both species in this interaction benefit from increased fitness. The plants produce floral and extrafloral nectar and fruits with pulp and seed arils that are consumed by the ants. In turn, the ants place the seeds of the plant in the walls of their carton nests. The seeds germinate and the plants grow and obtain nutrients from the detritus with which the ants construct their nests. The growth rate of plants that are not on ant nests is significantly less than the growth rate of plants on ant nests (Kleinfeldt 1978). Amazonian ant-garden ants are often parabiotic, with two or three species sharing a single nest (Davidson 1988; Benzing 1991; Orivel and Dejean 1999). The different ant species have bodies of different sizes, which may help the partitioning of a common living space, although it has been suggested that the most minute ant species may be parasites of the system (Benzing 1991). The association between epiphytes and ant-garden ants positively affects the fitness of the ants because epiphytes help the carton nest survive heavy rains. Moreover, compared to what they could do without the epiphytes, the ants can build larger nests in more exposed and light-richer environments, increasing the amount of food available to hemipteran-tending *Camponotus* and *Crematogaster* ants (Yu 1994).

The associations between epiphytes and ant-garden ants range from casual and facultative growth from ant cartons, to regular and obligate mutualisms between particular epiphyte and ant species under certain broad environmental conditions (Davidson and Epstein 1989). Preadaptations of plants and ants appear to have been very important to the origin of ant gardens. Although the evidence for evolutionary specialization, extreme specialization, and coadaptation is circumstantial, it is suggestive (Davidson 1988; Thompson 1994). Davidson and Epstein (1989) point out that the mutual and positive reinforcement of population dynamics should be strongest in associations in which ants disperse and plant the seeds of their epiphytes (chapter 4), thereby assuring constancy in selection pressures across ant and epiphyte life histories. These dynamics may have promoted both species specificity and abundance in ant-epiphyte associations even without evolution of mutualism (Davidson and Epstein 1989). Ant gardens represent a striking example of how a mutualism can dramatically increase the population density of the partners and alter community patterns (Yu 1994).

Conclusion

Despite the original considerations, not all examples of nutrient uptake from ant debris piles by plants involve epiphytes, and the origin of these ant-plant mutualistic associations is still unclear. Furthermore, the plants in all Southeast Asian examples are epiphytes and, with the notable exception of *Myrmecophila,* all American examples are nonepiphytes.

Nutrient uptake by plants has evolved independently in epiphytic and geophytic flowering plants as well as in ferns, both in the Old and the New World tropics. The only clear characteristic all these examples have in common is that the plants inhabit nutrient-poor environments, suggesting high levels of physical stress, and this component of their life histories may predispose them to mutualistic interactions (Roughgarden 1975; Thompson 1982). However, most of the ant-plant associations described above may also involve some degree of antiherbivore defense by the ants (e.g., Fisher 1992). Ant-derived plant protection seems to be strongly associated with environments rich in antagonistic associations, which is considered to be a very important trigger for the development of a mutualistic association (Thompson 1982). In general, the associated ants do not appear to nest obligatorily in the plants; and even though either party may survive without the other, very few of these plants are found in their natural habitat without ants. Indeed these ant-plants have specialized structures to attract and maintain ants and to take advantage of ant debris. The associations are not species-specific, and the same ants can be found elsewhere associated with other plants; however, specificity is high within a population, a given habitat, or a geographic area (Gay and Hensen 1992). Finally, the above generalizations about ant-fed plants need reevaluation since new examples are being studied, and we should fully quantify the costs and benefits inherent to each party, including the factors causing their variation.

APPENDIX 8.1 **Examples of ant-fed plants for which nutrient absorption from ant debris pile has been demonstrated or suggested**

Plant	Growth form	Distribution	Associated ant genera (number of species)	Source
Asclepiadaceae				
Dischidia major	epiphyte	Tropical, Malaysia	*Philidris* (1)	Treseder, Davidson, and Ehleringer 1995
Boraginaceae				
Cordia nodosa	shrub	Neotropical	*Allomerus* (1)	Solano and Dejean 2004
Cecropiaceae				
Cecropia peltata	tree	Neotropical	*Azteca* (> 1)	Sagers, Ginger, and Evans 2000
Chrysobalanaceae				
Hirtella physophora	shrub	Neotropical	*Allomerus* (1)	Solano and Dejean 2004
Gesneriaceae				
Codonanthe crassifolia	epiphyte	Neotropical	*Crematogaster* (1)	Kleinfeldt 1978
Melastomataceae				
Maieta guianensis	shrub	Neotropical	*Pheidole* (1)	Belin-Depoux and Bastien 2002; Solano and Dejean 2004
M. poeppigii	shrub	Neotropical	*Crematogaster* (1), *Pheidole* (1)	Christianini and Machado 2004
Tococa formicaria	shrub	Neotropical	*Azteca* (2), *Camponotus* (1), *Dolichoderus* (2), *Leptothorax* (1), *Linepithema* (1), *Oligomyrmex* (1)	Bizerril and Vieira 2002
T. guianensis	shrub	Neotropical	*Azteca* (1)	Cabrera and Jaffe 1994
Orchidaceae				
Myrmecophila christinae	epiphyte	Neotropical	*Camponotus* (3), *Crematogaster* (1), *Ectatomma* (1)	Rico-Gray et al. 1989

(*continued*)

APPENDIX 8.1 **(continued)**

Plant	Growth form	Distribution	Associated ant genera (number of species)	Source
Piperaceae				
Piper (2 spp.)	shrub	Neotropical	*Pheidole* (1)	Fischer et al. 2003
Polypodiaceae				
Lecanopteris (7 spp.)	epiphyte	Tropical, Malaysia	*Camponotus* (1), *Crematogaster* (3), *Iridomyrmex* (2)	Gay and Hensen 1992; Gay 1993
Rubiaceae				
Hydnophytum (> 40 spp.; data from *H. formicarium* and unidentified specimens in Huxley 1978)	epiphyte	Tropical, SE Asia	*Anoplolepis* (1), *Camponotus* (2), *Crematogaster* (> 1), *Iridomyrmex* (2), *Monomorium* (1), *Pheidole* (1), *Polyrachis* (1), *Tapinoma* (1), *Technomyrmex* (1), *Tetramorium* (1), *Turneria* (1), *Vollenhovia* (1)	Janzen 1974; Rickson 1979; Huxley 1978, 1986
Myrmecodia (5 spp.)	epiphyte	Tropical, SE Asia	*Camponotus* (1), *Crematogaster* (1), *Iridomyrmex* (4), *Pheidole* (1), *Technomyrmex* (1)	Janzen 1974; Huxley 1978, 1986

Note: See also Wheeler 1942; Benson 1985; Davidson and Epstein 1989; Hölldobler and Wilson 1990; Davidson and McKey 1993.

CHAPTER NINE

Canopy-Dwelling Ants, Plant and Insect Exudates, and Ant Mosaics

Studies of foliage-dwelling arthropods in tropical rain forests have shown that ants may represent 86% of the arthropod biomass and up to 94% of the arthropod individuals living in the canopy (Majer 1990; Tobin 1995). Although species richness of tropical arboreal ants is generally moderate, their numerical and behavioral dominance in the canopy environment has been repeatedly revealed in different tropical forest ecosystems (Adis, Lubin, and Montgomery 1984; Majer 1990; Stork 1991; Tobin 1991, 1994, 1995; Brühl, Gunsalam, and Linsenmair 1998; Dejean et al. 2000). Species of foliage-dwelling ants include both "true" canopy inhabitants (i.e., those that nest only in plant organs) and species that commonly nest on the ground but are also able to form colonies in hanging soil or are associated with the epiphytes and hemiepiphytes that abound in the canopy of a tropical forest (Davidson 1988; Davidson and Epstein 1989; Dejean, Olmsted, and Snelling 1995; Blüthgen et al. 2000). The remarkable ecological success of ants in the canopy can be attributed to, among other things, their ability to overcome some of the main constraints of this peculiar environment (as compared to the ground), including a more limited number of nest sites, decreased amounts and diversity of food resources, and low humidity and strong winds (Adis, Lubin, and Montgomery 1984; Wilson 1987a; Stork 1991; Davidson and Patrell-Kim 1996; Davidson 1997; Floren and Linsenmair 1997; Yanoviak, Dudley, and Kaspari 2005). Ant foraging on tropical foliage is promoted by the widespread occurrence of predictable and renewable liquid food sources such as extrafloral nectar, honeydew from sap-feeding hemipterans, and secretions from lepidopteran larvae (Way 1963; Bentley 1977a; Buckley 1987a, 1987b; Pierce et al. 2002). Extrafloral nectaries (EFNs), for example, are widespread among the flora of different vegetation physiognomies (chapter 10), and honeydew-producing hemipterans are thought to be far more abundant than currently estimated by canopy fogging techniques (Blüthgen et al. 2000; Dejean et al. 2000).

Ants are distributed over a wide variety of habitats and display a range of lifestyles; until recently they were regarded as fundamentally carnivorous (Sudd and Franks 1987). Although researchers recognized the role of some ant species as consumers of plant products, their function in food webs was considered to be primarily as predators and scavengers of animal matter (Wilson 1959; Carroll and Janzen 1973; Hölldobler and Wilson 1990). Because ants make up a major part of the arthropods living in the canopy of tropical forests, and because their biomass greatly surpasses that of their potential herbivore prey, this pattern poses a paradox and challenges our understanding of energy flow in the canopy. Tobin (1991, 1994) has explained the "ant-biomass paradox" by suggesting that most foliage-dwelling ants behave mainly as primary consumers and rely mostly on plant- and insect-derived exudates for nutrition.

Recent studies have provided strong evidence that the occurrence of liquid rewards on foliage plays a key role in mediating the foraging ecology of foliage-dwelling ants and may explain the predominance of ants in the canopy environment. Indeed, all dominant ants whose diets are well documented are typically exudate users (see below), and the few ant species that contribute disproportionately to the abundance of ants in canopy samples are known to feed extensively on exudates (Davidson 1997). In this chapter we present recent evidence supporting Tobin's suggestion and show how consumption of liquid foods and dominance status of particular ant species can be associated in the canopy environment. Competitive interactions among ant species at such food sources are shown to strongly affect the structure of arboreal ant communities, with important consequences for other associated arthropods and, ultimately, for the host plants.

Canopy Ants: Main Features and Trophic Role

A key feature of most canopy-dwelling ants is their ability to process large quantities of carbohydrate-rich plant and insect exudates, most especially extrafloral nectar and honeydew from sap-feeding Hemiptera. The proventriculus is an important element of the ant digestive system, and its morphological differentiation has been associated with increased dependence on liquid foods (Eisner and Wilson 1952). In all ants the proventriculus controls the flow of food from the crop (or "social stomach" [Wilson 1971]) to the midgut, where digestion occurs. Fifty years ago Eisner (1957) showed that species in the ant subfamilies Formicinae and Dolichoderinae are particularly well adapted for a diet of liquid foods due to some key innovations in their digestive systems. Associated with an expandable crop, alterations to the proventriculus in these two groups from the ancestral state in more predatory ants allows the storage of large liquid volumes and facilitates food sharing among nest mates (Eisner and Wilson 1952; Eisner 1957). These structural modifications connecting the crop to the worker's midgut most likely evolved independently in formicines and dolichoderines (Shattuck 1992; Chiotis, Jermiin, and Crozier 2000). The acquisition of more efficient modes of storing and processing liquid food may have prepared these ant taxa for further specialization on plant and insect exudates, which are rich in carbohydrates but very poor in protein and amino acids (Davidson 1997, 1998). Although the proventriculus in most of the Myrmicinae is reduced, functionally similar adaptations in some exudate-feeders (genera *Crematogaster* and *Cephalotes*) possibly allow them to benefit efficiently from liquid foods as well (Eisner 1957; Davidson 1997).

In an attempt to explain the extraordinary abundance of ants in tropical rain-forest canopies in Peru and Borneo, Davidson et al. (2003) investigated whether dominant arboreal ant species derive both carbohydrates and nitrogen (N) from plant and insect exudates and therefore act as herbivores as predicted by Tobin (1991). The ratio of N isotopes ($^{15}N/^{14}N$, formulated as $\delta^{15}N$) provides a tool for comparing N sources across ant species with different feeding habits. Relative to plant values, animal tissues are almost always enriched in ^{15}N, and there is a progressive enrichment of $\delta^{15}N$ at successive trophic levels (DeNiro and Epstein 1981; Post 2002). By comparing $\delta^{15}N$ values across ant taxa and with data for coexisting plants and other arthropod groups, Davidson et al. (2003) demonstrated that despite the very low concentrations of nitrogenous compounds in plant tissues, a number of specialized exudate-feeding ant species (especially formicines and dolichoderines) obtain their nitrogen lower in the trophic chain than do predominantly predacious species in the same habitat (fig. 9.1). A parallel work, carried out by Blüthgen, Gebauer, and Fiedler (2003) in an Australian rain forest produced

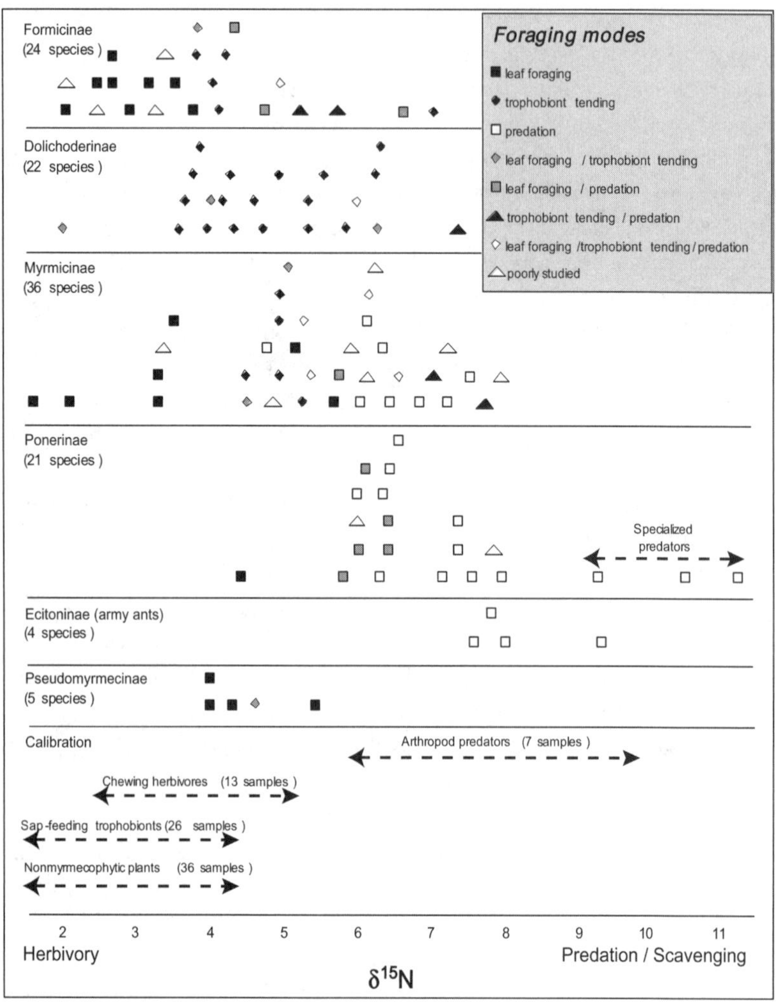

FIGURE 9.1. Mean δ^{15}N values (‰, by subfamily) for ant species in a Peruvian rain forest. Each symbol represents an ant species with a particular foraging mode; "leaf foraging" includes species that search leaves for dispersed food, such as extrafloral nectar, cast-off honeydew, plant wound secretions, isolated scale insects, vertebrate feces, pollen, microbes, and occasional dead or live prey. Data for plants and different arthropod groups in the same habitat calibrate the scales for the ants. Ratios of exudate-feeding ant species (mostly Formicinae and Dolichoderinae, but also some Myrmicinae and Pseudomyrmecinae) overlap extensively those of plants, trophobionts, and chewing herbivores but tend to segregate from those of typical predatory ant species (especially Ponerinae and Ecitoninae) and predatory arthropods (e.g., spiders and pseudoscorpions). The pattern is consistent with the hypothesis that many arboreal ant species obtain little nitrogen from animal prey and behave typically as herbivores. Modified from Davidson et al. 2003.

similar results (see also Davidson and Patrell-Kim 1996; Davidson, Cook, and Snelling 2004). These important findings support Tobin's (1991, 1994) hypothesis that the remarkable abundance of ants in tropical rainforest canopies occurs because the ants feed chiefly as herbivores, rather than as predators or scavengers. The hypothesis is further confirmed by the fact that the few dominant ant species in the canopy typically feed extensively on liquid foods (e.g., Blüthgen et al. 2000; Dejean et al. 2000; Blüthgen, Stork, and Fiedler 2004). Davidson (1997) has suggested that the high abundance of liquid food sources on foliage (extrafloral nectaries and honeydew-producing Hemiptera) plays an important role in shaping food-web structure (see below) in tropical forests by fueling costly prey-hunting activities by foliage-dwelling ants, especially if the ants are physiologically adapted to a diet of low nitrogen content. Moreover, microsymbionts of ants and their hemipteran trophobionts may help the ants upgrade their nitrogen from lower trophic levels (Sauer et al. 2000; Moran et al. 2003). However, because nitrogen can be a limiting factor, dominant arboreal species invest in nitrogen-free offensive and defensive chemical weaponry, including workers with a thin cuticle and nonproteinaceous venoms (Davidson 1997; Orivel and Dejean 1999). At high densities on foliage, exudate-fueled ant foragers would keep prey species at lower numbers than expected from an ant diet exclusively based on animal prey, with relevant consequences for some plant species (Davidson and Patrell-Kim 1996; Davidson, Cook, and Snelling 2004).

The ant taxa obtaining substantial N from live prey are the ones most likely to provide benefits to the plant through reduction of herbivore activity on foliage. However, because the most abundant canopy ants are frequently fueled by honeydew from large populations of hemipteran trophobionts, they not only act as parasites of plant N, carbohydrates, and water but also transmit plant pathogens via trophobiont tending (Davidson et al. 2003). Although honeydew-gathering ants may provide a net benefit to the plant if they also deter other herbivores (Messina 1981; Oliveira and Del-Claro 2005; see chapter 7), such situations are thought to be rarer in relation to net damage (Davidson, Cook, and Snelling 2004). Therefore it is expected that dominant exudate-feeding species that obtain their N mostly from prey (in the intermediate zone in the "herbivore-predation" spectrum shown in fig. 9.1) should have a greater potential to protect the plant by consuming and/or repelling herbivores (a mutualistic role) than other species feeding at lower trophic levels

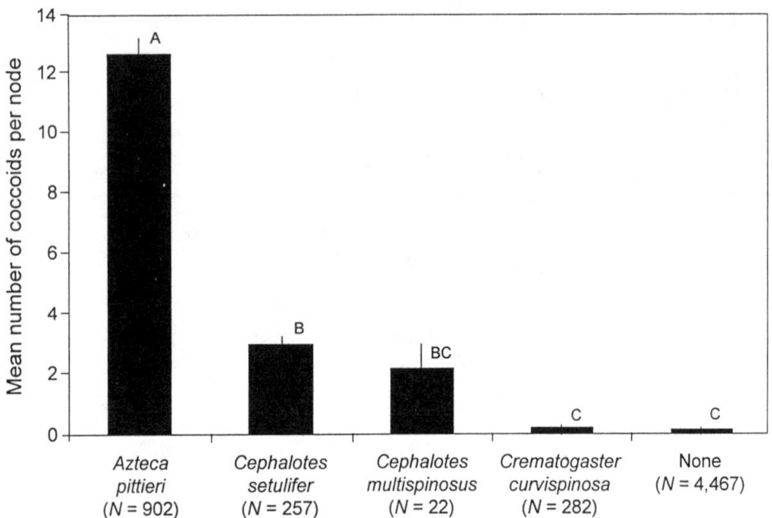

FIGURE 9.2. Mean number (± SE) of coccoid trophobionts per node of *Cordia alliodora* occupied by each ant species in a Costa Rican rain forest. Note that ant-inhabitant *Azteca pittieri* hosted by far the most coccoids. Different letters above the bars designate significant differences using Kruskal-Wallis tests. Modified from Tillberg 2004.

(a parasitic role) (Davidson et al. 2003). Working in the rain forest at La Selva Biological Station in Costa Rica, Tillberg recently investigated the behavior and diet of four ant inhabitants of *Cordia alliodora* (Boraginaceae), using stable-isotope analysis of nitrogen and carbon to infer the relative trophic position of the ant species. Tillberg showed that ant species differed in their rates of association with hemipteran trophobionts (fig. 9.2), performance of plant protective behaviors, and trophic role. These differences markedly affected the nature of the symbiotic relationship between the plant and its ant inhabitants, ranging from mutualism to parasitism. Although associated with the most coccoids, *Azteca pittieri* had the most beneficial behaviors on *Cordia alliodora,* including the most efficient bait-finding ability and the most vigorous recruitment response on foliage (Tillberg 2004). The positive or negative effect of each ant species was corroborated by stable-isotope data, indicating that the mutualistic *Azteca* and *Crematogaster* species have a more carnivorous diet than the two parasitic *Cephalotes* species (fig. 9.3) (see also Trimble and Sagers 2004).

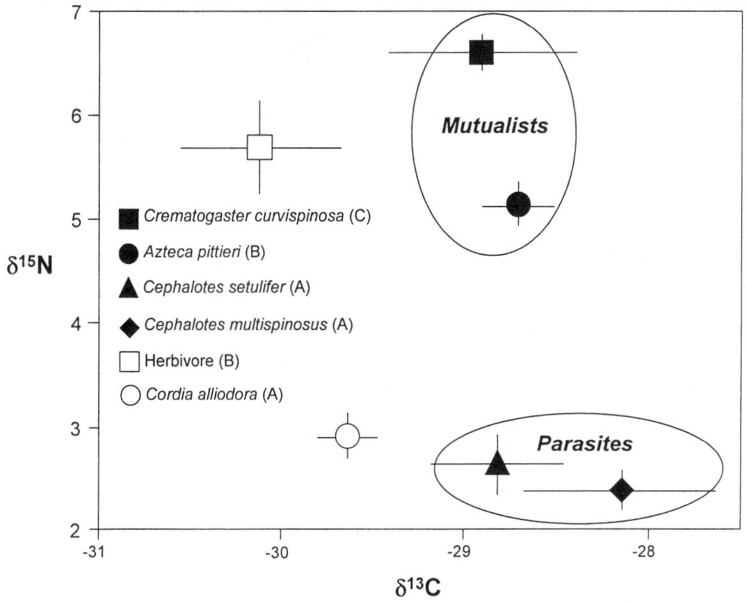

FIGURE 9.3. Graph of $\delta^{13}C$ by mean $\delta^{15}N$ (both ‰) showing clear dietary differentiation between mutualistic (*Azteca pittieri* and *Crematogaster curvispinosa*) and parasitic (*Cephalotes setulifer* and *C. multispinosus*) ants inhabiting *Cordia alliodora* trees in a Costa Rican rain forest. Results indicate that the two mutualistic ant species feed at a higher trophic level than the parasitic ones. Different letters after the ant species names indicate significant pairwise differences from K nearest-neighbors test after Bonferroni correction. Modified from Tillberg 2004.

The Competitive Environment

Interspecific competition has long been of interest to ecologists as a possible mechanism structuring natural communities and mediating phenomena such as resource partitioning within habitats (Schoener 1974, 1983). Although other forces such as predation and physical factors can be important for some animal communities (e.g., Lawton and Strong 1981), a considerable amount of data indicate that competition plays a key role in ant communities (Hölldobler 1987). Indeed, an array of factors ranging from overt aggression to overdispersion of colonies, character displacement, natural and experimental removal, and introduction of species have enhanced the relevance of competition in structuring ant communities (Hölldobler and Wilson 1990 and references therein).

Because ants are so abundant on both ground and foliage and frequently attend stationary food sources such as extrafloral nectaries, honeydew-producing hemipterans, and animal carcasses, interspecific contests at the same resource allow us to recognize regular behavioral types of the ant species. By observing ants at sugar baits, Wilson found three behavioral categories of ant species: (1) "opportunists" discover food quickly and consume it before competitors arrive, (2) "extirpators" recruit and aggressively dominate the food resource against competitors, and (3) "insinuators" rely on stealthy behavior to steal food from dominant ants. The three categories are not rigid, and Wilson (1971) mentions that a given ant species may behave differently depending on the context. While opportunists specialize in *exploitative competition,* extirpators specialize in *encounter competition,* which is a particular form of interference competition (Schoener 1983). Davidson (1998) subsequently equated Wilson's behavioral terminology by considering "encounter" species equivalent to extirpators and by including both opportunists and insinuators in the category of "submissives."

About two decades ago, Fellers (1987) proposed that an evolutionary trade-off between exploitative competitive ability (the capacity to find and use a resource before competitors) and interference competitive ability (here referred to only as behavioral dominance of a resource in encounter interactions) would mediate coexistence among species in a temperate community of woodland ants (see also Schoener 1983). By observing the activity of nine ant species at food baits, Fellers found an inverse correlation between exploitative and interference ability (table 9.1), which enabled the subordinate, opportunistic species to obtain resources. In most cases the speed of bait location was not correlated with distance of the bait from the colony, but rather with characteristics of particular species. Thus aggressive species used behavioral dominance (superior fighting and/or recruiting abilities) to ward off subordinates from resources. The subordinate species used their higher speed of food location to gain immediate access to resources and circumvent interference from dominant species (feeding time by subordinates was considerably reduced in the presence of other ant species). Some ant species exhibited intermediate performances in both discovering and defending food resources (table 9.1). Because interspecific interference in the form of aggression or avoidance occurred on more than half of the occasions when two or more species used the same food resource, Fellers's (1987) study emphasizes the importance of be-

TABLE 9.1 **Dominance rankings (behavioral performance during interspecific interactions at food baits) and discovery rankings (relative speed of food location) for nine species of ants in a temperate woodland community.**

Ant species	Dominance rank[a]	Discovery rank[b]
Camponotus ferrugineus	1	8
Prenolepis imparis	3	6
Lasius alienus	3	7
Formica subsericea	3	5
Myrmica spp. (2 species)	5	1
Aphaenogaster rudis	6	3
Tapinoma sessile	7	4
Leptothorax curvispinosa	8	2
Spearman's $r = -0.76$; $0.01 < p < 0.05$		

Source: Modified from Fellers 1987.

[a] Calculated as the percentage of time each species was dominant in its interspecific interactions; ranks range from 1 for the most dominant to 8 for the least dominant.

[b] Average percentage of baits attended by a species when it was the first to arrive; ranks range from 1 for the fastest (55% of the baits) to 8 for the slowest species (31% of the baits).

havioral interactions between species in mediating the structure of ant communities.

Davidson (1998) analyzed from a broader perspective the trade-off between exploitative competitive ability and behavioral dominance proposed by Fellers (1987) and examined mechanisms through which the trade-off could be broken. Davidson (1998) hypothesized that ecologically dominant ants (i.e., those that combine behavioral and numerical dominance) might use territoriality to break the trade-off. Studies with ant communities in the taiga biome have revealed three basic behavioral patterns with respect to the use of space by ant species (Vepsäläinen 1982; Savolainen and Vepsäläinen 1989): submissive species defend the nest site only (type I territoriality), extirpator species defend both nest site and food finds (type II territoriality), and territorial dominant species defend nest and foraging areas as absolute territories (type III territoriality). In tropical rain forests, dominant exudate-feeding ants defend exclusive territories that create a mosaic of different ant communities in the canopy zone (see below). Thus possession of absolute territories by dominant ant species, to the exclusion of other species, can be regarded as an amplified form of interference competition that adds to encounter interactions at food finds (Morrison 1996; Davidson 1998). This is likely to be the case for the dominant ant-garden ants—*Camponotus femoratus* and *Crematogaster limata* var. *parabiotica*—that are parabiotic (i.e., share carton nests and foraging trails) and are able to both discover and control honey baits in Amazonian rain forests (Davidson 1988). Similar results have been reported in African and Australian rain forests for the dominant exudate-feeders *Crematogaster depressa* and *Oecophylla smaragdina,* respectively (Dejean et al. 2000; Blüthgen and Fiedler 2002, 2004a), and in a Thai mangrove forest for *O. smaragdina* as well (Offenberg et al. 2004). By processing large volumes of exudates in order to upgrade nitrogen from lower trophic levels (Davidson et al. 2003), these ecologically dominant ants can use excess carbohydrates to invest in rapid motion and costly prey-hunting activities, as well as defense of absolute spatial territories. Davidson (1998) therefore suggests that modifications of the proventriculus for more efficient feeding on exudates (Eisner 1957) might have enabled dominant arboreal ants of tropical rain forests to break the discovery/dominance trade-off and invest in both exploitative competitive ability and interference ability.

Ant Mosaics

The idea that dominant ant species can generate a typical spatial distribution pattern in the ant community, known as "ant mosaic," was originally developed by Leston (1970) and further clarified by several authors working in less complex communities such as plantation systems, especially cocoa farms (Room 1971, 1975; Majer 1972, 1976a, 1976b; Taylor 1977; Jackson 1984). While studying canopy-dwelling organisms in the forests of West Africa, Leston (1973a, 1973b) noted that dominant ant species markedly affect the distribution and abundance of other local ant species. The persistent influence of the dominant species is such that they space out into mutually exclusive territories covering one to several trees each, so as to generate a three-dimensional mosaic pattern in the forest canopy. Leston further remarked that the effects (positive or negative) of dominants could be perceived not only on the distribution of other ant species, but on other arthropods and plants as well (see below). Dominant ant species typically have a potential for rapid population growth to attain extremely populous colonies (up to several million individuals) and also have the ability to build polydomous nests (i.e., a single colony occupies multiple nests) with carton, silk, and leaves (Davidson 1997, 1998). Polydomy allows the dominant species to economically allocate the worker force among different sites containing exudate-producing

Hemiptera and adjust worker numbers to monopolize these rich food resources (Hölldobler and Lumsden 1980; Blüthgen and Fiedler 2002). Finally, ecologically dominant ant species characteristically defend absolute spatial territories against both intraspecific and interspecific enemies (Vepsäläinen 1982; Morrison 1996; Davidson 1998).

The concept of an ant mosaic is of course better documented in more simplified communities (i.e., crop systems), where full visual access to the "canopy" permits rapid identification of the few local ant species and more accurate descriptions of the behavioral interactions taking place on the foliage. Indeed, as stated in the dominance-impoverishment rule of Hölldobler and Wilson (1990: 423), "the fewer the ant species in a local community, the more likely the community is to be dominated behaviorally by one or a few species with large, aggressive colonies that maintain absolute territories." Over the past three decades, however, the concept has been considerably expanded to include ant communities not only from species-rich forest and savanna ecosystems worldwide, but from tree crop plantations as well (e.g., Leston 1978; Hölldobler 1983; Majer 1990, 1993; Majer and Camer-Pesci 1991; Paulson and Akre 1991; Adams 1994; Dejean et al. 1994, 2000; Armbrecht et al. 2001; Dejean and Corbara 2003; Ribas and Schoereder 2004). In Asian pristine lowland forests, however, where ant communities are highly heterogeneous and their species composition in the canopy is unpredictable, the mosaic theory does not apply; the ant community organization seems to result from more complex dynamic processes (Floren and Linsenmair 1997, 2000). Indeed, after analyzing 14 studies on ant mosaics, Ribas and Schoereder (2002) have suggested that although competition may be an important process mediating assembly rules in ant communities, other biological processes and stochastic events also need to be considered to explain species distribution patterns.

Plant and Insect Exudates and Ant Community Structure

Although the abundance of plants with extrafloral nectaries is relatively well documented for a number of habitats (chapter 10), knockdown samples have apparently underestimated the relevance of honeydew-producing insects in the canopy of tropical environments (Dejean et al. 2000). Recent studies, however, have provided strong evidence that occurrence of such liquid foods on foliage plays a key role in mediating the

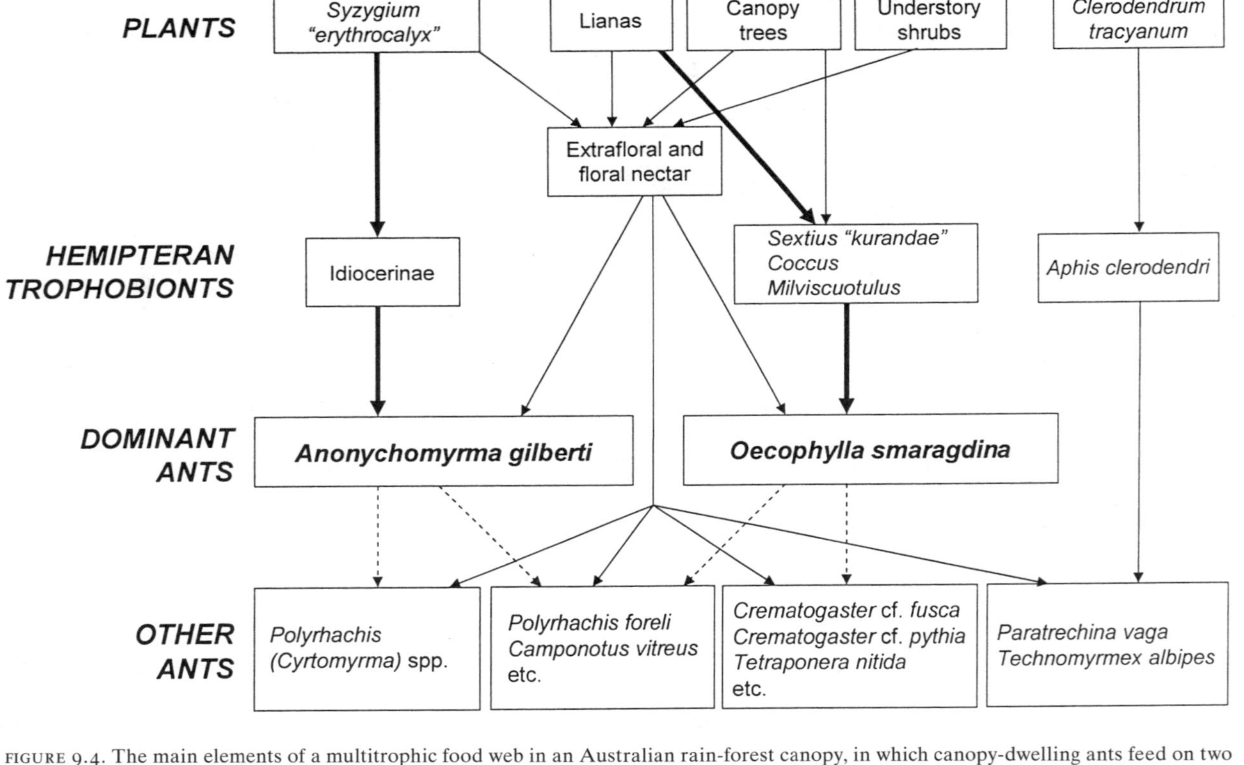

FIGURE 9.4. The main elements of a multitrophic food web in an Australian rain-forest canopy, in which canopy-dwelling ants feed on two main liquid food sources, hemipteran honeydew and/or nectar. Resource links (suggested bottom-up effects) are designated by continuous lines, whereas ant species co-occurrences are represented by broken lines. Thick continuous lines link the keystone species in the community. Note that some key plant species (*Syzygium* and lianas) hosting trophobionts ultimately shape the distribution of the two dominant ant species and, consequently, of the remainder of the ant community. Modified from Blüthgen, Stork, and Fiedler 2004.

foraging ecology and community organization of canopy-dwelling ants (e.g., Tobin 1994, 1995; Blüthgen et al. 2000; Davidson et al. 2003). By analyzing a diverse spectrum of sources of plant and insect exudates in the canopy and understory of an Australian tropical rain forest, Blüthgen, Stork, and Fiedler (2004) revealed that these resources produce important bottom-up effects on the ant community which are reflected in varying degrees of specialization and partitioning, as well as in the segregation and co-occurrence patterns among the ant species (fig. 9.4). On one hand, two dominant and abundant ant species—*Oecophylla smaragdina* (Formicinae) and *Anonychomyrma gilberti* (Dolichoderinae)—maintain exclusive territories and monopolize large aggregations of different trophobiont species. On the other hand, sources of extrafloral and floral nectar (48 plant species) are exploited by a dynamic and opportunistic ant assemblage (43 ant species), with frequent nonaggressive co-occurrences of nectar-gathering ant species on the same plant. Although the two dominant species are mutually exclusive on all types of liquid food sources, co-occurrence with nondominant species on nectaries is common (fig. 9.4).

The differential roles of honeydew (a specialized resource for dominant species) and nectar sources (an opportunistic resource for all ants) and their distinct distribution within the canopy provide strong evidence that some key plant species hosting trophobionts ultimately shape the distribution of dominant ants in this Australian canopy-ant community (Blüthgen, Stork, and Fiedler 2004). These findings strongly support the ant mosaic theory (Leston 1970; Room 1971; Majer 1972) and enhance the importance of interspecific competition and behavioral interactions in structuring ant communities as anticipated by other studies (Vepsäläinen 1982; Fellers 1987; Savolainen and Vepsäläinen 1989; Davidson 1998).

The Effect of Trophobiont Tenders on Associated Herbivores and on the Host Plant

Bottom-up effects from nectar and honeydew sources on foliage may go beyond the community of exudate-feeding ants. Visitation by ants to exudate sources may affect herbivore activity in diverse ways, with marked consequences for host plants. Although far better documented for plants bearing extrafloral nectaries (chapter 6), increased alertness and aggression by tending ants near trophobionts can also benefit the host plant if ant-derived gains from herbivore deterrence outweigh losses resulting

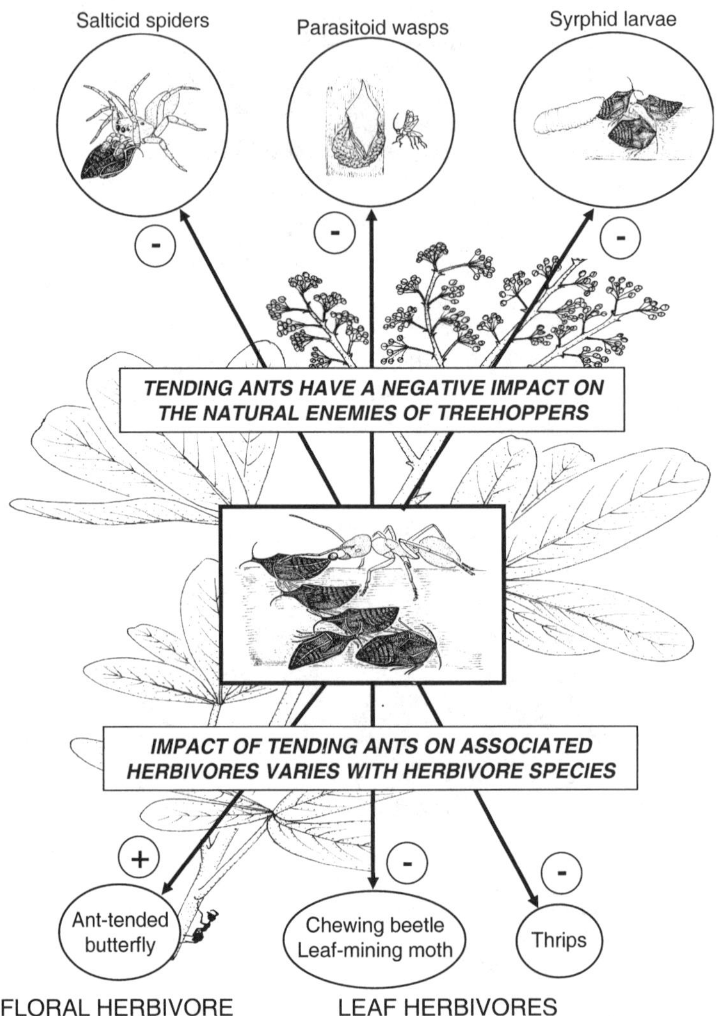

FIGURE 9.5. General outline of the multispecies system involving honeydew-gathering ants at aggregations of *Guayaquila xiphias* membracids on shrubs of *Didymopanax vinosum* in the Brazilian cerrado. The presence of tending ants increases membracid survival and significantly reduces the abundance of the main natural enemies of *G. xiphias* on the host plant. Patrolling activity by tending ants on leaves and flowers produces variable impacts on associated herbivore species. The variable character of the negative and positive effects among species in this complex multitrophic system likely results in temporal and spatial variation in the overall impact of the interactions on the host plant. Based on data in Del-Claro and Oliveira 2000; Oliveira and Del-Claro 2005.

from hemipteran feeding (Carroll and Janzen 1973; Messina 1981; chapter 7).

The multispecies system around ant–*Guayaquila xiphias* (Membracidae) associations on shrubs of *Didymopanax vinosum* (Araliaceae) in the Brazilian cerrado (fig. 9.5) illustrates well the multiple impacts that honeydew-gathering ants can have on associated arthropod species (Oliveira and Del-Claro 2005). Ant-exclusion experiments demonstrated that tending ants (21 species) have a negative impact on the abundance of natural enemies of *G. xiphias* on the host plant. Compared with other ants, the membracids receive better protection by dominant, aggressive *Camponotus rufipes* (Formicinae) that build satellite nests of dry grass to house trophobionts and monopolize entire aggregations day and night (Del-Claro and Oliveira 2000). However, field experiments with manipulated plants revealed that trophobiont-tending ants can have contrasting impacts on the abundance and damage caused by associated herbivores of *D. vinosum* shrubs (fig. 9.5). While three species of leaf herbivores were negatively affected by tending ants, one floral herbivore preferably infested plants with ants and trophobionts (see chapter 7, esp. figs. 7.4, 7.5). This multitrophic system is particularly complex: it involves four types of herbivores, each utilizing plant tissue in a different mode and all being affected by a focal ant-trophobiont interaction (Oliveira and Del-Claro 2005). As opposed to typical ant-plant systems, in which ant-derived benefits to plants result from plant traits (i.e., food and/or domatia) inducing ownership behavior by ants (chapter 6), the complex multitrophic/multispecies system around ant–*G. xiphias* associations is similar to others in which the analyses of pairwise interactions cannot predict the overall impact on the plant from all species involved (Price et al. 1980; Horvitz and Schemske 1984; Cushman 1991; Bronstein and Barbosa 2002; Del-Claro 2004).

* * *

In conclusion, the high incidence of plant and insect exudates on foliage likely accounts for the disproportionate abundance of ants compared to other arthropod groups in the canopy of tropical forests. The apparent biomass paradox (Tobin 1991) created by this extraordinary ant abundance on foliage (up to 86% of the arthropod biomass) is explained by data from stable-isotope analyses demonstrating that most canopy-dwelling ants behave chiefly as primary consumers and rely mostly on plant and insect exudates for nutrition (Davidson et al. 2003). Due to structural adaptations in the proventriculus, allowing them to efficiently process large quantities of liquid foods, exudate-feeding ants can use excess carbohydrates to invest in rapid motion and costly prey-hunting activities, as well as defense of absolute spatial territories (Davidson 1997). Ecological dominance (a combination of behavioral and numerical dominance [Davidson 1998]) by such ant species produces a typical distribution pattern in the ant community known as an ant mosaic, in which dominant species are spaced out into mutually exclusive three-dimensional territories within the forest canopy. Differential distribution of distinct liquid foods (plant nectar and hemipteran honeydew) within the canopy layer can produce bottom-up effects in the hierarchical structure of ant communities (Blüthgen and Fiedler 2004b). This in turn can be reflected on associated arthropods and plants as well (Oliveira and Del-Claro 2005) and thus can potentially affect food-web structure. Despite the extraordinary predominance of ants in tropical rain-forest canopies, we are only beginning to recognize the multiple peculiarities in the lifestyle of canopy-dwelling ants. Recent important findings include modified sticky tarsi that facilitate locomotion in vertical arboreal surfaces (Orivel, Malherbe, and Dejean 2001), gliding behavior that allows aerial return to home tree trunks (Yanoviak, Dudley, and Kaspari 2005), and the construction of elaborate traps to capture large prey that might otherwise escape (Dejean et al. 2005). Certainly much more remains to be discovered about the evolution of canopy-dwelling ants, their ecological function, and their importance.

CHAPTER TEN

Variation in Ant-Plant Interactions

The distribution of species is far from even, creating a spatial mosaic of species richness. In general, the tropics contain more species than equivalent areas at higher latitudes (Pielou 1979; Cox and Moore 1993; Huston 1994; Scheiner and Rey-Benayas 1994), although exceptions are numerous when specific taxa are reviewed (Price 1991; Huston 1994; Willing, Kaufman, and Stevens 2003). Moreover, the study of the distribution of species, which has long been a central focus of ecology and biogeography (Gaston 2000), is taking on new urgency as evidence increases on the global crisis of biological diversity (Reid 1998; Gaston 2000; McCann 2000; Tilman 2000). Several questions arise that are central to the design of effective conservation programs (McCann 2000; Purvis and Hector 2000; Tilman 2000). For example: How does the loss in biological diversity affect ecosystem functioning? Will declines in biological diversity accelerate the simplification of ecological communities? What geographic regions should be protected in order to maintain the most biological diversity? Thus, areas particularly rich in species, rare species, threatened species, or some combination of these (i.e., "hot spots") are increasingly being delineated to help set priorities for conservation (Reid 1998; Margules and Pressey 2000; and references therein). For the realization of conservation goals, whole landscapes should be studied and managed (Bronstein 1995), because reserves should sample or represent the biological diversity of each selected region, and they should also separate this diversity from processes that threaten its persistence (Margules and Pressey 2000). However, the focus in systematic conservation planning should be on the conservation not only of species but also of their interactions, because the diversity of life has resulted from the diversification of both species and the interactions among them (Thompson 1996). Moreover, changes in interactions, because they end up making changes in biological diversity, habitat fragmentation, and patchiness, alter the characteristics expressed by species and therefore the effect of species on ecosystem processes (Fagan, Cantrell, and Cosner 1999; Chapin et al. 2000). Thus, simply knowing that a species is present or absent is insufficient to predict its impact on ecosystems (Andersen 1993; Chapin et al. 2000).

Not only are species unevenly distributed, but their interactions also vary spatially and seasonally; variations occur in critical plant structures and/or food sources for interactions as well (e.g., Díaz-Castelazo et al. 2004; Dutra, Freitas, and Oliveira 2006). Consequently, the landscape approach to the study of ecological interactions should be emphasized much more than it has been (Bronstein 1995; Thompson 1996; Margules and Pressey 2000). Interactions vary in their probability of occurrence along environmental gradients (e.g., latitudinal or altitudinal) and under different disturbance regimes (e.g., Koptur 1991, 1992a; Rico-Gray, García-Franco, et al.1998), they vary in their outcome under different ecological conditions (Thompson 1982, 1988, 1994; Cushman and Addicott 1991) and between habitats (Barton 1986), and they vary between seasons (e.g., Rico-Gray and Castro 1996). This spatial heterogeneity in interactions may be a major factor maintaining the diversity of species and their interactions over both small and large geographic regions, by reducing the chance that one species can push another to extinction (Hassell and May 1974; Hassell, Comins, and May 1991; McLaughlin and Roughgarden 1993).

The structure of ant communities and of ant-plant interactions has been studied in a variety of habitats, and it is clear that spatial and temporal variation characterize ant communities (Herbers 1989; Fowler 1993; Morrison 1998; and references therein). Ant assemblages are very dynamic, and extrapolating from one ant community to another of superficially similar characteristics can lead to erroneous inferences (Herbers 1989); thus, the spatial and temporal dimensions of community structure in ant communities and their interactions with plants preclude broad generalizations (Feener and Schupp 1998).

The spatial variation in the richness and availability of energy-rich liquid food for ants (appendix 10.1) can be associated with interhabitat differences in ant diversity and abundance (Díaz-Castelazo et al. 2004) and thus generate variation in ant-plant interactions (Pemberton 1998; Rico-Gray, García-Franco, et al. 1998). Moreover, spatial scale may cause regional diversity gradients in ants, as processes that limit ant-species richness scale up with the size of the area sampled (Kaspari, Yuan, and Alonso 2003; Parr, Parr, and Chown 2003). In this chapter we present and discuss examples of temporal and geographic variation (latitudinal, altitudinal, and geographic variation) in ant-plant interactions, and we associate variation with community structure and coevolution.

Temporal Variation

Ant-plant interactions have been reported to change in number and in outcome among habitats, throughout the year and between years. However, most studies do not take into account temporal variation, or that ant colonies and their interactions with plants can change over time. For instance, ant response to elaiosome-bearing diaspores may differ seasonally (Thompson 1981a) as the need for elaiosome tissue as larval food varies, or with changes in the availability of alternative foods. Significant temporal variation has also been demonstrated for ant occupation of plants with domatia (Janzen 1975; McKey 1984; Davidson, Snelling, and Longino 1989; Vasconcelos 1993). For example, the percent occupation of ants associated with the facultative ant-plant *Conostegia setosa* (Melastomataceae) did not differ among censuses at two study sites—La Selva (Costa Rica) and Nusagandi (Panama)—but overall occupancy was lower in the dry season at La Selva (Alonso 1998).

Ant visitation to extrafloral and circumfloral nectaries also exhibits daily and seasonal changes (Bentley 1977a, 1977b; Rico-Gray 1989; Horvitz and Schemske 1990; Oliveira et al. 1999; Díaz-Castelazo et al. 2004). The number of ant associations with extrafloral nectaries (EFN) or with nectaries located in plant reproductive structures (floral or circumfloral) varied throughout the year in a tropical dry-deciduous forest on the central coast of Veracruz, Mexico (fig. 10.1). Most plant species in tropical dry forests produce a flush of new leaves at the onset of the rainy season (Bullock and Solís-Magallanes 1990). Extrafloral nectaries are associated with leaves, and many studies have shown that secretion

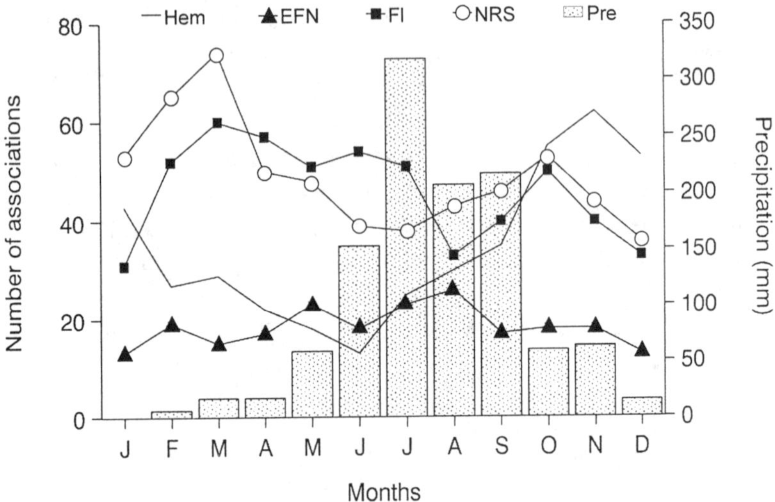

FIGURE 10.1. Number of ant-plant and ant-hemipteran associations registered per month per food resource on the coast of Veracruz, Mexico. Hem = hemipteran honeydew; EFN = extrafloral nectar; Fl = floral nectar; NRS = nectar from reproductive structures; Pre = monthly precipitation. Although ant-EFN associations increased significantly (Spearman $r = 0.552$, $p = 0.05$) with rainfall, they were more constant throughout the year than were those involving other plant resources. The numbers of ant visits to flowers and to hemipterans were negatively associated (Spearman $r = -0.664$, $p = 0.01$). Modified from Rico-Gray 1993.

of extrafloral nectar is greatest during periods of rapid vegetative growth such as the expansion of new leaves. Ant presence is highly correlated with these peaks of nectar flow (Bentley 1977a, 1977b), and an increase of ant-EFN cases would therefore be expected as new leaves appear. Ant associations with reproductive structures and flowers reached their peak during the dry season and decreased during the wet season (fig. 10.1). There are two main flowering peaks in the dry tropical lowlands of Central America, one in mid–dry season and one at the start of or during the wet season (Bullock and Solís-Magallanes 1990; Castillo and Carabias 1982). Those major periods of flowering are supplemented by erratic flowering of many species year-round, offering ants year-round energy-rich liquid sources. Furthermore, Díaz-Castelazo et al. found that the abundance of nectary-bearing plants differed among seasons, while ant density and richness differed seasonally in specific vegetation associations in the area. Seasonal patterns suggest higher nectar availability on vegetative structures during the rainy season and on reproductive structures during the dry season (Díaz-Castelazo et al. 2004).

Ant visitation to honeydew-producing Hemiptera also changed throughout the year (figs. 10.1 and 10.2). Rico-Gray and Castro have shown that the effect of an ant-aphid interaction on the reproductive fitness of *Paullinia fuscescens* (Sapindaceae) varied among years in coastal Mexico—plants with ant-tended aphids produced fewer seeds than plants without ants or aphids in two years, and in the third year there was no treatment effect (Rico-Gray and Castro 1996; see chapter 7).

Hemipterans usually spend the winter dry season in the egg stage; they hatch and become active as conditions improve (Borror, De Long, and Triplehorn 1981). The number of ant associations with honeydew-producing Hemiptera followed the above pattern in the tropical dry-deciduous forest of Veracruz: the number of associations increased after the onset of the rainy season and decreased abruptly once the dry season began (fig. 10.1). During the warm, humid months, plants produce new soft vegetative tissues, creating ideal feeding conditions for Hemiptera.

Because ant-Hemiptera associations sharply decrease in the dry season, however, ant-flower associations reach their peak during this period (fig. 10.1). This complementary pattern may result in the use of alternative resources with similar nutritional value, since 62.5% of the ant species using floral nectar also foraged for honeydew. The whole pattern indicates that the ants switch the use of plant-derived resources from dry

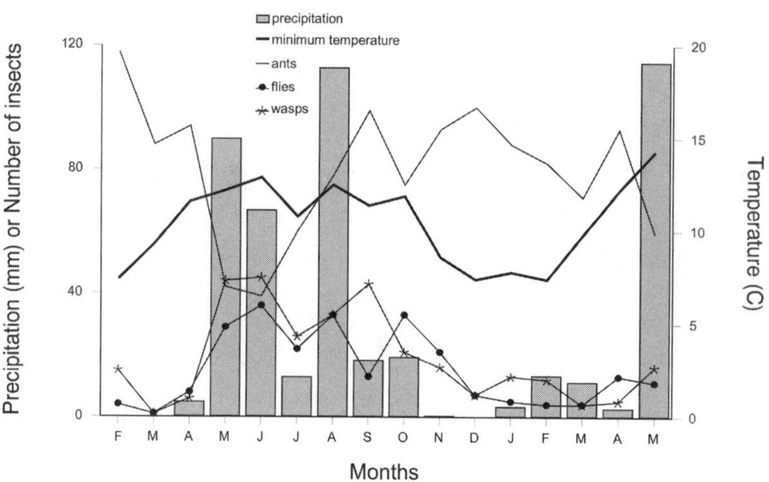

FIGURE 10.2. Precipitation, minimum temperature, and total number of insects (ants, flies, and wasps) per month (February 1994 to May 1995) that forage on the honeydew produced by hemipterans, which feed on the inner face of leaves of *Agave kerchovei* (Agavaceae) individuals in the Zapotitlán Valley, Puebla, Mexico. From Cuautle et al. 1999.

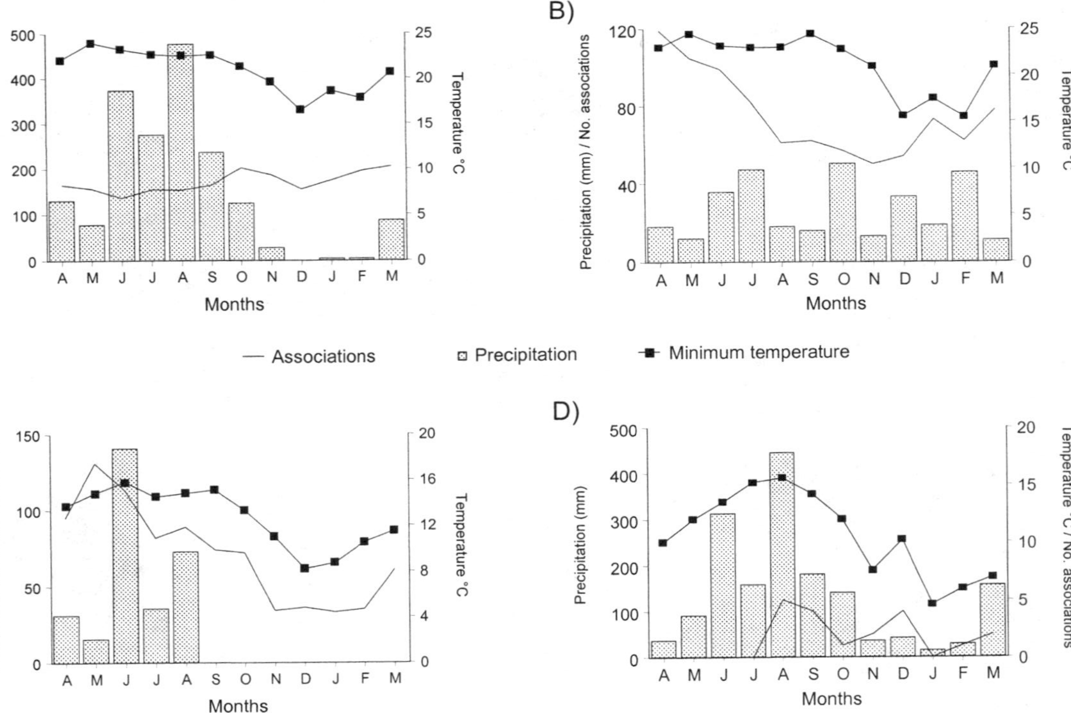

FIGURE 10.3. Number of ant-plant associations and precipitation and minimum temperature per month for the study sites: (A) La Mancha (tropical dry and deciduous forests, sand dune scrub, mangroves); (B) Zapotitlán (thorny scrub, columnar cacti, tropical dry forest); (C) San Benito (sand dune scrub, mangroves); (D) Xalapa (lower montane forest). Based on Rico-Gray, García-Franco, et al.1998.

to wet season, probably exploiting a greater variety in the latter period. Ants could also be feeding on a variety of insect prey during the wet season, when insects exhibit their peak activity in lowland tropical seasonal forests (Smythe 1982; Rico-Gray 1989). In summary, because more food resources are available during the wet season, ants are able to diversify their foraging activity at that time.

* * *

Using a different approach, Rico-Gray, García-Franco, et al. examined the seasonal variation of ant-plant interactions, expressed as the number of interactions per month, in four habitats with contrasting environmental characteristics. Two sites were at sea level with contrasting total annual precipitation (La Mancha in Veracruz, 1,500–1,700 mm; and San Benito in Yucatán, 300 mm) and two at higher elevations, also exhibiting contrasting total annual precipitation (Zapotitlán in Puebla, at 1,500 m and 300 mm; and Xalapa in Veracruz, at 1,300 m and 1,800–2,000 mm). The authors asked whether between-habitat differences in seasonality of ant-plant interactions could be explained by the differences in environmental parameters (for statistics, see Rico-Gray, García-Franco, et al. 1998).

Fewer ant species and interactions were registered in the drier and/or cooler habitats, regardless of altitude (fig. 10.3). La Mancha, the richest site in ant-plant interactions, was the most heterogeneous in terms of the number of different vegetation associations present, compared to the other three sites (see also Díaz-Castelazo et al. 2004). Even though rainfall was lower in Zapotitlán and San Benito, minimum temperature was the most important factor accounting for the seasonal distribution and number of ant-plant interactions. At La Mancha, with higher minimum temperatures and water availability, temperature alone did not account for the seasonal distribution and number of ant-plant interactions. The effect of the precipitation x temperature interaction was highly significant, suggesting that La Mancha is not as extreme a seasonal environment as Zapotitlán or San Benito. Xalapa exhibited the lowest temperatures and the highest precipitation of the habitats studied; their effect, however, was statistically marginal. It has been suggested that ants are less abundant in cool and humid, relatively aseasonal habitats (Janzen 1973c; Janzen et al. 1976; Greenberg, Macias-Caballero, and Bichier 1993; Samson, Rickart, and Gonzales 1997). However, Rico-Gray,

García-Franco, et al. (1998) suggest that the vegetation in Xalapa, a mixture of tropical and temperate floristic elements (Williams-Linera and Tolome 1996), constrains ant-plant interactions due to a limited presence of nectaries (Díaz-Castelazo and Rico-Gray 1998). In contrast, the other three habitats have tropical floristic elements (e.g., Fabaceae, Caesalpinaceae, Mimosaceae, Bignoniaceae, Cactaceae) that are abundant and frequently have nectar-producing structures (Oliveira and Leitão-Filho 1987; Koptur 1992a; Schupp and Feener 1991).

Besides evaluating among-site differences in abiotic factors (e.g., rainfall and minimum temperature), future studies involving multisite comparisons should take samples at several locations per site and should take into account physiognomic variation (vegetation associations) within sites.

Spatial Variation

Although species pairs and assemblages often occur across vast geographic areas, not much is known about the frequency of occurrence or the outcome of interactions across such vast spatial scales (Cushman et al. 1998; Rico-Gray, García-Franco, et al. 1998). Ant-plant interactions are geographically widespread, common in many plant communities, and have been shown to be important in plant defense against herbivores (Pemberton 1998; Oliveira and Freitas 2004; and references therein). Most studies of these relationships have analyzed the interactions between small subsets of ants and plants within communities (e.g., Rico-Gray and Thien 1989b; Deslippe and Savolainen 1995). Many ant-plant interactions, however, are highly facultative, and the diversity of ants involved in these interactions can vary considerably even over short geographic ranges (Horvitz and Schemske 1984; Barton 1986). Hence, our understanding of the ecology and evolution of these interactions and their effects on community organization requires a better understanding of how the diversity of ants and their use of plants varies geographically. Ant-plant interactions are suitable systems in which to study processes operating at the community level due to the parallels between the characteristics and features of both ant and plant communities (Andersen 1991a; López, Serrano, and Acosta 1994) and also because in some cases the number of plant species is the best predictor of the number of ant species (Morrison 1998). Although scarce, studies on the latitudinal, altitudinal, or geographic variation of ant-plant or ant-Hemiptera-plant interactions (or on their outcomes) have consistently shown that these associations are not evenly distributed.

Latitudinal Variation

Ant-plant interactions vary across latitudinal ranges. This variation could be the result of variation in ant density, abundance, or morphology (e.g., Kaspari, Yuan, and Alonso 2003; Parr, Parr, and Chown 2003) and/or of plant characteristics, diversity, and abundance (appendix 10.1). Here we will use three examples to illustrate latitudinal variation in ant or plant communities that may affect ant-plant interactions.

Ant assemblages in the British Isles and northern Europe change with latitude in two ways: (1) the number of ant species significantly decreases with increasing latitude, and (2) the body size of ant species significantly increases with increasing latitude (Cushman, Lawton, and Manly 1993; Porter and Hawkins 2001). Body-size pattern is present in the subfamily Formicinae and, to a lesser extent, in the Myrmicinae, the two subfamilies that comprise the main groups of ant species (95%) in the region. However, formicines increase in size with latitude faster than myrmecines do. The pattern of increasing body size was due mainly to the fact that ranges of ant species shift to higher latitudes as their body size increases, with larger formicines becoming less represented at southern latitudes and larger myrmicines becoming more represented at northern latitudes. Cushman, Lawton, and Manly (1993) suggest that the above pattern is most likely explained by the starvation-resistance hypothesis, in which harsher climates at higher latitudes would lead to more rapid starvation of ant species with small-bodied workers and thus favor species with larger-bodied workers. Related studies have not shown this association to be a general rule (e.g., Garrido et al. 2002) and suggest that these approaches may fail to control for the possible confounding effects of phylogeny and also that they do not consider latitudinal overlap of geographic ranges (Cardillo 2002). Furthermore, studies testing for the relationship between body size of animals and a gradient (e.g., latitude) are complicated by the facts that typically there is a single average size for each species and that each species occurs at several sample stations over the gradient. Manly (1998) has suggested that this problem can be overcome by using a randomization test, a general method for relating body size to latitude and subfamily differences.

Pemberton (1998) showed a significant increase in the abundance of EFN-bearing plants as a function of total plant cover (from 10.25% to 40.18%), when moving from tundra to subtropical vegetation (70°–26° N) in East Asia. The pattern is similar for the number of species per sampled area (from 0.11 to 1.13/100 m) and the proportion of species within regional floras (from 0.32% to 7.46%). Ants, the primary mutualists associated with plants bearing EFNs, exhibit a similar pattern of increasing abundance (species richness, nest density, and colony size) along the same north-south latitudinal gradient (Pemberton 1998 and references therein; see also Kusnezov 1957; Jeanne 1979 for latitudinal patterns in species richness and rates of ant predation, respectively, in the Americas).

Leaf-cutting ants (*Acromyrmex* and *Atta*) are commonly considered to be characteristic of tropical forests (Wirth et al. 2003). However, leaf-cutters exhibit rich species assemblies (abundance and diversity) in the subtropical habitats (particularly open areas) of South America (Fowler and Claver 1991). Farji-Brener and Ruggiero report that species richness of leaf-cutting ants also decreases as latitude and longitude increase in Argentina (Patagonia and the Andes lack leaf-cutters), where mean annual precipitation, intra-annual temperature variation, and the minimum mean winter temperature are the principal determinants of leaf-cutting-ant richness. Species harvesting different types of plant material showed differences in the size of their geographic distributions. While grass-dicot cutters were the most widespread species, grass-cutter species were strongly affected by the change from woody to grass vegetation. Farji-Brener and Ruggiero (1994) thus show that the size of the geographic ranges and the patterns of leaf-cutter species richness were strongly dependent on both climatic/vegetational features of the environment and the biological attributes of the different foraging groups.

Altitudinal Variation

Not only latitudinal changes affect the distribution patterns of ants and their interactions. The generalized distribution pattern indicates that abundance, richness, and activity levels of ants in tropical habitats increase in lowlands and up to mid-elevations and decline sharply with altitude (Janzen 1973c; Janzen et al. 1976; Bentley 1977b; Koptur 1985; Hölldobler and Wilson 1990; Olmstead and Wood 1990; Stein 1992; Samson, Rickart, and Gonzales 1997). Moreover, this pattern is not peculiar to tropical regions or to ant-angiosperm interactions (Lawton,

MacGarvin, and Heads 1987). In the mid-Atlantic region of the United States, the total abundance of ants shows an overall decline with increasing elevation (100–1,700 m) (McCoy 1990), and in southeastern Arizona ant distribution and behavioral dominance change over an elevational gradient, with site species richness ranging from 4 in a Douglas fir forest (2,600 m) to 33 in an oak-juniper woodland (1,400 m) (Andersen 1997a).

It has been suggested that the above pattern in the distribution of ants affects the distribution of ant-tended hemipterans, the presence of EFNs, and the prevalence of different defense mechanisms (e.g., ants vs. chemicals). The proportion of obligate ant-tended treehopper species that are dependent upon ants for defense declines with increasing altitude in the neotropics (Colombia); those species that have parental care and do not rely on ants for defense are more common at higher elevations (Olmstead and Wood 1990). In Jamaica, Keeler (1979a) has shown that the cover of plants with EFNs decreased from 28% at sea level to 0% at 1,310 m; ant abundance was also greater at the lower-elevation site. Extrafloral nectaries in *Centropogon* (Lobeliaceae) are more prevalent in species found at lower elevations; below 1,500 m approximately 80% of the species have that trait, according to Stein. Moreover, in species of *Centropogon* with wide elevational ranges, the glands are often degenerate or nonfunctional at higher elevations (Stein 1992). A similar reduction or loss of extrafloral nectaries with increasing altitude was found in *Bixa orellana* (Bixaceae) in Costa Rica (Bentley 1977b). Koptur (1985), studying two *Inga* species, showed that nectarivorous ants were less active at higher than at lower elevations in Costa Rica and that, even though herbivory damage by caterpillars to leaves was greater at higher elevations, ant defense was less effective in the upland populations of *I. punctata* and *I. densiflora*. She also showed that ant defense at high elevations is compensated by alternative defenses, such as increased concentrations of phenolics or visitation to EFNs by adult fly and wasp parasitoid species.

Geographic Variation

The study of the geographic distribution of interactions is essential for understanding the evolutionary and coevolutionary consequences of species interactions, as well as the influence of these interactions on population dynamics and community structure (Alonso 1998). Thus, even caste structure among ant populations can vary geographically (Yan, Martin, and Nijhout 2004). Three studies using different approaches and methodology are considered here to exemplify geographic variation in ant-plant interactions. The first two focus on extensive geographic ranges, one in southern Africa analyzing the occurrence and outcome of an association (Cushman et al. 1998) and the other, in Mexico, comparing the richness of ant-plant interactions (Rico-Gray, García-Franco, et al.1998). The third study explores the geographic variation in the interactions between elaiosome-bearing seeds and ant dispersers in Spain (Garrido et al. 2002).

Cushman et al. (1998) studied the geographic distribution and taxonomic diversity of a positive interaction involving ant-tended hemipterans and *Ficus* trees across a large geographic region (southern Africa and Madagascar). The system is based on the indirect benefits that *Ficus sur* trees receive from the presence of an ant-tended hemipteran species (*Hilda patruelis,* Tettigometridae)—attracted ants (mainly *Pheidole megacephala*) reduce the effects of both predispersal seed predators and parasitoids of pollinating wasps (Compton and Robertson 1988, 1991; Zachariades 1994). Cushman et al. used a method for obtaining geographic-scale data on species interactions based on (1) experimentally documenting the outcome of an interaction on a local scale, (2) the use of these results to identify conditions that must be met in order to show nonexperimentally that the interaction occurs, and (3) the evaluation of these conditions on a geographic scale to estimate the distribution of the interaction. Their data on 429 trees distributed throughout southern Africa and Madagascar indicates that 20 of 38 *Ficus* species, and 46% of all the trees they sampled, had ants on their fruits. Moreover, members of the *Ficus* subgenus *Sycomorus* were significantly more likely to attract ants than those in the subgenus *Urostigma,* and ant-colonization levels in *Sycomorus* were significantly greater than for species of *Urostigma.* The honeydew-producing hemipteran *H. patruelis* was the most common food source for the ants, although a range of other ant-tended hemipterans also attacked the plant. On average each ant-occupied *F. sur* individual had 37% of its fruit crop colonized by ants, compared to 24% for the other *Ficus* species. *Pheidole megacephala* was the most common ant species on the *Ficus* trees, occupying 58% of the sampled trees. Ant densities usually exceeded 4.5 per fruit, which was enough to provide protection against ovule gallers and parasitoids of pollinators. Cushman et al. (1998) concluded that the positive indirect effects in the system are geographically widespread, involving at least 20 host plant species.

* * *

In an extensive field study about variation of interactions on a geographic scale, Rico-Gray, García-Franco, et al.(1998) compared the richness of ant-plant interactions in four habitats with contrasting environmental characteristics (fig. 10.3). In particular, they asked if there were similarities among the habitats in the frequency of use by ants of plant-based food resources. However, they did not account for the outcome of those interactions. Their main results are presented below.

The ant genera with more species registered were *Camponotus* and *Pseudomyrmex;* and only six ant species overlapped among habitats. Species in the genera *Azteca, Camponotus, Crematogaster, Forelius, Monomorium,* and *Pseudomyrmex* used a wider variety of plant-based food resources (appendix 10.2). The latter finding is not surprising, since ants in the genera *Azteca, Camponotus,* and *Crematogaster,* in particular, have been reported as being able to tap the high productivity of canopy foliage by feeding on plant and insect exudates (Davidson 1997; see chapter 9).

Floristic similarity between sites was relatively low, particularly when comparisons were done at the species level. The two seaside sites (La Mancha and San Benito) were the most similar, whereas Xalapa was the most distinctive one (table 10.1). The ants more commonly used trees, shrubs, and herbs than any other life form such as woody or herbaceous vines, epiphytes, parasites, cactus, or treelike species (e.g., *Yucca, Beaucarnea*).

The diversity of resource use by ants differed significantly between La Mancha and the other three habitats, but not among Zapotitlán, Xalapa, and San Benito (table 10.2). Diversity of resource use was higher at La Mancha, where more plant species were involved, even though the total number of flowering plants per site is roughly similar (ca. 300–350 species). In La Mancha, San Benito, and Zapotitlán, nectar from plant reproductive structures and floral nectar were used more, whereas extrafloral nectar (nectar associated with leaves only) was used less (table 10.2).

TABLE 10.1 **Floristic similarity (Sørensen, %) between sites, at the levels of family, genera,and species**

	La Mancha (LM)			San Benito (SB)			Zapotitlán (ZA)			Xalapa (XA)		
Site	Fam	Gen	Sp	Fam	Gen	Sp	Fam	Gen	Sp	Fam	Gen	Sp
LM	—	—	—	33.0	22.9	11.3	26.3	12.5	2.6	16.7	4.2	0
SB	—	—	—	—	—	—	30.5	15.2	0	14.6	4.0	0
ZA	—	—	—	—	—	—	—	—	—	6.3	0	0
XA	—	—	—	—	—	—	—	—	—	—	—	—

Source: Rico-Gray, García-Franco, et al. 1998.

TABLE 10.2 **Number of cases in which each food resource was recorded being used by ants, and diversity of resource use per site**

Resource	La Mancha (LM)	San Benito (SB)	Zapotitlán (ZA)	Xalapa (XA)
Floral nectar	41	24	33	1
Nectar of reproductive structures	41	35	86	0
Hemipteran honeydew	46	9	20	2
Extrafloral nectar	13	2	6	13
Diversity Index (H')	0.57	0.47	0.46	0.26

LM vs. ZA, $t = 3.7898$, df = 210, $p = 0.00$* ZA vs. XA, $t = 2.0769$, df = 18, $p = 0.052$ ns
LM vs. XA, $t = 3.3367$, df = 16, $p = 0.004$* ZA vs. SB, $t = -0.2886$, df = 160, $p = 0.774$ ns
LM vs. SB, $t = 2.8616$, df = 93, $p = 0.006$* XA vs. SB, $t = -2.1623$, df = 20, $p = 0.042$ ns

Source: Rico-Gray, García-Franco, et al. 1998.
Notes: * Significant difference: global $\alpha = 0.10$, individual $\alpha = 0.0083$
ns = not significant (all Bonferroni corrected)

Food resources are patchily distributed for most animals, regulating their feeding behavior, population dynamics, and ultimately their evolution (Bronstein 1995). Plant-based food resources change seasonally, and ants are forced to undergo diet changes (see fig. 10.1; Rico-Gray and Sternberg 1991). For instance, plant-derived exudates, the main food resources utilized by ants, are unevenly distributed in space and among plant species and plant life forms in coastal Mexican communities (Rico-Gray et al. 2004). Moreover, many studies have demonstrated differences in the taxonomic distribution and abundance of plants bearing EFNs in a variety of habitats worldwide, supporting the idea that such glands are less common in temperate than in tropical regions (Bentley 1977a; Coley and Aide 1991; appendix 10.1).

Other studies have revealed the uneven distribution of other resources used by ants that may affect their within-habitat distribution (Torres 1984a, 1984b; Fowler and Claver 1991; Perfecto and Sediles 1992; Andersen and Patel 1994; Perfecto and Vandermeer 1994; Kaspari 1996). Ant size, abundance, and diversity and the abundance of ant-plant and ant-aphid interactions are known to decline with increasing altitude and/or latitude (e.g., Janzen 1973c; Janzen et al. 1976; Torres 1984a, 1984b; Koptur 1985; Olmstead and Wood 1990; Cushman, Lawton, and Manly 1993; Andersen 1997a; Samson, Rickart, and Gonzales 1997). The use of certain plant-based food resources may not only reflect ant preference; it may also reflect the relative abundance of plant life forms with nectaries per habitat. For instance, in a humid tropical forest Schupp and Feener

(1991) showed that lianas and vines were more likely to have nectaries than were trees and shrubs, and for a Mexican sand dune matorral Rico-Gray and Castro (1996) suggested that the presence and/or abundance of certain plant life forms might depend on the successional stage of the vegetation. Indeed, EFN-bearing plants (especially vines) are more common in secondary-growth areas, such as forest clearings and edges, where potential ant-derived protection is also increased (Bentley 1976, 1981).

* * *

In another ant-plant geographic variation study, Garrido et al. explored the interaction between the elaiosome-bearing seeds of *Helleborus foetidus* (Ranunculaceae) and its ant dispersers, searching for correlations between diaspore traits and functional characteristics of ant communities across nine sites in Spain. According to their results, mean diaspore traits did not show a distance-dependent pattern, but seed size-related traits (length, width, diaspore mass, and seed mass) varied mostly among sites, and ant-reward-related traits (elaiosome mass and elaiosome-to-seed mass ratio) varied mainly within plants. The ant communities studied showed distance-dependent patterns in composition and abundance, and almost all ant species responded positively to diaspore offerings and preferred seeds with elaiosomes. The extent of the preference differed among species and was affected by ant size. However, ant size was similar in almost all localities and the quality of the disperser guild, estimated as the ant response to seed offerings, did not vary. Except for *Messor capitatus,* all ant species that responded positively to the offerings were significantly more likely to respond to seeds with elaiosomes (table 10.3). Seed traits could not be predicted from ant community composition or geographic distance, nor could structural and functional ant community parameters explain variation in seed traits. Finally, comparison of ant sizes, position on a Principal Components Analysis of seed traits, and indirect estimates of dispersal success, suggests that there is a mosaic of well-matched and mismatched situations that probably obscures the overall relationships among seed traits and ant assemblages (Garrido et al. 2002). This is consistent with the role proposed for the geographic structure of interactions in recent coevolutionary theories (Thompson 1994).

* * *

TABLE 10.3 **Ant species-specific response to seed offerings**

Ant species	SW	SWO	Wald's X^2	*P*
Aphaenogaster iberica	0.85	0.44	8.30	0.0039
A. senilis	0.85	0.10	25.44	0.0001
Cataglyphis velox	0.92	0.18	38.68	0.0001
Crematogaster scutellaris	1.00	0.00	55.45	0.0001
C. sordidula	0.00	0.00	—	—
Formica cunicularia	0.78	0.05	27.00	0.0001
F. decipiens	0.56	0.10	10.92	0.0009
F. lugubris	0.80	0.25	14.17	0.0001
F. rufibarbis	0.20	0.025	9.53	0.002
Lasius niger	0.75	0.00	30.43	0.0001
Leptothorax unifasciatus	0.00	0.00	—	—
Messor capitatus	0.85	0.76	0.55	0.4562
Myrmica aloba	0.95	0.67	6.55	0.0104
M. scabrinodis	1.00	0.00	31.77	0.0001
M. sabuleti	1.00	0.00	55.45	0.0001
Pheidole pallidula	0.64	0.04	23.03	0.0001
Plagiolepis pygmaea	0.00	0.00	—	—
Tapinoma erraticum	0.67	0.00	29.11	0.0001
T. nigerrimun	0.50	0.00	20.92	0.0001
Tetramorium caespitum	0.90	0.00	40.83	0.0001

Source: From Garrido et al. 2002.
Note: The proportions of positive responses to seeds with (SW) and without (SWO) elaiosome are shown, together with the results of the Generalized Linear Model to test for differences in the response of each species to seeds with and without elaiosomes.

Despite the differences among habitats and studies, ant communities seem to be similar in their use of four plant-based food resources, which suggests a guild (i.e., a group of species that are similar in some way that is ecologically relevant) with a similar spatial evolutionary history and ecology (Wilson 1999). Community-wide assembly patterns have provided evidence that some communities are competitively structured (Denno, McClure, and Ott 1995; Stone, Dayan, and Simberloff 1996; but see Farrell and Mitter 1993; Wilson 1999). Mosaics of exclusive foraging territories appear to be a recurring feature of community organization for many arboreal ant communities throughout the tropics, but not in relatively untouched lowland rain forests (chapter 9). There are regional to local effects. However, although competition occurs among ground-dwelling ants (Jackson 1984), species composition and foraging performances (rates of resource discovery) may not differ between gaps and the forest understory (Feener and Schupp 1998).

Rico-Gray, García-Franco, et al. (1998) show considerable variation among habitats in the number, diversity, and seasonal distribution of ant-plant interactions, and they suggest that interhabitat variation of ant-plant interactions is the effect of variation in environmental parameters, the richness of plants with nectaries in the vegetation, and the richness in habitat heterogeneity (i.e., more than one vegetation association present). They also suggest that the abiotic environment drives vegetational diversity, which determines the nature of the ant community to a certain extent. Coupled with results showing geographic and temporal variation in the benefits of these interactions to plants (Thompson 1998; Beattie 1991; Koptur 1991; Rico-Gray and Castro 1996), this variation may explain, in part, geographic variation in species interactions, because it suggests a link between species diversity and interaction diversity (Thompson 1994, 1996, 1997; Travis 1996). The geographic mosaic of interactions between species may often influence the nature of the coevolutionary process and the evolution of interactions in general, since natural selection integrates geographic differences both in the species involved in an interaction and in the outcomes of an interaction (Thompson 1994). To generate such processes, species need to overlap, and the results by Rico-Gray, García-Franco, et al. (1998) show that very few species overlapped at the broad scale used. Of course, local specialization may be taking place in the different species/populations (Thompson 1994, 1997), and complete overlap is not a requirement. However, the positive outcome of the interactive system studied by Cushman et al. (1998) is present throughout a relatively large geographic range.

Few studies have addressed seasonal or geographic variation in ant-plant interactions, whether in their outcome or in their richness. In particular, we should explore the possibility that the studied variation is a response of the ant-plant populations not only to the environmental characteristics of the habitat (e.g., Rico-Gray, García-Franco, et al. 1998) but also to the genetic variation within and/or between populations. Local populations are genetically differentiated groups within a species, and the outcome of a particular interaction may vary among the populations of a given species (Thompson 1997). Moreover, in many instances not all individuals of a given plant species within a population interact with ants (Rico-Gray 1993), and this variation could be genetically based. For instance, in many polyploid species, the polyploids often have different suites of floral traits and different flowering periods than their diploid progenitor species, and such differences may subsequently affect their interactions with pollinating and other insect visitors (Segraves and Thompson 1999; Segraves et al. 1999). Polyploidy, therefore, may result in a geographic mosaic of interspecific interactions across a species' range, contributing to diversification in both plant and insect groups (Segraves and Thompson 1999).

APPENDIX 10.1 **Occurrence of plants with extrafloral nectaries (EFNs) in different vegetation types in tropical, subtropical, and temperate habitats worldwide**

Geographic region and type of vegetation	Species with EFNs (%)	Cover of plants with EFNs (%)	Source
Tropical and subtropical			
Brazilian cerrado			
Southeast Brazil, São Paulo (5 sites; woody species)	15.4–21.9	7.6–20.3	Oliveira and Leitão-Filho 1987
Western Brazil, Mato Grosso (5 sites; woody species)	16.5–25.5	14.2–31.2	Oliveira and Oliveira-Filho 1991
Brazilian amazon			
Terra firme forest (woody species)	17.6	19.1	Morellato and Oliveira 1991
Successional forest (woody species)	18.5	42.6	Morellato and Oliveira 1991
Buritirana (palm) vegetation (woody species)	33.3	29.7	Morellato and Oliveira 1991
Shrub canga (woody species)	53.3	50	Morellato and Oliveira 1991
Mexico			
Coastal vegetation (forest, sand dune, grassland, water marsh)	14.8	—	Díaz-Castelazo et al. 2004
Costa Rica			
Tropical dry forest hillside	—	30–80	Bentley 1976
Tropical riparian forest	—	10–40	Bentley 1976
Lowland rain forest	—	1–8	Koptur 1992a
Lower montane cloud forest	—	3–22	Koptur 1992a
High montane oak forest	—	0–3	Koptur 1992a

Panama			
Rain forest (shrubs and trees)	14–34	—	Schupp and Feener 1991
Rain forest (climbing plants)	44	—	Schupp and Feener 1991
Venezuela			
Rain forest (epiphytes)	19	28	Blüthgen et al. 2000
Jamaica			
Second-growth forest (sea level)	—	28	Keeler 1979a
Second-growth forest (montane)	—	0	Keeler 1979a
Florida (USA)			
Sawgrass prairie	—	2	Koptur 1992b
Rockledge pinelands	—	34	Koptur 1992b
Hardwood hammock	—	23	Koptur 1992b
Cameroon			
Rain forest (trees)	41.8	55.7	Dejean et al. 2000
Rain forest (climbing plants)	44.4	70	Dejean et al. 2000
East Asia			
Bonin islands (forest, sclerophyll shrub)	7.5	40.2	Pemberton 1998
Malaysia rain forest (woody species)	12.3	19.3	Fiala and Linsenmair 1995
Australia			
Rain forest (trees)	16.9	14.4	Blüthgen and Reifenrath 2003
Rain forest (climbing plants)	21.3	19.2	Blüthgen and Reifenrath 2003
Polynesia			
Hawaii (native species)	1.2	—	Keeler 1985

(*continued*)

APPENDIX 10.1 **(continued)**

Geographic region and type of vegetation	Species with EFNs (%)	Cover of plants with EFNs (%)	Source
Temperate and cold			
North America			
Southern California (desert communities)	—	0–27.7	Pemberton 1988
Northern California (grassland, forest, chaparral)	0	0	Keeler 1981a
Nebraska (forest, prairie)	3.8	0–14.2	Keeler 1979b, 1980a
Nebraska (forest, prairie) (woody plants only)	2.3	—	Keeler 1979b
Arizona (forest, desert, chaparral)	—	0–39	Keeler 1981d
Russia			
Tundra	0.3	10.3	Pemberton 1998
Cool-cold temperate forest	0.6–1.2	12.3–12.5	Pemberton 1998
Korea			
Deciduous forest	4.1	0.5–55	Pemberton 1990
Warm temperate forest	3–4	14.6	Pemberton 1998

Source: Modified from Oliveira and Freitas 2004.
Notes: Frequency of plants is expressed as a percentage of species and/or individuals (cover).
Unless otherwise indicated, surveys include all plant life forms.

APPENDIX 10.2 **Ant species using plant-derived food resources in four Mexican habitats**

Ant species	Site	Number of plant species visited	Resource
Dolichoderinae			
Azteca sp.	LA	24	fl, hem, nrs
Dorymyrmex flavus	ZA	10	fl, hem, nrs
Dorymyrmex insanus	ZA	10	arl, fl, hem, nrs
Dorymyrmex sp. 1	LA	19	efn, fl, fls, hem, lep, nrs
Dorymyrmex sp. 2	SB	7	fl, hem, nrs
Forelius sp. 1	ZA	10	arl, fl, hem, nrs
Forelius sp. 2	LA	20	efn, fl, hem, lep, nrs
Forelius sp. 3	SB	6	fl, nrs
Ectatomminae			
Ectatomma tuberculatum	SB	5	fl, hem, nrs
Formicinae			
Brachymyrmex sp. 1	ZA	10	fl, nrs
Brachymyrmex sp. 2	LA	1	hem, nrs
Camponotus abdominalis	LA, SB	6, 7	hem, nrs/arl, fl, nrs
Camponotus mucronatus	LA	10	efn, fl, hem, nrs
Camponotus planatus	LA,SB	72, 21	sa, efn, fl, fls, hem, lep, nrs/arl, fl, nrs
Camponotus rectangularis	SB	2	fl, nrs
Camponotus rubrithorax	ZA	28	arl, efn, fl, hem, nrs
Camponotus sericeiventris	LA	7	efn, fl, hem, nrs
Camponotus (Myrmobrachys) sp.	LA	11	efn, fl, hem, nrs
Camponotus (Tanaemyrmex) sp.	ZA	4	fl, nrs
Camponotus sp. 1	LA	10	efn, fl, hem, nrs
Camponotus sp. 2	XA	2	efn
Camponotus sp. 3	XA	1	efn
Paratrechina longicornis	LA, XA	5, 3	efn, fl, hem, nrs/efn, fl, hem
Myrmicinae			
Atta texana	ZA	11	fl, le, lgl, nrs
Atta sp.	SB	4	arl, fl, nrs, pe
Cephalotes aztecus	ZA	2	fl, nrs
Cephalotes sp.	LA	4	fl, hem, nrs
Crematogaster brevispinosa	LA, SB	42, 18	sa, efn, fl, fls, hem, nrs/fl, hem, nrs
Crematogaster opaca	ZA	16	arl, efn, fl, hem, nrs
Monomorium sp.	LA	21	efn, fl, hem, lep, nrs
Pheidole sp. 1	LA	3	fl, hem, nrs
Pheidole sp. 2	XA	3	efn, fl
Pogonomyrmex barbatus	ZA	4	fl, hem, nrs
Solenopsis geminata	LA, XA	3, 4	efn, fl, hem/efn
Xenomyrmex stolli	ZA	6	arl, efn, fl, nrs
Pseudomyrmecinae			
Pseudomyrmex ejectus	LA	5	efn, fl, nrs
Pseudomyrmex ferrugineus	LA	1	efn

(*continued*)

APPENDIX 10.2 **(continued)**

Ant species	Site	Number of plant species visited	Resource
Pseudomyrmex filiformis	LA	4	efn, fl, hem, nrs
Pseudomyrmex gracilis	LA	5	efn, fl, hem, nrs
Pseudomyrmex ita	LA	1	fl
Pseudomyrmex major	ZA	10	efn, fl, hem, nrs
Pseudomyrmex pallidus	ZA, LA	10, 4	fl, hem, nrs/efn, hem, nrs
Pseudomyrmex simplex	LA	4	hem, nrs
Pseudomyrmex sp.	SB	16	arl, fl, hem, nrs
Unknown 1	LA	1	fl
Unknown 2	LA	1	nrs
Unknown 3	LA	2	fl
Unknown 4	LA	1	fls
Unknown 5	LA	1	frs
Unknown 6	LA	3	fls
Unknown 7	LA	2	fls
Unknown 8	SB	7	fl, hem, nrs

Sources: Data from Rico-Gray 1989, 1993; Díaz-Castelazo and Rico-Gray 1998; Rico-Gray, García-Franco, et al. 1998; Rico-Gray, Palacios-Rios, et al. 1998.

Notes: Sites: LA = La Mancha; SB = San Benito; ZA = Zapotitlán; XA = Xalapa.

Resources: arl = nectaries on areole; sa = seed aril; efn = extrafloral nectar; fl = floral nectar; fls = liquids from flowers on the ground; frs = liquids from fruits on the ground; hem = hemipteran honeydew; le = chewing a leaf; lep = lepidopteran honeydew; lgl = liquids from ligula; nrs = nectar from reproductive structures other than floral; pe = liquids from petals.

CHAPTER ELEVEN

Ant-Plant Interactions in Agriculture

Data obtained from studies of various topics in ant-plant interactions (e.g., chapters 6 and 7) could potentially be applied in insect pest-management programs of agricultural systems (Vander Meer, Jaffe, and Cedeno 1990; Way and Khoo 1992). Ants possess many characteristics that are associated with the potential to act as biological control agents, especially in tropical agroecosystems (Risch and Carrol 1982a, 1982b), and an economically beneficial role has been associated with ants used for such purposes (e.g., Atsatt and O'Dowd 1976; Gotwald 1986; Majer 1986; Fernandes et al. 1994). Nevertheless, the positive effects of several ant attributes (e.g., predation of herbivores, pollination, soil improvement, and nutrient cycling) must be weighed against possible disadvantages (e.g., leaf-cutter ants and seed predators). Some ants feed on or disturb plants, act as vectors of plant diseases, benefit damaging Hemiptera, and may attack humans, domestic animals, or other beneficial animals. In short, virtually all ant species that prey on pests also possess some potential disadvantages. However, because ants play a crucial role in the regulation of certain insect populations, they should be taken into account when chemical control programs are planned (Gotwald 1986; Majer 1986; Perfecto 1990, 1991). Even though not all of the examples described below are strictly ant-plant interactions (e.g., no reward is offered by the plant in most cases; see chapter 6), the predatory activity by ants is shown to benefit the yield of crops, and as such one could say that it indirectly increases plant fitness. In this chapter we review some general characteristics of agricultural systems, the herbivore-ant relationship, the role of ants as biological control agents (describing two case studies: maize and coffee), and the relationship between biological control and the study of interspecific interactions.

Agricultural Systems, Herbivores, and Ants

Agricultural systems are areas of land where humans manipulate physical, chemical, and biological processes using a group of practices to produce desired products for their own use (e.g., food, fiber, and wood). As a result of such practices, natural ecosystems have been radically transformed over the years, and agriculture has extensively reshaped landscapes and affected ecological processes at the population, community, and ecosystem levels (Bentley 1983; Chapin et al. 2000; Klink and Moreira 2002). When land is cleared and plowed, there is a dramatic change in vegetation structure, which has a profound impact on above-ground plant and animal distributions and interactions, on soil-inhabiting organisms, on the physical and chemical properties of soil, and on the microclimatic conditions above and below ground (Stinner and Stinner 1989). Overall, agriculture generally reduces species composition and diversity, modifying the number and types of interspecific interactions relative to those present in natural ecosystems (Perfecto and Sediles 1992; Roth, Perfecto, and Rathcke 1994; Perfecto and Vandermeer 1996). Moreover, insecticide applications negatively affect organisms of higher trophic levels, such as predators and parasites, which help regulate herbivore populations (Price et al. 1980, 1986; Chapin et al. 2000). Because interacting species can differ widely in their responses to the details of habitat structure, it is becoming increasingly apparent that habitat fragmentation and patchiness have at least as much potential to affect species interactions and communities as they do to affect population dynamics (Fagan, Cantrell, and Cosner 1999 and references therein). Although there is not much direct evidence for how agriculture affects the relative proportions of the different types of interactions (e.g., antagonistic and mutualistic), it has been hypothesized that there is a shift toward antagonistic interactions (e.g., crop-weeds, plant-pathogens, and plant-herbivores) in agricultural versus natural ecosystems (Crossley et al. 1984; Stinner and Stinner 1989; Chapin et al. 2000; but see Olofsson, Moen, and Oksanen 1999). It is precisely in these particular antagonistic interactions that humans spend a great deal of energy (mechanical and/or chemical) trying to control damage to their crops.

Phytophagous insects damage nearly all crop species and can be broadly divided into tissue consumers and sap feeders. Most herbivorous arthropods (e.g., lepidopteran caterpillars) belong to the former group; however, the sap feeders include important herbivores that are also vectors of plant pathogens (e.g., Hemiptera) (Schowalter 2000). Chemicals have been used in insect pest management since the indiscriminate use of insecticides in the 1940s and 1950s. First inorganic substances (e.g., arsenic, copper, and cyanide compounds) were used, followed by DDT, and currently organophosphates, carbamate, and pyrethroid compounds are in use (Stinner and Stinner 1989). These toxic compounds exert high environmental damaging effects and sooner or later lose efficiency. In addition, many insects and other arthropods have evolved genetic resistance against them. Thus, it is quite apparent that eradication, or complete control of insect pests, will probably never be accomplished. Furthermore, and more dramatically, we may have accomplished the extinction of nonpest species (e.g., insect pollinators) more than we have eliminated pest species (Bentley 1983). We have reached a stage where all individual methods and strategies for insect pest management can be drawn together and used in a system that combines the most suitable characteristics of each to produce an ecologically sound, economically viable package called integrated pest management (Speight, Hunter, and Watt 1999). The use of natural enemies that may have a role in biological control, such as pathogens (bacteria and fungi), predatory invertebrates (spiders and ants), and parasitoids (flies and wasps) should therefore be amplified toward more environmentally friendly pest-control strategies (Jervis and Kidd 1996).

Some of the most important attributes of ants that make them potentially useful biological control agents are these (Finnegan 1974; Risch and Carroll 1982a, 1982b; Majer 1982a, 1986; Wilson 1987b):

1. They are diverse and abundant in most tropical and some temperate ecosystems, and most can be considered predators.
2. They respond to changes in the density of prey.
3. They can remain abundant even when prey is scarce, because they can cannibalize their brood and/or use plant and insect exudates as stable sources of energy.
4. They can store food and hence continue to capture prey even if it is not

immediately needed; i.e., predator satiation is not likely to limit the effectiveness of ants.

5. Besides killing pests, they can deter many others including some too large to be successfully captured.
6. Ants can be managed to enhance their abundance, distribution, and contacts with prey.

The use of ants in pest management has long been in practice in traditional management systems. In ancient times Chinese orchard growers observed the activities of the red tree ant (*Oecophylla smaragdina*) and began to manipulate ant nests to culture the ant, to encourage the ant to reside in their orchards, and to sell ant nests in markets. This type of biological control is still practiced after 1,700 years (Gotwald 1986; Huang and Yang 1987; Way and Khoo 1992). Experimental studies have also been developed. Early in the 1900s, the U.S. Department of Agriculture tested in Texas the effect of *Ectatomma tuberculatum* (Ectatomminae), an ant imported from Guatemala, on the herbivores of cotton (*Gossypium* spp.: Malvaceae), a plant bearing extrafloral nectaries (EFNs) (Cook 1904, 1905). Unfortunately, not much was heard of the success of this attempt, which was later reevaluated using *Ectatomma ruidum* (Weber 1946). With the development and wide use of DDT, the support for this kind of research decreased abruptly. It has also been demonstrated that planting EFN-bearing plants among crops lacking nectaries successfully recruits biological control agents in agroecosystems (Atsatt and O'Dowd 1976; Rudgers and Gardener 2004). Most of the studies on the protective effect of ants on plant fitness ("ant-guard studies"), whether they focus on disrupting herbivore activity or on eliminating herbivores directly (chapter 6), form an excellent basis to evaluate the management of insect herbivores using predatory ant species. For instance, manipulations of extrafloral nectar (regardless of the plant species) would help determine whether this resource can affect the diversity and abundance of associated arthropods. If this is the case, in addition to the ant-exclusion approach, extrafloral nectar traits (volume, composition, and volatiles) should also be analyzed in order to evaluate their effects on species interactions in agroecosystems (Rudgers 2004; Rudgers and Gardener 2004). Finally, besides direct deterrence or elimination of insect herbivores, foraging ants may also benefit crop plants by preying on eggs of pest species (Risch 1981; Letourneau 1983; Kluge 1991; Way, Cammell, and Paiva 1992; Goebel et al. 1999; Mansfield, Elias, and Lytton-Hitchins 2003).

Ants as Biological Control Agents

Predatory ants can significantly affect the behavior of prey and depress the size of potential pest populations (Green and Sullivan 1950; Gotwald 1986; Perfecto 1990, 1991; Cole et al. 1992; Sanders and Pang 1992; Mahdi and Whittaker 1993; Karhu and Neuvonen 1998). Published work has emphasized seven genera of dominant ant species that are either beneficial or potentially beneficial predatory ants: *Oecophylla, Dolichoderus, Anoplolepis, Wasmannia,* and *Azteca* in the tropics; *Solenopsis* in the tropics and subtropics; and *Formica* in temperate environments (reviewed by Way and Khoo 1992). Species in other ant genera, e.g., *Ectatomma, Pheidole, Dorymyrmex, Leptothorax,* and *Tetramorium,* have also been mentioned as possible biological control agents due to their foraging activity (e.g., Taylor 1977; Lachaud 1990; Perfecto 1990, 1991; Fernandes et al. 1994). Although most predatory ants can eliminate insect herbivores from plants, some species associated with honeydew-producing Hemiptera could potentially decrease plant fitness by transmitting plant pathogens (Buckley 1987a, 1987b; see chapter 7). It may also occur that a potentially useful ant species in a particular agricultural system is driven away by a not-so-helpful dominant ant (Andersen and Patel 1994), or that diversity and abundance of ants are associated with particular forest clearance procedures and methods of plantation establishment (Watt, Stork, and Bolton 2002).

Other factors that should be considered are the complexity of interactions within an ant mosaic or ant community (Perfecto 1994; Kaspari 1996; Bruhl, Gunsalam, and Linsenmair 1998; Blüthgen, Stork, and Fiedler 2004), the diversity of the vegetation, and the kinds of agricultural management systems, because all these factors can affect ant diversity and, in turn, the effect of ants on herbivorous pests (Perfecto and Sediles 1992; Roth, Perfecto, and Rathcke 1994; Perfecto and Vandermeer 1994, 1996). In order to assess which ant species will eventually be the most active and important as biological control agents and to manipulate ant distribution, one needs to know the dominant ants in the community and how the spatial pattern of the community is influenced and maintained (Majer 1986; Ryti and Case 1992; Ewuim, Badejo, and Ajayi 1997). Unfortunately, there is a general lack of published case studies, and more experimental data is needed before one can evaluate the role of ants as biological control agents in crop systems. Some specific examples are reviewed below, followed by two case studies.

Workers of *Oecophylla longinoda* and *O. smaragdina* (Formicinae) attack and kill a wide range of arthropods for food (Way 1953, 1954; Hölldobler 1983). Their highly organized aggressive predatory behavior (which may constitute a constraint for their use by humans), combined with extensive foraging throughout the area occupied by a colony, may explain the success of *Oecophylla* species in killing or driving away many pests or potential pests, particularly Hemiptera (e.g., Coreidae, Miridae) and foliar-feeding Coleoptera, in coconut, oil palm, cacao, coffee, citrus, eucalyptus, mango, and timber plantations (e.g., Way 1953; Brown 1959; Majer and Camer-Pesci 1991; Majer, Delabie, and Smith 1994). Moreover, they can also protect cacao plantations against rodents, and oil palm plantations against some lepidopterous defoliators (Way and Khoo 1992).

Dolichoderus thoracicus (Dolichoderinae) is used in the humid Southeast Asian tropics in biological control of coconut palms and cacao trees; these ants are not aggressive and thus pose no threat to the staff of the plantation (Way and Khoo 1991, 1992). They form dense colonies that may cover an area of many hectares, and they need suitable nesting sites (including artificial nests) to play a role in biological control (Way and Khoo 1991). In Malaysia this ant is associated with mealybugs (Hemiptera) but does not appear to decrease yield; *D. thoracicus* is particularly important in protecting cacao against Miridae (Hemiptera) (Way and Khoo 1992).

Two species of *Anoplolepis* (Formicinae) have biological control attributes. *A. longipes* (= *A. gracilipes*) probably originated in Africa (Haynes and Haynes 1978a) but occurs throughout the tropics and has important biological control attributes that can usefully be encouraged in places where it is clearly beneficial. Even though it can kill or disturb domestic animals and harm plants directly and indirectly, it is not conspicuously aggressive to humans (Haynes and Haynes 1978b). Although in some areas it has displaced the beneficial *Oecophylla longinoda* or is a constraint to the establishment of *Dolichoderus thoracicus, A. longipes* can provide some protection to cacao trees but especially to coconut palms from Coleoptera (Way and Khoo 1992). *A. custodiens* is limited to certain habitats in Africa (Way 1953), and it is considered a pest in South Africa because its attended Hemiptera cause serious damage to crops like citrus (Steyn 1954). In Tanzania dense populations may protect coconut palms from Coreidae, but when extremely abundant their associated Hemiptera might be damaging to crops (Way and Khoo 1992).

Wasmannia auropunctata (Myrmicinae) is native to tropical America and is sometimes regarded as a pest in some areas of its original distribution (Ulloa-Chalon and Cherix 1990); it also may affect the native insect community when introduced to other areas (Lubin 1984). The species has been shown to protect cacao plantations against Miridae (Hemiptera) in Cameroon, and coconut plantations against Coreidae in the Solomon Islands (Way and Khoo 1992). In the latter, *W. auropunctata* has also displaced the dominant ants *Iridomyrmex cordatus* and *Pheidole megacephala,* which do not protect coconut palms.

Species in the tree-nesting neotropical genus *Azteca* (Dolichoderinae) have already been discussed as important antiherbivore agents of tree species in the genus *Cecropia* (Cecropiaceae) (chapter 6). The Kayapo Indians in Brazil recognized the value of these ants and used them against leaf-cutting ants (Attini) (Overal and Posey 1984). It has been shown that *Azteca*-occupied citrus trees in Trinidad are damaged much less by *Atta cephalotes* ants than unoccupied trees are (Jutsum, Cherrett, and Fisher 1981) and that cacao plants hosting *Azteca* colonies had higher yields than neighboring unoccupied plants (Delabie 1990). However, the aggressiveness of *Azteca* ants may cause discomfort to people, and their attended Hemiptera may outweigh the benefits that the ants confer (Delabie 1990; Catling 1997). Thorough research is needed on the predatory behavior (Morais 1994) and on the role and use of *Azteca* ants as biological control agents (Leston 1978).

Three neotropical species in the myrmicine genus *Solenopsis* (*S. geminata* in the hotter climates and *S. invicta* and *S. richteri* in subtropical South America; *S. richteri* has also been introduced to the southern United States) have obvious biological control characteristics, particularly the fire ant *S. invicta.* The introduced fire ant opportunistically exploits and thrives in disturbed agricultural habitats, and in the United States it has turned into a serious pest that creates severe medical, veterinary, and agricultural problems (Pereira 2003). However, fire ants are not regarded as major pests for crops. *S. invicta* is an important predator of pests of sugar cane, cotton, and other crops and usually will not harm other predatory species in cotton fields (Sterling 1978; Jones and Sterling 1979; Adams et al. 1981; McDaniel and Sterling 1982; Fillman and Sterling 1985; Reagan 1986; Fuller and Reagan 1988). Chemical control of the ants has sometimes made pests worse, so emphasis is currently on preservation and enhancement through cultural practices and selective use of chemicals where the benefits of *S. invicta* outweigh the disadvantages (Way and Khoo 1992).

Pheidole (Myrmicinae) is a hyperdiverse cosmopolitan ant genus that is extremely abundant in most of the warm regions of the world, especially in soil and ground litter (Wilson 2003). In cotton fields of Brazil, Fernandes et al. (1994) have shown that *Pheidole* ants are very effective in suppressing overwintering adult boll weevils (*Anthonomus grandis*) on the ground during the between-season period. Predation by ants as a whole destroyed 20% of the weevils and involved four species in each of four genera: *Pheidole, Solenopsis, Dorymyrmex,* and *Tapinoma*. Foraging *Pheidole* accounted for 94% of the predation due to ants in the study period. In Texas, United States, Sterling (1978) showed that *Solenopsis invicta* is efficient at suppressing larval stages of *Anthonomus,* consuming up to 85% of the weevils. Moreover, Agnew and Sterling (1981) demonstrated that *S. invicta* also kills large numbers of pupae and adults of the weevil when infested mature bolls open and expose the insects. While the benefit of weevil suppression by *Solenopsis* late in the season is the reduction of the weevil population that will enter diapause and survive the winter in Texas, predation by *Pheidole* of overwintering adult boll weevils in Brazil is mainly that of reducing the risk of high infestation in the next cropping cycle (Fernandes et al. 1994). The presence of extrafloral nectaries in cotton further promotes ant visitation to the plants and tends to increase potential ant-derived benefits afforded through predation on boll weevils (Agnew and Sterling 1981; Rudgers 2004).

The importance of species of *Formica* (Formicinae) in reducing defoliation by outbreak pests in temperate forests has long been recognized in Germany (Gosswald 1990 and references therein). The pests controlled by *Formica* spp. are high-density epidemic species, particularly lepidopteran caterpillars, whose periodic increase in abundance needs fast suppression (Adlung 1966). As facultative predators that are active both day and night over a long season, *F. polyctena* and *F. lugubris* have been favored due to their high population densities at all levels of the forest and their capacity to kill both active and quiescent stages of different prey species, particularly pest caterpillars during outbreaks (Way and Khoo 1992). When prey is scarce, *Formica* ants maintain their large colonies on the honeydew from tended Hemiptera, which at times may make these ants undesirable (Way 1963).

The research conducted by Ivette Perfecto and co-workers contains some of the best field experimental examples of how ants can be useful biological pest control agents in agricultural systems (maize, maize-beans, and coffee), of how insecticides modify ant densities and the outcome of their effect, and of how vegetation diversity and its modification affect ant diversity (Perfecto 1990, 1991, 1994; Perfecto and Sediles 1992; Perfecto and Vandermeer 1994, 1996, 2002; Roth, Perfecto, and Rathcke 1994; Perfecto and Armbrecht 2003; Philpott, Greenberg, et al. 2004; Philpott, Maldonado, et al. 2004). Some of the important findings from these works are reviewed below.

Case Studies

The Maize-Pest-Ant System in Nicaragua

Working in a traditional small-scale agroecosystem involving tropical maize (*Zea mays*) in Nicaragua, Perfecto (1990) examined and monitored ant foraging activity, in particular by *Pheidole radowszkoskii* (Myrmicinae) and *Ectatomma ruidum* (Ectatomminae), and the population levels of two important herbivores, the fall armyworm (*Spodoptera frugiperda,* Lepidoptera: Noctuidae) and the corn leafhopper (*Dalbulus maidis,* Hemiptera: Cicadellidae). Monitoring was done in four insecticide treatment scenarios: (1) the systemic carbofuran, (2) the broad-spectrum chlorpyrifos, (3) a combination of carbofuran and chlorpyrifos, and (4) a control (no insecticide).

Her results show that (1) the application of carbofuran reduced ant foraging activity (fig. 11.1) and increased population levels of *S. frugiperda* (fig. 11.2); (2) the application of chlorpyrifos significantly reduced ant foraging activity as well as densities of *S. frugiperda,* while the density of *D. maidis* significantly increased (fig. 11.3); and (3) ant predation was apparently responsible for a higher larval removal rate in insecticide-free plots (fig. 11.4) (Perfecto 1991).

These results by Perfecto (1990) show the indirect effect resulting from the application of insecticides. The application of carbofuran significantly increased the numbers of *S. frugiperda* larvae, representing probably the effect of the dramatic reduction of predatory ants. The application of chlorpyrifos, however, significantly reduced the abundance of *S. frugiperda* but increased leafhopper abundance, probably due to a negative correlation between *S. frugiperda* and *D. maidis* and the fact that chlorpyrifos did kill *S. frugiperda*. Furthermore, leafhoppers apparently preferred plants with the least damage from *S. frugiperda*.

Interestingly, at the time when this research was carried out, carbofu-

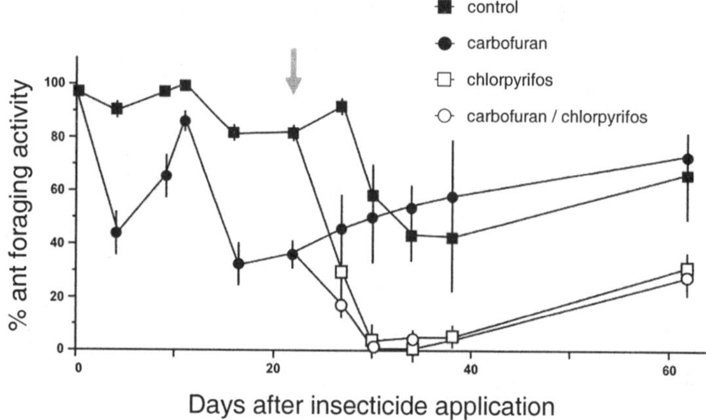

FIGURE 11.1. Percentage of ant foraging activity for the carbofuran and control treatments, before and after the application of chlorpyrifos in a maize agroecosystem. The arrow represents the last sampling date before the application of chlorpyrifos, after which ant foraging activity for all treatments is shown. Error bars indicate ± 1 SE. Modified from Perfecto 1990.

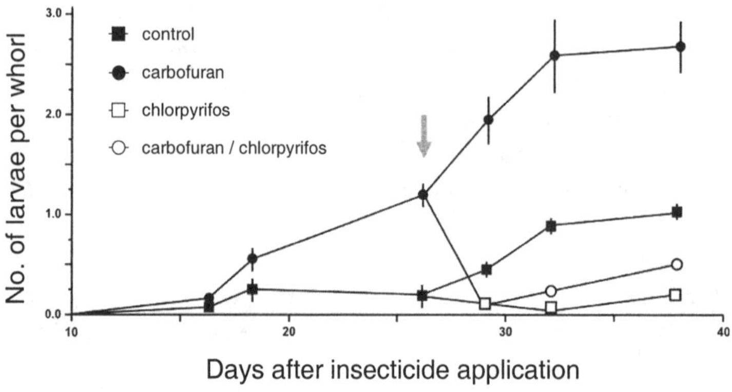

FIGURE 11.2. Numbers of armyworm larvae (*Spodoptera frugiperda*) per maize plant during the growing season for the carbofuran and the control treatments before and after the application of chlorpyrifos. The arrow represents the last sampling date before the application of chlorpyrifos, after which numbers of *S. frugiperda* are shown for all four treatments. Error bars indicate ± 1 SE. Modified from Perfecto 1990.

ran was highly recommended in Nicaragua as an effective control against both *S. frugiperda* and *D. maidis* (Perfecto 1990). The ineffectiveness as a control mechanism for both maize pests was probably the effect of the evolution of resistance. In contrast to their effects on herbivorous pests, both carbofuran and chlorpyrifos had a strong negative effect on ant foraging activity. This effect is not surprising because resistance among predators is unusual, and ants are especially susceptible to insecticides (Perfecto 1990 and references therein).

In a different field experiment, using a large-scale production system of irrigated maize in Nicaragua, Perfecto (1991) showed specifically that ants markedly reduced the abundance of *S. frugiperda* and *D. maidis* (fig. 11.4, table 11.1), with significantly higher numbers of both pests in plots where ant-foraging activity was reduced with insecticide (Perfecto and Sediles 1992).

These results suggest that ants can play a significant role as agents of biological control in integrated pest-management programs and that ant susceptibility to chemical insecticides should be taken into consideration (Perfecto 1990, 1991; Perfecto and Sediles 1992). Additional work has shown that although conservation of ant diversity may be pos-

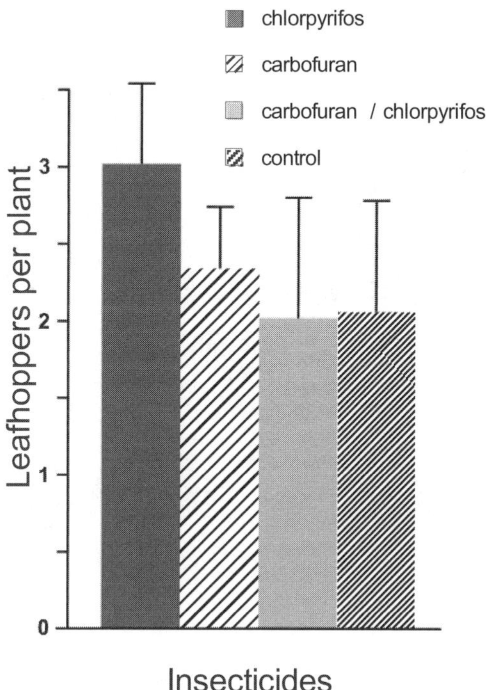

FIGURE 11.3. Number of corn leafhoppers (*Dalbulus maidis*) per maize plant at four levels of insecticide: chlorpyrifos, carbofuran, chlorpyrifos with carbofuran, and a control (with no insecticide). Leafhoppers were significantly more abundant in the chlorpyrifos treatment than in all other treatments ($p < 0.001$; Duncan's multiple range test). Error bars indicate ± 1 SE. Modified from Perfecto 1990.

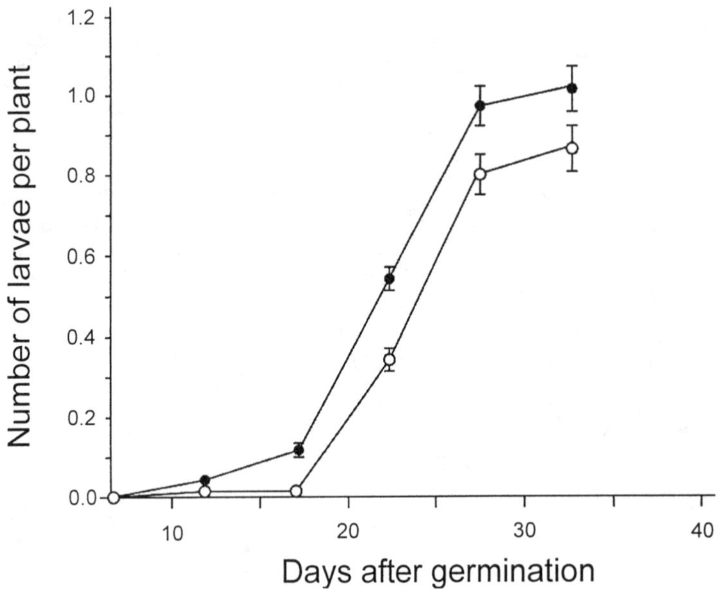

FIGURE 11.4. Numbers of fifth and sixth instar larvae of the fall armyworm (*Spodoptera frugiperda*) per maize plant throughout the growing season, for treatments with normal (open circles) and reduced ant-foraging activity (black circles). Error bars indicate ± 1 SE. Modified from Perfecto 1991.

TABLE 11.1 **Numbers of the corn leafhopper (*Dalbulus maidis*) per maize plant throughout the season, in plots with reduced (treatment) and normal (control) ant-foraging activity**

Days after plant germination	Treatment	Control	*F*
7	0.12 ± 0.05	0.27 ± 0.05	5.740*
12	0.56 ± 0.05	0.51 ± 0.13	1.027*
17	1.29 ± 0.16	0.90 ± 0.05	4.803*
22	0.39 ± 0.14	0.13 ± 0.05	10.754**
27	0.19 ± 0.09	0.02 ± 0.01	6.082*
32	0	0	—
37	0	0	—
42	0	0	—

Source: Modified from Perfecto 1991.
Note: Data are means ± 1 SE.
* $p < 0.05$; ** $p < 0.01$ (ANOVAs, df = 1, 3)

sible in a mosaic of different land uses (Perfecto and Vandermeer 1994; Roth, Perfecto, and Rathcke 1994), physical factors (e.g., light and temperature) may change the nature of competitive interactions between the most abundant species (Perfecto 1994; Perfecto and Vandermeer 1994, 1996).

Ants in Coffee Plantations in Mexico

Traditional production techniques for coffee include an overstory of shade trees, usually dominated by legumes in the genera *Inga* or *Erythrina,* but in many cases including a variety of fruit and timber species as well (Vandermeer et al. 2002; Perfecto and Armbrecht 2003). Although arboreal ants are commonly found nesting in these trees (Perfecto and Vandermeer 2002), their role in the system remains largely unknown. Many of these ant species are known to tend hemipterans that may reach pest proportions, thus making the ant a pest species as well. However, ants are predators and may have a net positive effect as biocontrol agents (Vandermeer et al. 2002; Philpott, Greenberg, et al.2004). Indeed, their effectiveness in shaping arthropod assemblages has been tested in a Mexican coffee farm. For instance, Vandermeer et al. (2002) showed that ant presence resulted in the virtual absence of lepidopteran caterpillars. They were either preyed upon by ants or harassed to the point of dropping off the bush. Furthermore, Philpott, Greenberg, et al. (2004) found differences in the effect of the two ant species studied. *Azteca instabilis* ants had no effect on arthropods (either total number, whether small or large, or of specific taxonomic orders) and additionally had the potential to be pests through their positive effect on scale insects (Hemiptera) (Vandermeer et al. 2002). In contrast, *Camponotus senex* tended to remove arthropods (total, small), especially in the dry season, and affected arthropod densities of some taxonomic orders both positively and negatively (Philpott, Greenberg, et al. 2004).

Birds were also present and exerted a significant pressure on herbivore populations. They reduced densities of total and large arthropods and of some arthropod orders (e.g. spiders, beetles, and roaches), according to Philpott, Greenberg, et al. Interestingly, birds and ants differed in their effects on total and large arthropods, indicating that these different taxa should not be treated as one entity (Philpott, Greenberg, et al. 2004). In summary, birds exerted stronger negative effects on arthropods than ants did, and the two dominant ant species (*A. instabilis* and *C. senex*) had strong effects on arthropods in different seasons.

TABLE 11.2 **Numbers of spiders (or webs) found in a sample of 18 pairs of trees in an organic coffee farm in Chiapas, Mexico**

	Bushes without ants	Bushes with ants	Significance[a]
Number of nymphs	103	74	0.024
Number of adults	10	36	0.010
Number of webs	55	42	0.195
Foraging adults[b]	4	8	0.300
Foraging nymphs	26	8	0.005
Linyphiid adults	2	20	0.059

Source: From Vandermeer et al. 2002.
[a] Significance based on a paired *t*-test of the logs of the numbers.
[b] Foraging adults or nymphs refer to those individuals found not in a web.

Coffee plantations are a more complex system than just ants, Hemiptera, and other herbivores, implying a more complicated food-web structure. Indeed, a census of spiders on coffee plants with and without foraging *Azteca* resulted in a negative relationship between ants and spiders (table 11.2), suggesting a complicated relationship between these two predatory organisms (Vandermeer et al. 2002).

The presence of other organisms in coffee agroecosystems, such as parasitic phorid flies (*Pseudacteon* sp., Diptera: Phoridae) can have a strong, indirect effect on ant prey through their interaction with ants. In the same coffee plantations studied above, the presence of phorid flies significantly decreased recruitment by the ant *Azteca instabilis,* to the point of having no effect on larvae (*Spodoptera frugiperda*) experimentally introduced. Furthermore, ants were significantly slower at killing or removing larvae from coffee plants when phorids were present. Philpott, Maldonado, et al. (2004) suggest that phorid flies may have much wider influences, affecting interactions of *A. instabilis* ants with herbivore communities as well as with other plant species. They propose the existence of a behaviorally mediated species-level trophic cascade in the coffee agroecosystems, with considerable effects in ant and herbivore communities, as well as for coffee production.

These studies show that (1) certain dominant ant species have potential as biological control agents; (2) ants have the potential to be pests through their positive effect on hemipterans; (3) there is an assemblage of different possible predator species, ants being part of it; and (4) aggregating taxonomically related and unrelated predators into trophic levels without previous experimental data evaluating the sign and strengths of effects may lead to a misinterpretation of food-web interactions in coffee plantations (Vandermeer et al. 2002; Perfecto and Armbrecht 2003; Philpott, Greenberg, et al. 2004; Philpott, Maldonado, et al. 2004).

Biological Control and Interspecific Interactions

Even though the use of ants as a biological pest control may have certain drawbacks (e.g., tending Hemiptera, killing other predatory species, and irritating humans), ants in general have proved to be beneficial in controlling a number of insect herbivores (Agnew and Sterling 1981; Gotwald 1986; Majer 1986; Way and Khoo 1992; Perfecto 1990, 1991; Fernandes et al. 1994; Rickson and Rickson 1998). An added benefit from the use of ants to control herbivorous pests is the decrease in the use of pesticides, which in turn helps to promote a better and more healthful environment (e.g., Andersen 1997b; Longino and Colwell 1997). Pesticides have to be constantly tested, and/or new compounds must be developed because herbivores become resistant to their effect, but ants continue to evolve, just as herbivores do, and they are also generalist predators capable of controlling a wide range of prey.

More research is needed for a better understanding of the biology and management of ants (Rickson and Rickson 1998), especially when the introduction of exotic predatory ants is planned (Majer 1986). Unfortunately, support for research on these topics is not easy to obtain. The use of pesticides is promoted by manufacturing companies and/or by governments that have special programs in which pesticide-fertilizer packages are offered to farmers under special deals. The latter is particularly true in developing countries (e.g., Mexico, India, and Brazil), which are the natural habitat for many of the ant species and interactions described above, and where traditional practices of agricultural management have developed over hundreds of years. This is the case for the Yucatecan Maya in Mexico, whose knowledge is rapidly being lost to the point of no recovery (Rico-Gray, García-Franco, and Chemás 1988; Rico-Gray et al. 1988, 1990; Chemás and Rico-Gray 1991; Rico-Gray, Chemás, and Mandujano 1991; Rico-Gray and García-Franco 1991, 1992). Furthermore, in some of these countries, large-scale agricultural production, planned to attain food self-sufficiency and make the economy less dependent on international economic conditions, is resulting in severe pest and general environmental problems (e.g., rice fields in the Yohaltún valley in Campeche or the Uxpanapa valley in Veracruz, Mexico). The use of more ecologically rational agriculture, such as the acceptance of biological and cultural pest control to create integrated pest-management programs (which include biological control agents) is one way to solve some of the problems created by large-scale monocultures (Rickson and Rickson 1998). The latter are more prone to herbivore attack than a temporal and spatial mosaic of vegetation (Whitham, Williams, and Robinson 1984).

As the reduction of natural ecosystems dramatically increases, our possibilities to study untouched natural populations are also decreasing fast. Thus, we are facing large areas of modified ecosystems that, in most cases, ecology and evolutionary biologists have neglected to study, always looking for "untouched" systems. Modified ecosystems will be the norm in coming years, and access to funding has to change in these circumstances; we should put together basic ecological and evolutionary studies with applied ecological studies (e.g., Vander Meer, Jaffe, and Cedeno 1990). Ecology and evolution continue to take place in disturbed habitats, and we have to face this fact and study with a different scope, combining natural selection and artificial selection, even though these processes add complications to our research. This approach creates an opportunity for joint research in evolutionary and applied ecology and vice versa. The knowledge we have from studies in natural populations is crucial to our long-term success in manipulating those communities as we attempt to preserve species and their interactions and genetic diversity, minimize the impact of introduced species through biological control programs, and fight against emerging diseases. Many questions in the biology of interspecific interactions have the potential to make evolutionary ecology one of the most important of the applied ecological sciences (Thompson 1999a, 1999c). Even though our lack of data is partly due to not taking up the study of more coevolutionary relationships within natural communities (Thompson 1999c), we believe it is time to take disturbed communities more seriously into account, in addition to conducting research in "untouched" systems. All of these have to change fast because we are faced every day with the increase of disturbed systems.

CHAPTER TWELVE

Overview and Perspectives

Ants are probably the most dominant insect family on earth, and flowering plants have been the dominant plant group on land for more than 100 million years. The evolutionary success of angiosperms cannot be ascribed solely to benefits conferred by possessing flowers; it is also the result of benefits conferred by an array of interspecific interactions (e.g., pollination, herbivory, and seed dispersal) that have helped shape their great diversity. On those bases alone, the results of studies on the ecology and evolution of ant-plant interactions are crucial to an understanding of the ecology of terrestrial biological communities. Moreover, ant-plant interactions are suitable systems in which to study processes operating at the community level because of the parallels between the characteristics of ant and plant communities and because in some cases the number of plant species is the best predictor of the number of ant species. Here we will address the importance of studies on ant-plant interactions for evolutionary ecology and present an overview of what has been learned by studying ant-plant interactions. We will finish by suggesting perspectives on what needs to be studied and how these studies should be approached, and by reporting on research that is currently in development.

Even though fossil evidence is extremely scarce, it is inferred that the evolutionary history of ant-plant interactions may have developed as early as the mid-Cretaceous (Grimaldi and Engel 2005; Wilson and Hölldobler 2005; Moreau et al. 2006). The current understanding of the joint evolutionary history of the two groups has recently been summarized by Wilson and Hölldobler (2005) in three key events (see chapter 1), following especially the origin and changes in dominance of the four most diverse, abundant, and geographically widespread living subfamilies (Ponerinae, Myrmicinae, Formicinae, and Dolichoderinae). Although one of the first ant-plant interactions probably involved ants acting as seed predators (Labandeira 2002), the first fossil evidence suggesting ant-plant interactions is the oldest known fossil of an extrafloral nectaried leaf (Pemberton 1992). This fossil comes from a site that also harbored 32 ant species, as well as predatory and parasitic insects whose modern relatives visit extrafloral nectaries (EFNs). Of these 32 ant species, 10 belong to extant genera that have species acting as ant-guard mutualists of plants bearing EFNs. This finding, together with the fact that EFNs are visited not only by ants but also by a variety of arthropod groups, indicates that ant-plant interactions and the relative effects of these groups on the origin and evolution of nectaries and on plant fitness need to be comparatively analyzed using a multispecies approach (Cuautle and Rico-Gray 2003; Heil and McKey 2003; Del-Claro 2004).

The study of ant-plant interactions has permeated the literature since Janzen's (1966, 1967a, 1967b, 1969b) experimental studies with the *Acacia-Pseudomyrmex* system. These crucial studies were seminal in stimulating further research on ant-plant interactions, mutualism, and coevolution in many contexts and on an array of topics of general interest relating to animal-plant interactions. The research accomplished during the past 40 years has shown, for instance, that ant-plant interactions are geographically widespread, that they are common in different types of plant communities (Bentley 1977a; Oliveira and Oliveira-Filho 1991; Schupp and Feener 1991; Koptur 1992a; Fiala and Linsenmair 1995; Fonseca and Ganade 1996; Pemberton 1998; Rico-Gray, García-Franco, et al.1998; Oliveira and Freitas 2004), that they play an important role in plant defense against herbivores (Coley and Aide 1991; Davidson and McKey 1993; Heil and McKey 2003; and references therein), and that the effect of ants as seed predators can be considerable (chapter 2).

Studies on foliage-dwelling arthropods in tropical rain forests have shown that ants may represent 86% of the arthropod biomass and up to 94% of the arthropod individuals living in the canopy (Majer 1990; Tobin 1995). Although researchers generally acknowledge consumption of plant products by ants, these have been regarded mostly as predators and scavengers of animal matter (Wilson 1959; Carroll and Janzen 1973; Hölldobler and Wilson 1990). However, because the ants' biomass greatly exceeds that of herbivore prey, our understanding of energy flow in the canopy is challenged by this apparent paradox. Tobin (1991, 1994) has solved this "ant-biomass paradox" by suggesting that most foliage-dwelling ants behave mainly as primary consumers and rely mostly on plant- and insect-derived exudates for nutrition (chapter 9). This view is supported by the frequent incidence on foliage of renewable liquid foods such as extrafloral nectar, hemipteran honeydew, and secretions from lepidopteran larvae (Blüthgen et al. 2000; Dejean et al. 2000; Pierce et al. 2002) and by the fact that those few species that disproportionately account for the extraordinary ant abundance in canopy samples are known to feed extensively on exudates (Davidson 1997). A key feature of canopy ants, especially formicines and dolichoderines, is their ability to process liquid foods due to innovations in their digestive systems, which include an expandable crop and a differentiated proventriculus that allows the storage of large liquid volumes and facilitates food sharing among nest mates (Eisner 1957). A major step toward explaining the abundance of tropical arboreal ants was taken by Davidson et al. (2003) in the rain forests of Peru and Borneo using data from stable-isotope analyses (see also Blüthgen, Gebauer, and Fiedler 2003). Despite the very low concentrations of nitrogenous compounds in plant tissues, Davidson et al. (2003) showed that a number of specialized exudate-feeding ant species (particularly formicines and dolichoderines) acquire their nitrogen lower in the trophic chain than predominantly predacious species do, and therefore they act as herbivores as originally predicted by Tobin (1991). Thus the high incidence of exudates on foliage plays a key role in shaping food-web structure in tropical forests by fueling costly prey-hunting activities by canopy ants, which keep prey species at lower numbers than expected from an ant diet exclusively based on animal prey, with relevant consequences for some plant species (Davidson 1997). Dominant exudate-feeding species that obtain N mostly from prey should therefore have a greater potential to protect plants by attacking herbivores (a mutualistic role) than other species feeding at lower trophic levels (a parasitic role) (Davidson et al. 2003). This was recently confirmed by stable-isotope data from ants inhabiting *Cordia alliodora* (Boraginaceae) in Costa Rica—mutualistic ant species have a more carnivorous diet than parasitic ones (Tillberg 2004).

The results of studies on ant-plant interactions have played a major role in shaping our broad understanding of mutualism, by developing approaches to measure benefits, costs, and net outcomes and by their explicit consideration of variability (Bronstein 1998; Holland et al. 2005). Ant-plant-herbivore associations offer an excellent opportunity to analyze the effects of both historical and ecological factors on the evolution of mutualisms (McKey and Davidson 1993) and have served with increasing frequency as models for examining conditional outcomes in interspecific interactions (Cushman and Beattie 1991; Del-Claro and Oliveira 2000; Billick and Tonkel 2003) and adaptive specialization toward more stable symbiotic mutualisms (Heil, Rattke, and Boland 2005). Moreover, such interaction systems have helped us understand the evolutionary stability of mutualisms in the face of potentially destabilizing conflicts, and species coexistence (Palmer, Young, and Straton 2002; Yu, Wilson, and Pierce 2001).

The associations between ants and plants can be used as model systems in ecological and evolutionary research. Although we have rarely tested the beneficial effects of mutualism on the ant partners (see below), as antiherbivore defense agents, ants pose few barriers to experimentation and make protective ant-plant interactions tractable models to study resource allocation to defense and its regulatory mechanisms (Heil and McKey 2003). Ant-plant-herbivore systems have aided understanding of the structure of food webs and many central issues of community ecology (Letourneau and Dyer 1998a, 1998b, 2005; Schmitz, Hamback, and Beckerman 2000; Heil and McKey 2003; Blüthgen, Stork, and Fiedler 2004; Oliveira and Del-Claro 2005; McKey et al. 2005). These associations commonly consist of complex multitrophic-multispecies interactions varying in species specificity and in the impact on the fitness of participants (Bronstein and Barbosa 2002). For instance, the complex multitrophic system formed by several interconnected pairwise relationships surrounding *Cnidoscolus aconitifolius* (Euphorbiaceae) has a significant impact on the fitness of the plant. This plant has a considerable chemical arsenal for defense (Barrientos-Benítez and Gutíerrez-Lugo 1994; Arredondo-Ramírez and Castorena-Adame 1996) and is also visited by ants that forage for extrafloral nectar and deter herbivores (mostly lepidopteran caterpillars). The plant is pollinated mainly by butterflies (and also by a variety of bees) that, in turn, are consumed by predatory spiders that cause a decrease in plant fitness (lower seed set) and have a negative effect when their activity deters pollinators, but mostly the butterflies, the bees, and the spiders do not interact. The ants apparently

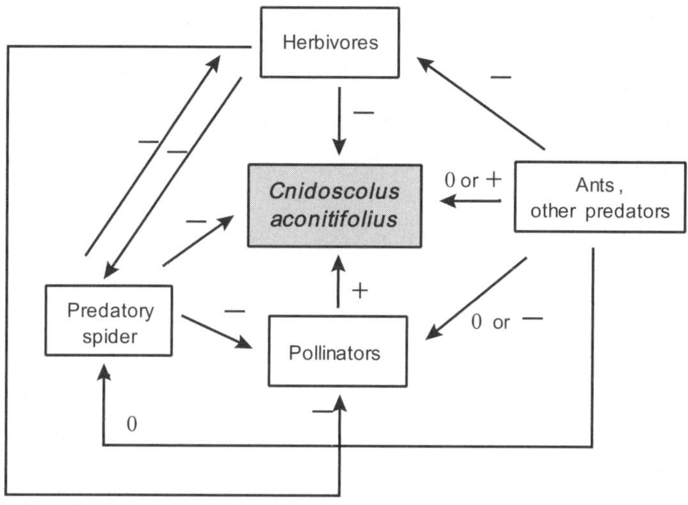

FIGURE 12.1. The complex multitrophic-multispecies associations that influence the fitness of *Cnidoscolus aconitifolius* (Euphorbiaceae) in Yucatán, Mexico. The arrows represent the direction of the interactions, and the signs the outcome of the interaction for the affected participant: positive (+), negative (−), or no recorded change in outcome (0). Based on Arango, Rico-Gray, and Parra-Tabla 2000; Parra-Tabla, Rico-Gray, and Carbajal 2004; V. Parra-Tabla and A. Canto, unpubl. data; A. M. Arango and V. Rico-Gray, unpubl. data).

do not interact with the spider, the main predator in the system (Arango, Rico-Gray, and Parra-Tabla 2000; Parra-Tabla, Rico-Gray, and Carbajal 2004; V. Parra-Tabla and A. Canto, unpubl. data; A. M. Arango and V. Rico-Gray, unpubl. data) (fig. 12.1).

Furthermore, because ant-plant mutualisms are horizontally transmitted and vulnerable to parasites and "cheaters," their study should help to assess the evolutionary dynamics of interspecific interactions (Currie, Mueller, and Malloch 1999; Currie 2001; Izzo and Vasconcelos 2002). These associations offer rich material to study ant social evolution in novel contexts, in settings where colony limits, resource supply, and nest-size availability are all easily quantifiable (Heil and McKey 2003). Blüthgen, Stork, and Fiedler (2004) recently showed that plant and insect exudates produce important bottom-up effects on the ant community that are reflected in varying degrees of specialization and partitioning, as well as in the segregation and co-occurrence patterns among the ant species in an Australian tropical rain forest. The differential roles of honeydew (a specialized resource for dominant species) and nectar sources (an opportunistic resource for all ants), and their distinct distribution within the canopy, indicate that some key plant species hosting trophobionts may ultimately shape the distribution of dominant ants in the canopy (Blüthgen, Stork, and Fiedler 2004) and provide strong support for the ant mosaic theory (Leston 1978).

Finally, the study of ant-plant interactions may help in the planning of conservation efforts (e.g., Gove, Majer, and Rico-Gray 2005) because the community dynamics of ants can be a robust ecological indicator of disturbance and of habitat recovery (Carvalho and Vasconcelos 1999; Vasconcelos 1999; Andersen and Majer 2004; Majer et al. 2004; Vinson, O'Keefe, and Frankie 2004), and structured inventories of ants can be used in biodiversity assessments (Longino and Colwell 1997; Vasconcelos 1999; Vinson, O'Keefe, and Frankie 2004). Indeed, ant-plant associations have served to highlight many promising topics in evolutionary ecology, expanding our current frontiers in the study of interspecific interactions (see also Beattie and Hughes 2002; Heil and McKey 2003). We suggest below some ideas that we consider worthwhile examining in future studies of ant-plant relationships.

SPATIAL AND TEMPORAL VARIATION Studies on ant-plant interactions have usually been accomplished within a single population or habitat, leaving aside or neglecting the possible existence of interpopulation variability in the interactions, or even the role of abiotic factors (De la Fuente and Marquis 1999; Kersch and Fonseca 2005). Recent research has stressed the importance of studying in several populations and/or habitats in order to assess whether interspecific interactions vary geographically (Thompson 1994) or among different habitats (Alonso 1998; Cogni, Freitas, and Oliveira 2003). This approach should make it possible to study coevolution in a geographic mosaic context (Thompson 2005) and to compare the variation in outcome in the same interacting species among populations (Cushman et al. 1998; Garrido et al. 2002). In order to elucidate the complexity of ant-plant interactions and search for patterns, future studies should analyze the existence of variable outcomes and specificity, the effect of seasonality, and the effect of individual variation within a population (Rico-Gray 1993; Díaz-Castelazo et al. 2005). Moreover the "study site" should comprise several populations and/or habitats. For instance, a study on ant-plant associations conducted in one site on the coast of Veracruz, Mexico, an area no larger than 100 ha, revealed significant variation among vegetation types and seasons for the mean value of the ant-density index and ant-species richness (fig. 12.2; Díaz-Castelazo et al. 2004). Thus, even a relatively

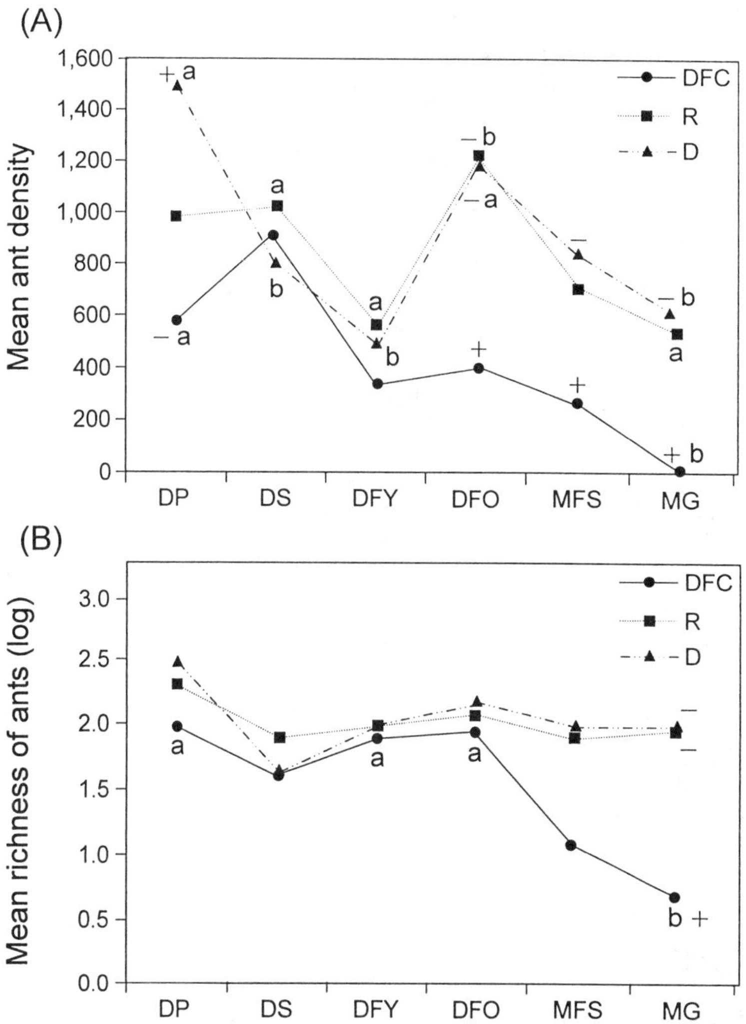

FIGURE 12.2. Mean values of ant density index (A) and ant species richness (B) for vegetation types and seasons. Significant differences at $p < 0.05$ (Tukey HSD test) for factor interactions are shown among vegetation types within a season as (a) and (b) and among seasons within vegetation types as (+) and (−). Vegetation types: DP = dune pioneer; DS = dune scrub; DFY = tropical dry forest; DFO = tropical deciduous forest; MFS = ecotone; MG = mangroves. Seasons: DFC = cold fronts with winds; R = wet season; D = dry season. Modified from Díaz-Castelazo et al. 2004.

small area can generate considerable variation that has to be accounted for to really obtain a clear picture of the study system.

ALTERNATE DEFENSE STRATEGIES Many studies have proposed trade-offs or a lack of redundancy of defenses acting over the same temporal, spatial, and/or herbivore scales (e.g., McKey 1979; Davidson and Fisher 1991; Ågren and Schemske 1993). However, an absolute trade-off between defense systems is not always found (Steward and Keeler 1988). What happens when a plant species exhibits two defensive tactics, such as chemical defense to deter herbivores and EFNs to attract ants and/or other organisms?

Before designing an experiment on plant protection by EFN-gathering ants, for example, one needs to assess whether a plant species exhibits important physical and/or chemical defenses besides harboring EFNs. Furthermore, it is becoming critically important to consider the presence and the role of induced responses to herbivory (Koptur 1989; Karban and Baldwin 1997; Agrawal and Rutter 1998; Heil 2004; Romero and Izzo 2004). For example, herbivory can induce glandular trichome production in *Cnidoscolus aconitifolius,* and it has been suggested by Abdala-Roberts and Parra-Tabla that it apparently represents a relevant selective force determining trichome abundance in natural populations of this species. However, the importance of herbivory as a primary factor determining trichome production will depend on its own intensity and duration over time and on the plant part under study (Abdala-Roberts and Parra-Tabla 2005).

PLANTS AND ANT-TENDED INSECTS Ant-plant systems harboring ant-tended insects such as exudate-producing hemipterans and lepidopteran larvae represent complex multispecies interaction systems (Messina 1981; Horvitz and Schemske 1984; Gaume, McKey, and Terrin 1998; Oliveira and Del-Claro 2005). Ant-tended hemipterans and lepidopterans are herbivores whose damage can severely decrease plant fitness, and the extent of this damage is counteracted only if deterrence of other herbivores by exudate-feeding ants is enough to increase plant fitness above the level of damage caused by the ant-tended insects (Carroll and Janzen 1973).

Future research should consider the association, if any, between ant-tended insects (sap-feeding hemipterans in particular) and the evolution of extrafloral nectaries (Becerra and Venable 1991). Ant-Hemiptera-plant interactions increase in complexity and uncertainty when plants have EFNs. The presence of hemipterans may disrupt the ant-EFN interaction and amplify the negative effects of the hemipterans on the plant, which may be due to the increased attractiveness of insect exudates to ants. We need to determine whether ants switch food sources based on quality and/or quantity and not merely based on their presence or absence (e.g., Rico-Gray and Morais 2006), whether there is spatial and/or seasonal variation in ant foraging patterns and levels of defense, whether there is a potential for coevolution, and what determines conditionality in ant-Hemiptera interactions (e.g., Gove and Rico-Gray 2006). Plant-hemipteran-ant associations are interesting not so much because of the direct interactions between different pairs of organisms but rather because of the diverse indirect effects that these complex multitrophic interactions can have on the plants (McKey et al. 2005; Moreira and Del-Claro 2005; Oliveira and Del-Claro 2005).

PHYLOGENY Apart from a few exceptions (e.g., McKey 1989, 1991; Ward 1991; Davidson and McKey 1993; Chapela et al. 1994; Hinkle et al. 1994; Feldhaar, Fiala, Gadau, et al.2003; Brouat, McKey, and Douzery 2004), there is a general lack of a phylogenetic historical approach to the study of ant-plant interactions. Because evolutionary history leaves a significant imprint on the composition of the component communities that must exert selection on plant characteristics, and vice versa, phylogenetic history should help to explain certain aspects of ant-plant associations. For instance, this approach may help us understand why certain species possess both a powerful chemical arsenal and EFNs that attract ants offering no protection to the plant (as in *Cnidoscolus;* see above), or it may help determine the possible evolutionary association between ant-plant-Hemiptera systems and the origin and evolution of EFNs. Moreover, a phylogenetic approach will help us explore the interrelationships between the history of ants and that of plants (see also Grimaldi and Engel 2005; Wilson and Hölldobler 2005; Moreau et al. 2006). Why is it that nearly all the ants involved in "pure" antagonistic interactions with plants evolved in one ant subfamily (Myrmicinae, with a few in the Ponerinae and Formicinae)? Is this the result of ants being the earliest, perhaps the first, insect predators on the ground? And why is it that the best example of a coevolved mutualism between ants and plants (*Pseudomyrmex-Acacia*) is in the ant subfamily Pseudomyrmecinae? Finally, the evolutionary history of the interactions between ants and ferns (e.g., Koptur, Rico-Gray, and Palacios-Rios 1998) has been poorly studied and could be elucidated with a phylogenetic approach.

ANT-FED PLANTS For many years, ant-fed plants were considered to be primarily tropical epiphytes living in nutrient-poor soils in the families Rubiaceae, Orchidaceae, Polypodiaceae, and Asclepiadaceae (e.g., Huxley 1980; Rico-Gray et al. 1989; Treseder, Davidson, and Ehleringer 1995). However, the discovery that shrubs and trees in a variety of plant families (Melastomataceae, Boraginaceae, Chrysobalanaceae, Piperaceae, and Cecropiaceae) can also obtain nutrients from debris deposited by ant inhabitants (e.g., Sagers, Ginger, and Evans 2000; Fischer et al. 2002, 2003; Dejean et al. 2004; Solano and Dejean 2004) is changing the general view on ant-fed plants (see chapter 8). Thus, contrary to the earlier discoveries, not all examples of nutrient uptake from ant debris piles by plants involve epiphytes. All Southeast Asian examples are epiphytes and, with the notable exception of *Myrmecophila,* all American examples are geophytes. The characteristic they have in common is that they inhabit nutrient-poor environments, suggesting high levels of physical stress, a component of life histories that may predispose mutualistic interactions (Thompson 1982). It is also clear that this type of association has arisen independently in several angiosperm families and in the ferns. These associations may also involve some degree of antiherbivore defense by the ants (e.g. Rico-Gray and Thien 1989b). The associated ants do not appear to nest obligatorily in the plants, and even though either party may survive without the other, very few of these plants are found in their natural habitat without ants. The associations are not species-specific; however, specificity is high within a population, a given habitat, or a geographic area (Gay and Hensen 1992; Fiala et al. 1999). The traditional generalizations about ant-fed plants, for instance, that they are largely tropical epiphytes in Southeast Asia living in nutrient-poor soils, need reevaluation under the current information that many species in a wide array of plant families in South America that are not epiphytes exhibit this type of interaction. Also, we should fully quantify the costs and benefits inherent to each party, including the factors causing their variation.

PLANT-DERIVED BENEFITS TO ANTS Most studies on "mutualism" between ants and plants have failed to demonstrate that both the ant and the plant benefit significantly from the association, that is, that a gain in fitness is experienced by both. In such cases it is not possible to conclude that a given interaction is mutualistic; one can only infer a mutualism (Cushman and Beattie 1991). The vast majority of the studies on ant-plant interactions have assessed only the increase in fitness for the plant, the better-known and/or easier-to-study partner. The benefits for the ants in these interactions are poorly known or have seldom been analyzed. Probably the general assumption is that if ants are obtaining food and/or domatia, there should be a gain in fitness. Almost all we know has to do with the effect of ants on herbivores (either positive or negative) and the consequences for plant fitness. For instance, the effect of sugar sources (EFNs and/or honeydew) and fruits and seeds (pulp and aril) on the diet and demography of an ant colony have hardly been explored (e.g., Fisher, Sternberg, and Price 1990; Cushman, Rashbrook, and Beattie 1994; Morales and Heithaus 1998). The consumption of elaiosomes has recently been shown to significantly affect the dynamics of an ant colony by shifting the mass and numerical investment ratio in colony reproductive output toward gynes (virgin queens) (Morales and Heithaus 1998). Elaiosomes were also shown to have a quantitative effect on larval development because larvae that accumulated more radio-label from elaiosomes tended to develop into virgin queens, whereas other female larvae developed into workers (Bono and Heithaus 2002). Thus, more effort should be made to measure fitness gains (if any) by ant colonies caused by their interaction with plants.

ASSOCIATED ARTHROPOD ASSEMBLAGES Most studies addressing plant defense by ants have not considered the possibility of estimating the effect of more than one group of nectar-gathering insects (fig. 12.3) or that EFNs may be attracting ants that may affect other services (e.g., seed dispersal) rather than defense (table 3.2). Figure 12.3 shows that wasps constitute an even better defensive system than ants. Even though both groups of organisms have a significant effect on plant fitness, their effect is not additive (Cuautle and Rico-Gray 2003).

Although spiders and insects other than ants (wasps, bees, and flies) can visit EFNs together with ants, the joint effect of this assemblage on plant fitness has rarely been considered (Ruhren and Handel 1999; Cuautle and Rico-Gray 2003; Cuautle, Rico-Gray, and Díaz-Castelazo 2005). Does the assemblage exert a positive and significant effect on plant fitness? Are the effects of the different components of the assemblage on plant fitness additive? Do the components of the visiting assemblage

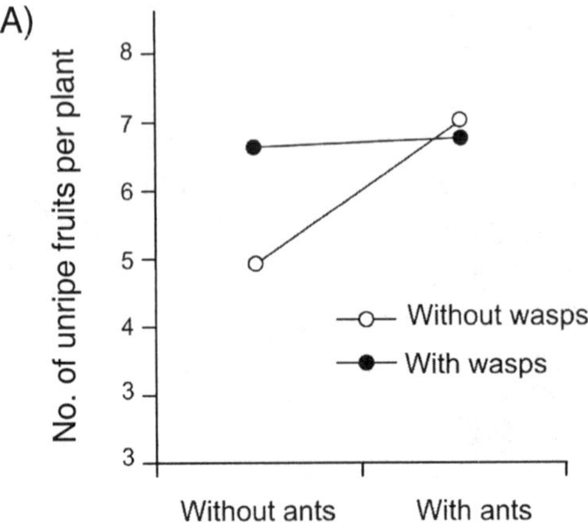

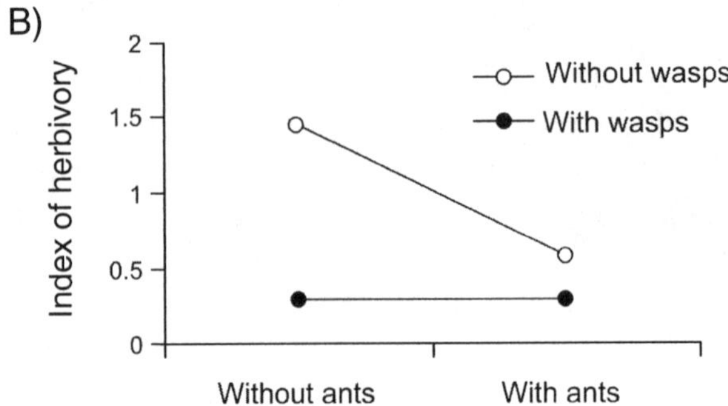

FIGURE 12.3. (A) Significant effects of wasp-ant interaction on the number of unripe fruits per plant. (B) Significant effects of wasp-ant interaction on the index of herbivory. Modified from Cuautle and Rico-Gray 2003.

compete for the nectar source? And if so, does such competition affect plant fitness? Moreover, cases in which a given herbivore species can have both antagonistic (herbivore) and mutualistic roles (e.g., by attracting ant guards or by acting as pollinators) relative to a plant species are particularly suitable for investigation of net effects on host plants within multispecies systems (Horvitz and Schemske 1984; Gaume, McKey, and Terrin 1998; Oliveira and Del-Claro 2005).

BEHAVIORAL INTERACTIONS Surprisingly, aspects of behavioral ecology at the ant-plant-herbivore interface have rarely been explored; they should offer a wide avenue for research. Although ants provide a consistent defense system that is relatively immune to evolutionary changes by the herbivore (Schemske 1980), it has already been shown that immature and adult herbivores present morphological and behavioral traits that markedly reduce ant interference on foliage (Heads and Lawton 1984, 1985; Freitas and Oliveira 1996; Eubanks et al. 1997; Oliveira and Freitas 2004). In contrast, ant-tended herbivores receiving protective services from ants may promote contact with ant partners by ovipositing on ant-occupied plants (Pierce and Elgar 1985; Oliveira and Del-Claro 2005), by producing substrate-borne vibrational calls that attract nearby ants (DeVries 1990; Cocroft 1999; Travassos and Pierce 2000), or by flicking honeydew beneath the host plant (Del-Claro and Oliveira 1996). Thus, the behavioral traits of herbivores when confronted with ants on the host plant can be crucial for understanding the outcomes of multitrophic interactions, either in antagonistic or in mutualistic ant-herbivore interactions (Oliveira, Freitas, and Del-Claro 2002). Furthermore, it has been shown for certain Amazonian myrmecophytes (*Tococa bullifera* and *Maieta guianensis*) that hemipteran abundance can vary significantly between species but that the nature of the association is mediated by the ant partner; because hemipterans are herbivores, the costs and benefits to the host plant of the different ant partners may vary in many ways (Lapola et al. 2005).

Likewise, the "quality" component of ants as seed-dispersers should necessarily include key behavioral traits of ants related to their capacity for transporting and/or cleaning seeds, as well as to their effects on seedling recruitment in both myrmecochorous and nonmyrmecochorous plant species (Horvitz 1981; Gorb, Gorb, and Punttila, 2000; Passos and Oliveira 2002, 2004; Cuautle, Rico-Gray, and Díaz-Castelazo 2005). Current results show that the chemical contents (lipids and sugars) in the fleshy portion of diaspores and the relative size of the diaspores are important factors determining the patterns of ant-seed/fruit interactions (Hughes, Westoby, and Jurado 1994; Ness et al. 2004; Pizo, Passos, and Oliveira 2005). Finally, the interactions between ants and fleshy seeds, including those primarily dispersed by vertebrates, carry implications for the conservation and regeneration of natural systems. This too is an open field for research.

APPLICATIONS The study of leaf-cutter ants (tribe Attini) should assume a unique importance in future research on ant-plant interactions because they (1) may interact with fruits and seeds in a beneficial way (Farji-Brener and Silva 1996; Leal and Oliveira 1998), (2) may become severe pests as defoliators (Vasconcelos and Cherrett 1997; Wirth et al. 2003), and (3) can have a significant effect on the regeneration of tropical forests (Farji-Brener and Illes 2000; Moutinho, Nepstad, and Davidson 2003; Farji-Brener 2005). The joint effect of these three aspects needs further investigation.

Although the use of ants in programs of pest control has certain drawbacks (e.g., tending Hemiptera, killing other predatory species, and irritating humans), ants have proved to be beneficial in suppressing many insect herbivores (Risch and Carroll 1982a, 1982b; Perfecto 1990, 1991; Fernandes et al. 1994) and thereby may promote a better and more healthful environment (e.g., Andersen 1997b; Longino and Colwell 1997). Moreover, as opposed to pesticides that need to be regularly tested, ants are generalist predators capable of controlling a wide range of prey (chapter 11). More research is needed on the management of ants, especially in developing countries (e.g., Mexico, India, and Brazil) that are the natural habitats of many ant species participating in protective ant-plant systems (chapter 6), and where traditional practices of agricultural management have developed over hundreds of years (chapter 11). The use of more ecologically rational agriculture, such as the acceptance of biological and cultural pest control to create integrated pest-management programs is one way to solve some of the problems created by large-scale monocultures (Rickson and Rickson 1998), which are more prone to herbivore attack than is a temporal and spatial mosaic of vegetation (Whitham, Williams, and Robinson 1984). Although there is already a great body of work on plant defense by ants in nature, we have not been able yet to show how this large data set can be effectively used for pest control in agricultural systems. This issue is closely related with the preservation of the vast biodiversity of natural systems in the tropics. If the huge diversity of ants inhabiting tropical habitats could offer help "for free" in pest control, then the preservation of forests and savannas will assume a very important applied role. We think that ecologists have not taken advantage of the link between studies on interspecific interactions in natural habitats and the applications that may result from the use of these data.

THE SEARCH FOR PATTERNS Ecological research in general, and research on interspecific interactions in particular, is mostly based on hypotheses testing. However, a lot can be gained just by searching for patterns over a certain geographic area or within an interaction (Lawton 1996). For instance, the study of patterns of biological diversity has demonstrated that the distribution of species is far from even, creating a spatial mosaic of species richness (Pielou 1979; Cox and Moore 1993; Huston 1994; Scheiner and Rey-Benayas 1994). Not only are species unevenly distributed, but their interactions, as well as critical plant structures and/or food sources for interactions, also vary spatially and seasonally (e.g., appendix 10.1). Consequently, a landscape approach to the study of ecological interactions should be used much more (Bronstein 1995; Thompson 1996; Margules and Pressey 2000).

Interactions vary in probability of occurrence along environmental gradients (e.g., latitudinal or altitudinal) and under different disturbance regimes (e.g., Koptur 1991, 1992a; Rico-Gray, García-Franco, et al.1998), they vary in their outcome under different ecological conditions (Thompson 1994; Cushman and Addicott 1991) and in different habitats (Barton 1986; Cogni, Freitas, and Oliveira 2003), and they also vary between seasons (e.g., Rico-Gray and Castro 1996). The structure of ant communities and of ant-plant interactions has been studied in a variety of habitats, and it is clear that spatial and temporal variation characterizes ant communities (e.g., Herbers 1989; Fowler 1993; Morrison 1998). Ant assemblages are very dynamic, and extrapolating from one ant community to another with superficially similar characteristics can lead to erroneous inferences (Herbers 1989). Thus, the spatial and temporal patterns of community structure in ant communities and their interactions with plants should be studied in order to assess broad generalizations (chapter 10). For instance, the spatial patterns in the richness and availability of energy-rich liquid food for ants can be associated with interhabitat differences in ant diversity and abundance and thus generate variation in ant-plant interactions (Díaz-Castelazo et al. 2004). Furthermore, ants exhibit a differential use of the different types of nectaries (fig. 12.4; Díaz-Castelazo et al. 2005), generating clusters of association between certain types of nectaries and their ant species (C. Díaz-Castelazo and V. Rico-Gray, unpubl. data).

INTERACTION NETWORKS Species interact within communities as networks, with each species connected to one or more other species (Jordano 1987). A network of interacting species may have a small number of links among species, indicating an assemblage of ecological specialists, or many links, indicating ecological generalists (Bascompte and Jordano 2006). One of the central problems to solve in community ecol-

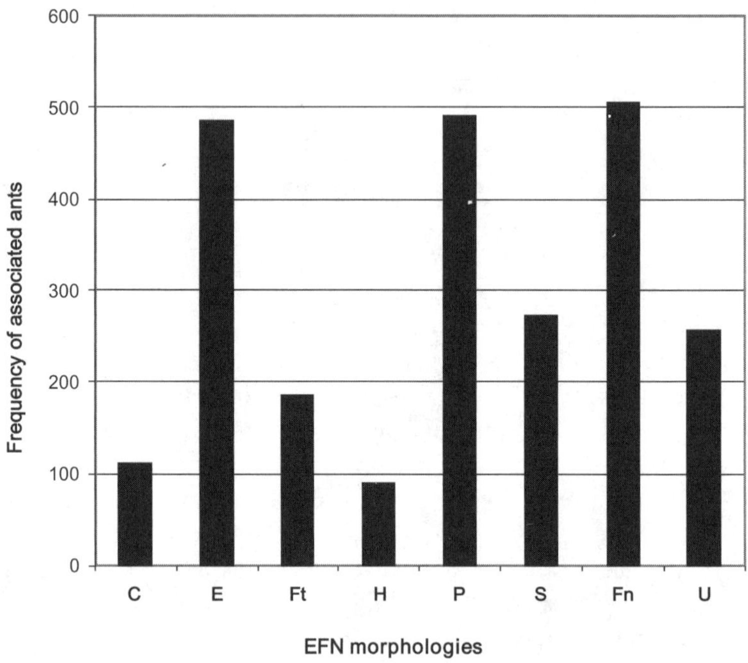

FIGURE 12.4. Frequency of ant visits (all species included) to different morphotypes of extrafloral nectaries (EFNs). C = Capitated; E = Elevated; Ft = Flattened; H = Hollow; P = Peltate; S = Scale-like; Fn = Functional; U = Unicelular. From Díaz-Castelazo et al. 2005.

ogy is whether different forms of interaction or different ecological conditions favor alternative structures in these networks of interacting species (Guimarães et al. 2006).

Networks are usually not homogeneous, so we can find among them groups of species more connected with one another than with the other species. Certain general characteristics of networks are maintained independently of the type of community studied (Jordano, Bascompte, and Olesen 2003). The comprehension of such properties is fundamental to our understanding of the evolution of ecosystems and to the development of viable conservation measures (Bascompte et al. 2003). For instance, the biodiversity of a community should include the list of species names but also their interactions. There is a small but significant body of research on how mutualistic communities ("mutualistic networks") are structured (e.g., Bascompte et al. 2003; Jordano, Bascompte, and Olesen 2003; Bascompte and Jordano 2006), which is now permeating ant-plant interactions studies (Guimarães et al. 2006). Such an approach can lead to specific questions on community organization, geographic variation, and coevolution (Thompson 2005).

Mutualisms between free-living species often form multispecific networks in which individual species differ in the number of species with which they interact, creating complex webs of interaction that often appear idiosyncratic. Studies of pollinator-plant and frugivore-fruit interactions have suggested how specialization is distributed among interacting species (Bascompte et al. 2003; Jordano, Bascompte, and Olesen 2003; Vazquez and Aizen 2004; Bascompte and Jordano 2006). Bascompte et al. (2003) found that pollination and seed-dispersal networks often show a specific type of asymmetrical specialization called nestedness. Nested networks are characterized by (1) generalists that all interact with one another, forming a core of interacting species; (2) specialist species that commonly interact only with generalists; and (3) the absence of specialists that interact only with other specialists (see also Vazquez and Aizen 2004).

We do not really know whether asymmetries in specialization are common in other forms of mutualism or how different ecological conditions may shape the extent of asymmetry (the degree of nestedness). We know that species in mutualisms commonly differ geographically in the species with which they interact (Anderson and Majer 2004; Rudgers and Strauss 2004; Herrera 2005) and that some interactions coevolve as a geographic mosaic in which populations differ across landscapes in their adaptation and specialization with regard to other species (Thompson 1994, 2005). The problem to solve is whether interaction networks show similar patterns of specialization in different communities regardless of the particular species involved. By exploring geographic variation in macroscopic patterns of mutualistic networks, we will be able to relate the two main approaches to exploring the organization of multispecies mutualisms: geographic mosaic theory and complex network theory (Bascompte and Jordano 2006).

Using one of the most studied types of animal-plant mutualisms (Bronstein 1998), Guimarães et al. explored whether interactions between plants with EFNs and ants (EFN networks) showed predictable patterns of asymmetry in specialization and whether those patterns varied geographically in four contrasting sites in Mexico. They addressed the following questions (Guimarães et al. 2006): Do ant-plant networks show a predictable pattern of specialization within and among communities? To what extent is the pattern of specialization similar to that found in studies of other forms of mutualism?

Guimarães et al. reported that the three large, tropical networks analyzed were highly nested (those in Zapotitlán, Puebla; San Benito, Yucatán; and La Mancha, Veracruz), but the single small network in mixed vegetation (in Xalapa, Veracruz) was not (fig. 12.5). The observed nested values are similar to those of pollination and seed-dispersal networks (fig. 12.6A), and nestedness is not explained by random processes (fig. 12.6B). After a control for network size effects is imposed, each tropical network shows a different value of nestedness, suggesting that there is an idiosyncratic component to the nested pattern. The results suggest that nestedness is a common feature of nonspecific, species-rich mutualisms but that size, vegetation structure, and idiosyncratic aspects of each community also affect the degree of nestedness in mutualistic assemblages. The results also suggest that loss of biodiversity within biological communities may have strong effects on the organization of mutualistic assemblages (Guimarães et al. 2006).

Further questions for future testing are these: How are the number of interactions of one species in a community and the macroscopic pattern (the extent of distribution of the community) related to geographic variation? Are the most widespread species or genera (in geographic terms) also the most connected species in the plant-animal community

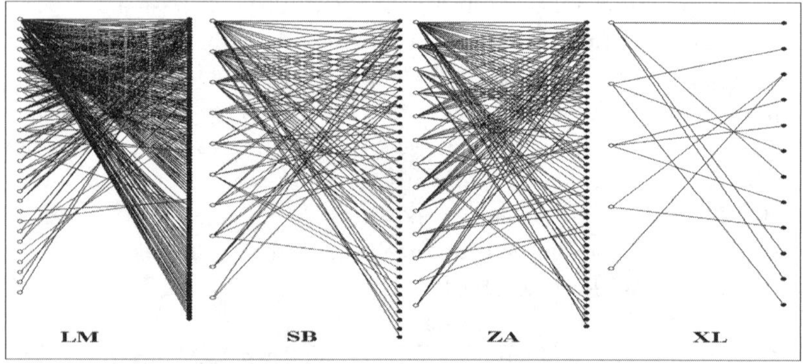

FIGURE 12.5. Networks based on the interactions between ants and extrafloral nectaries. LM = La Mancha; SB = San Benito; ZA = Zapotitlán; XL = Xalapa. Open nodes are ant species, and closed nodes are extrafloral-nectary-bearing plant species. The first three networks show high degrees of nestedness; that is, specialists interact with a subset of the species that generalists interact with. In contrast, XL shows nonsignificant nestedness. From Guimarães et al. 2006.

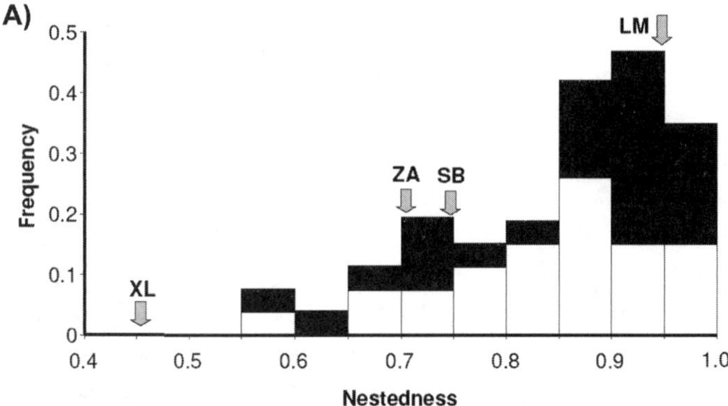

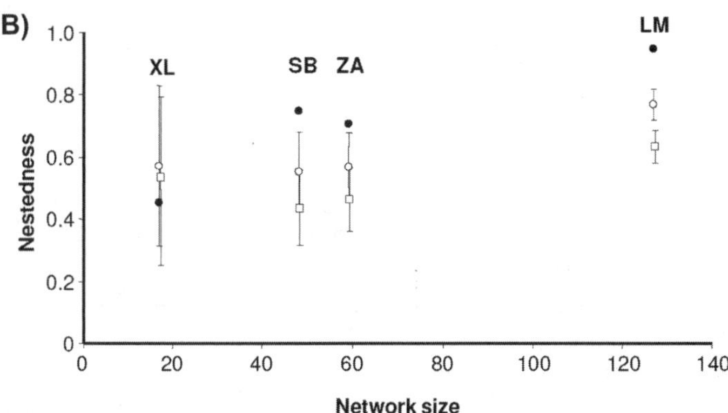

FIGURE 12.6. (A) Frequency of different degrees of nestedness observed in published studies on pollination, seed dispersal, and EFN networks (black bars = pollination networks; white bars = frugivory networks; gray arrows = EFN networks). (B) Relationship between degree of nestedness and network size (black circles = observed values of nestedness for each EFN network; white squares = average nested value predicted by null model I; white circles = average value predicted by null model II). Bars indicate 95% confidence interval (10,000 replicates for each null model). See Guimarães et al. 2006.

network? How do different mixes of species that vary in their geographic distribution lead to the emergence of distinct network topologies (P. R. Guimarães, V. Rico-Gray, S. F. dos Reis, J. N. Thompson, and C. Díaz-Castelazo, unpubl. data)?

Finally, the tools and the possibilities offered by network analysis to study the organization of biotic communities are still largely unexplored (e.g., Guimarães et al. 2005; Vázquez, Morris, and Jordano 2005; Blüthgen, Menzel, and Blüthgen 2006). They represent an open avenue for research using large sets of species interactions (e.g., Bascompte et al. 2003), which will improve our understanding of community organization and, in particular, enable us to explore how different types of interactions shape communities (Olesen and Jordano 2002; Bascompte and Jordano 2006). The nested pattern obtained for highly structured mutualistic networks is important for understanding coevolution in species-rich communities, which can neither be reduced to pairs of coevolving species nor to diffuse, randomly interacting assemblages (Jordano, Bascompte, and Olesen 2003; Thompson 2005; Bascompte and Jordano 2006). The implications of this dynamic approach to network structure are important for studies in coevolution, community ecology, and conservation ecology (Bascompte and Jordano 2006). We hope that more studies on species interactions in general, and ant-plant interactions in particular, will soon use this community approach so that it will be possible to compare different types of communities, including both their internal and their geographic organization.

CONSERVATION OF INTERACTIONS Several ecological concerns (e.g., loss of important interactions from communities such as keystone species, invasion of alien species, or spread of other species) have been the focus of many studies designed to evaluate the ways in which changed interspecific interactions affect biodiversity and conservation efforts. Also, changes in ecological conditions affect not only the diversity and evolution of species but also the diversity and evolution of the interactions themselves (Thompson 1996). As mentioned above, biological communities are networks of species (i.e., species and their interactions) that maintain independent characteristics, and understanding them is fundamental to understanding their evolution and the development of viable conservation measures (Bascompte et al. 2003). Thus, the biodiversity of a community should include not only a list of species, but their interactions as well. Just as species are reservoirs of genetic diversity, interspecific interactions in communities are storehouses of information on how evolution has shaped associations between species under different ecological conditions (Thompson 1994). Consequently, conservation should have a landscape rather than a species approach; in this way both species and their interactions can be preserved.

One of the first studies to consider the effect of ants in food-web organization and conservation of diversity was that of Gilbert (1980); since then several studies have assessed ant diversity in different communities (e.g., Andersen 1997b; Longino and Colwell 1997; Agosti et al. 2000). For example, Vasconcelos and Vilhena (2006) have shown for the Brazilian Amazon that there are as twice as many species of ants in the forest as in the savanna; within habitats, there were more species on the ground than in the vegetation. Thus, they conclude that ant species richness is affected by both habitat and strata. Studies of this kind are needed to support a variety of conservation and management projects.

Different studies have shown that forest disturbance and vegetation management practices affect the conservation of ant communities and of their interactions (e.g., Vinson, O'Keefe, and Frankie 2004). Anthropogenic disturbances (Vasconcelos 1999) and forest fragmentation (Carvalho and Vasconcelos 1999) affect ground-foraging and litter-dwelling ant communities in central Amazonia, respectively. Vasconcelos found that there were more ant species in mature and old regrowth forest than in abandoned pastures; however, ant abundance tended to decrease with forest maturity. Moreover, the ant fauna in pastures and early regrowth differed from that in old regrowth and mature forest. Thus, different land-management practices result in different rates of recovery of the ant forest fauna (Vasconcelos 1999).

Similarly, the distance from forest edges significantly affects ant species composition. This effect is partly the result of variation in litter depth, Carvalho and Vasconcelos explain. Although no significant changes in ant densities or species richness were recorded with increasing distance away from the forest edge, species richness of ants was greater in continuous forest than in fragments. Thus, both edge and isolation effects are significant in structuring litter-dwelling ant communities in central Amazonia (Carvalho and Vasconcelos 1999). Furthermore, it has been suggested that although communities of ant-plant mutualists are likely to persist in fragmented tropical landscapes 25 years after fragment isolation, most species become rare and population sizes in fragments decrease dramatically (Bruna, Vasconcelos, and Heredia 2005). Future studies should focus on evaluating how fragmentation has altered herbivore pressure and the dispersal of ants and plants to fragments (see also Guimarães and Cogni 2002), since the interactions of these factors are likely to have the greatest impact on long-term patterns of population persistence (Bruna, Vasconcelos, and Heredia 2005).

Another factor that affects the conservation of ant communities and their interactions is biological invasions (Morrison 2006), because they can disrupt, for example, important seed-dispersal mutualisms that can have community-level consequences (Christian 2001; Carney, Byerley, and Holway 2003; Wetterer et al. 2006). For instance, the invasion of South African shrublands by the ant *Linepithema humile* led to a shift in composition of the plant community as a result of a disproportionate reduction in the densities of large-seeded plants; thus, the preservation of mutualistic interactions may be essential for maintaining natural communities (Christian 2001; Christian and Stanton 2004). Also, ants may compete with other animals, such as land hermit crabs, for food (Morrison 2006), and these interactions also affect ants' role within the community. The conservation of ant-plant interactions, as well as of ant and plant communities, has to be based on studies that consider both a landscape approach and the nature of the interactions (Vinson, O'Keefe, and Frankie 2004).

* * *

With the exception of two antagonistic interactions (seed predation, and grazing by leaf-cutter ants), most of the ant-plant interactions discussed in this book are basically mutualisms; i.e., the interacting species receive benefits that significantly increase their fitness. However, we should not forget that there is a very close, almost intangible, association between antagonism and mutualism (see chapter 1), to the point where some authors consider most mutualisms as resource-based (Holland et al. 2005). The outcomes of interactions are based on a cost-benefit model, such that outcomes can vary among populations of the same interacting species. It is thus of great importance to study the outcomes of interactions under a geographic or multiple-site design to assess variation among populations of interacting species and to consider such processes under the theory of the geographic mosaic of coevolution.

Literature Cited

Abdala-Roberts, L., and V. Parra-Tabla. 2005. Artificial defoliation induces trichome production in the tropical shrub *Cnidoscolus aconitifolius* (Euphorbiaceae). *Biotropica* 37:251–57.

Adams, C. T., T. E. Summers, C. S. Lofgren, D. A. Focks, and J. C. Prewitt. 1981. Interrelationships of ants and the sugarcane borer in Florida sugarcane fields. *Environmental Entomology* 10:415–18.

Adams, E. S. 1994. Territory defense by the ant *Azteca trigona:* Maintenance of an arboreal ant mosaic. *Oecologia* 97:202–8.

Addicott, J. F. 1978. Competition for mutualists: Aphids and ants. *Canadian Journal of Zoology* 56:2093–96.

———. 1979. A multispecies aphid-ant association: Density dependence and species-specific effects. *Canadian Journal of Zoology* 57:558–69.

Adis, J., Y. D. Lubin, and G. C Montgomery. 1984. Arthropods from the canopy of inundated terra firme forests near Manaus, Brazil, with critical considerations on the pyrethrum fogging technique. *Studies on Neotropical Fauna and Environment* 19:223–36.

Adler, L. S. 2000. Alkaloid uptake increases fitness in a hemiparasitic plant via reduced herbivory and increased pollination. *American Naturalist* 156:92–99.

Adlung, K. G. 1966. A critical evaluation of the European research on the use of red wood ants (*Formica rufa* group) for the protection of forests against harmful insects. *Zeitschrift für Angewandte Entomologie* 57:167–69.

Agnew, C. W., and W. L. Sterling. 1981. Predation of boll weevils in partially-open cotton bolls by the red imported fire ant. *Southwestern Naturalist* 6: 215–19.

Agosti, D., J. D. Majer, L. E. Alonso, and T. R. Schultz, eds. 2000. *Ants: Standard methods for measuring and monitoring biodiversity.* Washington, DC: Smithsonian Institution Press.

Agrawal, A. A. 1998a. Induced responses to herbivory and increased plant performance. *Science* 279:1201–2.

———. 1998b. Leaf damage and associated cues induce aggressive ant recruitment in a neotropical ant-plant. *Ecology* 79:2100–2112.

Agrawal, A. A., and M. T. Rutter. 1998. Dynamic anti-herbivore defense in ant-plants: The role of induced responses. *Oikos* 83:227–36.

Ågren, J., and D. W. Schemske. 1993. The cost of defense against herbivores: An experimental study of trichome production in *Brassica rapa. American Naturalist* 141:338–50.

Alonso, L. E. 1998. Spatial and temporal variation in the ant occupants of a facultative ant-plant. *Biotropica* 30:201–13.

Altshuler, D. L. 1999. Novel interactions of non-pollinating ants with pollinators and fruit consumers in a tropical forest. *Oecologia* 119:600–606.

Alvarado, A., C. W. Berish, and F. Peralta. 1981. Leaf-cutter ant (*Atta cephalotes*) influence on the morphology of Andepts in Costa Rica. *Soil Science Society of America Journal* 45:790–94.

Alvarez, G., I. Armbrecht, E. Jiménez, H. Armbrecht, and P. Ulloa-Chacón. 2001. Ant-plant association in two *Tococa* species from a primary rain forest of Colombian Choco (Hymenoptera: Formicidae). *Sociobiology* 38:585–602.

Andersen, A. N. 1988. Dispersal distance as a benefit of myrmecochory. *Oecologia* 75:507–11.

———. 1991a. Parallels between ants and plants: implications for community ecology. In Huxley and Cutler 1991, 539–57.

———. 1991b. Seed harvesting by ants in Australia. In Huxley and Cutler 1991, 493–503.

———. 1993. Ant communities in the gulf region of Australia's semi-arid tropics: Species composition, patterns of organization, and biogeography. *Australian Journal of Zoology* 41:399–414.

———. 1997a. Functional groups and patterns of organization in North American ant communities: A comparison with Australia. *Journal of Biogeography* 24:433–60.

———. 1997b. Using ants as bioindicators: Multiscale issues in ant community ecology. *Conservation Ecology* 1:8.

Andersen, A. N., and J. D. Majer. 2004. Ants show the way down under: Invertebrates as bioindicators in land management. *Frontiers in Ecology and the Environment* 2:291–98.

Andersen, A. N., and S. C. Morrison. 1998. Myrmecochory in Australia's seasonal tropics: Effects of disturbance on distance dispersal. *Australian Journal of Ecology* 23:483–91.

Andersen, A. N., and A. D. Patel. 1994. Meat ants as dominant members of Australian ant communities: An experimental test of their influence on the foraging success and forager abundance of other species. *Oecologia* 98:15–24.

Andersen, M. 1991. An ant-aphid interaction: *Formica fusca* and *Aphthargelia symphoricarpi* on Mount St. Helens. *American Midland Naturalist* 125:29–36.

Anderson, G. J., and D. E. Symon. 1985. Extrafloral nectaries in *Solanum. Biotropica* 17:40–45.

Andresen, E. 1999. Seed dispersal by monkeys and the fate of dispersed seeds in a Peruvian rain forest. *Biotropica* 31:145–58.

Angiosperm Phylogeny Group (APG). 1998. An ordinal classification for the families of flowering plants. *Annals of the Missouri Botanical Garden* 85:531–53.

———. 2003. An update of the Angiosperm Phylogeny Group classification for the orders and families of flowering plants: APG II. *Botanical Journal of the Linnean Society* 141:399–436.

Apple, J. L., and D. H. Feener 2001. Ant visitation of extrafloral nectaries of *Passiflora:* The effects of nectary attributes and ant behavior on patterns in facultative ant-plant mutualisms. *Oecologia* 127:409–16.

Arango, A. M., V. Rico-Gray, and V. Parra-Tabla. 2000. Population structure, seasonality, and habitat use by the green lynx spider *Peucetia viridans* (Oxyopidae) inhabiting *Cnidoscolus aconitifolius* (Euphorbiaceae). *Journal of Arachnology* 28:185–94.

Armbrecht, I., E. Jiménez, G. Alvarez, P. Ulloa-Chacon, and H. Armbrecht. 2001. An ant mosaic in the Colombian rain forest of Chocó (Hymenoptera: Formicidae). *Sociobiology* 37:491–509.

Aronne, G., and C. C. Wilcock. 1992. Reproductive characteristics and breeding systems of shrubs of the Mediterranean region. In *Plant-animal interactions in Mediterranean type ecosystems,* ed. C. A. Thanos, 223–27. Athens, Greece: University of Athens Press.

———. 1994a. First evidence of myrmecochory in fleshy-fruited shrubs of the Mediterranean region. *New Phytologist* 127:781–88.

———. 1994b. Reproductive characteristics and breeding systems of shrubs of the Mediterranean region. *Functional Ecology* 8:69–76.

Aronne, G., C. C. Wilcock, and P. Pizzolongo. 1993. Pollination biology and sexual differentiation of *Osyris alba* L. (Santalaceae) in the Mediterranean region. *Plant Systematics and Evolution* 188:1–16.

Arredondo-Ramírez, E., and M. G. Castorena-Adame. 1996. Flavonoides y triterpenos de *Cnidoscolus tehuacanensis* Breckon (Euphorbiaceae). B.Sc. thesis, Universidad Nacional Autónoma de México.

Atsatt, P. R. 1981a. Ant-dependent food plant selection by the mistletoe butterfly *Ogyris amaryllis* (Lycaenidae). *Oecologia* 48:60–63.

———. 1981b. Lycaenid butterflies and ants: Selection for enemy-free space. *American Naturalist* 118:638–54.

Atsatt, P. R., and D. J. O'Dowd. 1976. Plant defense guilds. *Science* 193:24–29.

Auclair, J. L. 1963. Aphid feeding and nutrition. *Annual Review of Entomology* 8:439–90.

Augspurger, C. K. 1984. Seedling survival of tropical tree species: Interactions of dispersal distance, light-gaps, and pathogens. *Ecology* 65:1705–12.

———. 1990. The potential impact of fungal pathogens on tropical plant reproductive biology. In *Reproductive ecology of tropical forest plants,* ed. K. S. Bawa and M. Hadley, 237–46. Carnforth, UK: Parthenon.

Azcárate, F. M., and B. Peco. 2003. Spatial patterns of seed predation by harvester ants (*Messor* Forel) in Mediterranean grassland and scrubland. *Insectes Sociaux* 50:1–7.

Azuma, H., L. B. Thien, M. Toyota, Y. Asakawa, and S. Kawano. 1997. Distribution and differential expression of (*E*)-4,8-dimethyl-1,3,7-nonatriene in leaf and floral volatiles of *Magnolia* and *Liriodendron* taxa. *Journal of Chemical Ecology* 23:2467–78.

Bach, C. E. 1991. Direct and indirect interactions between ants (*Pheidole megacephala*), scales (*Coccus viridis*), and plants (*Pluchea indica*). *Oecologia* 87:233–39.

Baiges, J. C., X. Espadaler, and C. Blanché. 1991. Seed dispersal by ants in west Mediterranean *Euphorbia* species. *Botanika Chronika* 10:697–705.

Baker, H. G., and I. Baker. 1978. Ants and flowers. *Biotropica* 10:80.

———. 1979. Starch in angiosperm pollen grains and its evoutionary significance. *American Journal of Botany* 66:591–600.

Balick, M., D. Furth, and G. A. Cooper-Driver. 1978. Biochemical and evolutionary aspects of arthropod predation on ferns. *Oecologia* 35:55–89.

Baroni-Urbani, C., B. Bolton, and P. W. Ward. 1992. The internal phylogeny of ants (Hymenoptera: Formicidae). *Systematics Entomology* 17:301–29.

Barrientos-Benítez, T., and M. T. Gutiérrez-Lugo. 1994. Determinación de la actividad antimicrobiana y citotóxica potencial de extractos derivados de treinta especies vegetales utilizados en la medicina tradicional mexicana. B.Sc. thesis, Universidad Nacional Autónoma de México.

Barton, A. M. 1986. Spatial variation in the effect of ants on an extrafloral nectary plant. *Ecology* 67:495–504.

Bascompte, J., and P. Jordano. 2006. The structure of plant-animal mutualistic networks. In *Ecological networks: Linking structure to dynamics in food webs,* ed. M. Pascual and J. A. Dunne, 143–59. Oxford: Oxford University Press.

Bascompte, J., P. Jordano, C. J. Melián, and J. M. Olesen. 2003. The nested assembly of plant-animal mutualistic networks. *Proceedings of the National Academy of Sciences* 100:9383–87.

Beattie, A. J. 1978. Plant-animal interactions affecting gene flow in *Viola.* In *The pollination of flowers by insects,* ed. A. J. Richards, 151–64. Linnean Society Symposium Series, no. 6. London: Academic Press.

———. 1985. *The evolutionary ecology of ant-plant mutualisms.* Cambridge: Cambridge University Press.

———. 1989. Myrmecotrophy: Plants fed by ants. *Trends in Ecology and Evolution* 4:172–76.

———. 1991. Problems outstanding in ant-plant interaction research. In Huxley and Cutler 1991, 559–76.

Beattie, A. J., and D. C. Culver. 1977. Effects of the mound nests of the ant, *Formica obscuripes,* on the surrounding vegetation. *American Midland Naturalist* 97:390–99.

———. 1981. The guild of myrmecochores in the herbaceous flora of West Virginia forest. *Ecology* 62:107–15.

———. 1982. Inhumation: How ants and other invertebrates help seeds. *Nature* 297:627.

———. 1983. The nest chemistry of two seed-dispersing ant species. *Oecologia* 56:99–103.

Beattie, A. J., and L. Hughes. 2002. Ant-plant interactions. In Herrera and Pellmyr 2002, 211–35.

Beattie, A. J., and N. Lyons. 1975. Seed dispersal in *Viola* (Violaceae): Adaptations and strategies. *American Journal of Botany* 62:714–22.

Beattie, A. J., C. Turnbull, T. Hough, S. Jobson, and R. B. Knox. 1985. The vulnerability of pollen and fungal spores to ant secretions: Evidence and some evolutionary implications. *American Journal of Botany* 72:606–14.

Beattie, A. J., C. Turnbull, R. B. Knox, and E. G. Williams. 1984. Ant inhibition of pollen functions: A possible reason why ant pollination is rare. *American Journal of Botany* 71:421–26.

Becerra, J. X., and D. L. Venable. 1989. Extrafloral nectaries: A defense against ant-Homoptera mutualisms? *Oikos* 55:276–80.

———. 1991. The role of ant-Homoptera mutualisms in the evolution of extrafloral nectaries. *Oikos* 60:105–6.

Beckmann, R. L., and J. M. Stucky. 1981. Extrafloral nectaries and plant guarding in *Ipomoea pandurata* (L.) G.F.W. Mey. (Convolvulaceae). *American Journal of Botany* 68:72–79.

Begon, M., J. L. Harper, and C. R. Townsend. 1990. *Ecology: Individuals, populations, and communities.* Oxford: Blackwell Scientific Publications.

Belin-Depoux, M., and D. Bastien. 2002. Regards sur la myrmécophilie en Guyane française: Les dispositifs d'absorption de *Maieta guianensis* et la triple association *Philodendron*-fourmis-*Aleurodes*. *Acta Botanica Gallica* 149:-299–318.

Belt, T. 1874. *The naturalist in Nicaragua.* London: John Murray.

Ben-Dov, Y., and C. J. Hodgson, eds. 1997. *Soft scale insects: Their biology, natural enemies, and control.* Amsterdam: Elsevier Science.

Benkman, C. W. 1999. The selection mosaic and diversifying coevolution between crossbills and lodgepole pine. *American Naturalist* 153:S75–S91.

Bennett, B., and M. D. Breed. 1985. On the association between *Pentaclethra macroloba* (Mimosaceae) and *Paraponera clavata* (Hymenoptera: Formicidae) colonies. *Biotropica* 17:253–55.

Benson, W. W. 1985. Amazon ant-plants. In *Key environments in Amazonia,* ed. G. T. Prance and T. E. Lovejoy, 239–66. Oxford: Pergamon Press.

Bentley, B. L. 1976. Plants bearing extrafloral nectaries and the associated ant community: Interhabitat differences in the reduction of herbivore damage. *Ecology* 57:815–20.

———. 1977a. Extrafloral nectaries and pugnacious bodyguards. *Annual Review of Ecology and Systematics* 8:407–27.

———. 1977b. The protective function of ants visiting the extrafloral nectaries of *Bixa orellana* (Bixaceae). *Journal of Ecology* 65:27–38.

———. 1981. Ants, extrafloral nectaries, and the vine life-form: An interaction. *Tropical Ecology* 22:127–33.

———. 1983. Nectaries in agriculture, with an emphasis on the tropics. In *The biology of nectaries,* ed. B. Bentley and T. S. Elias, 204–22. New York: Columbia University Press.

Bentley, B. L., and W. W. Benson. 1988. The influence of ant foraging patterns on the behavior of herbivores. In *Advances in myrmecology,* ed. J. C. Trager, 297–306. New York: E. J. Brill.

Benzing, D. H. 1991. Myrmecotrophy: Origins, operation, and importance. In Huxley and Cutler 1991, 353–73.

Bequaert, J. 1922. Ants in their diverse relations to the plant world. *Bulletin of the American Museum of Natural History* 45:333–621.

Berg, R. Y. 1958. Seed dispersal, morphology, and phylogeny of *Trillium*. *Skrifer Utgitt av Det Norske Videnskaps-Akademi* 1:1–36.

———. 1966. Seed dispersal of *Dendromecon:* Its ecologic, evolutionary, and taxonomic significance. *American Journal of Botany* 53:61–73.

———. 1972. Dispersal ecology of *Vancouveria* (Berberidaceae). *American Journal of Botany* 59:109–22.

———. 1975. Myrmecochorous plants in Australia and their dispersal by ants. *Australian Journal of Botany* 23:475–508.

Bergstrom, B., I. Groth, O. Pellmyr, P. K. Endress, L. B. Thien, A. Hubener, and W. Francke. 1991. Chemical basis of a highly specific mutualism: Chiral esters attract pollinating beetles in Eupomatiaceae. *Phytochemistry* 30:3221–25.

Bernhardt, P., and L. B. Thien. 1987. Self-isolation and insect pollination in the primitive angiosperms: New evaluations of older hypotheses. *Plant Systematics and Evolution* 156:159–76.

Billick, I., and K. Tonkel. 2003. The relative importance of spatial vs. temporal variability in generating a conditional mutualism. *Ecology* 84:289–95.

Bizerril, M. X. A., and E. M. Vieira. 2002. *Azteca* ants as antiherbivore agents of *Tococa formicaria* (Melastomataceae) in Brazilian cerrado. *Studies on Neotropical Fauna and Environment* 37:145–49.

Blom, P. E., and W. H. Clark. 1980. Observations of ants (Hymenoptera: Formicidae) visiting extrafloral nectaries of the barrel cactus, *Ferocactus gracilis* Gates (Cactaceae), in Baja California, Mexico. *Southwestern Naturalist* 25:181–96.

Blüthgen, N., and K. Fiedler. 2002. Interactions between weaver ants *Oecophylla smaragdina,* homopterans, trees, and lianas in an Australian rain forest canopy. *Journal of Animal Ecology* 5:793–801.

———. 2004a. Competition for composition: Lessons from nectar-feeding ant communities. *Ecology* 85:1479–85.

———. 2004b. Preferences for sugars and amino acids and their conditionality in a diverse nectar-feeding ant community. *Journal of Animal Ecology* 73:155–66.

Blüthgen, N., G. Gebauer, and K. Fiedler. 2003. Disentangling a rainforest food web using stable isotopes: Dietary diversity in a species-rich ant community. *Oecologia* 137:426–35.

Blüthgen, N., G. Gottsberger, and K. Fiedler. 2004. Sugar and amino acid composition of ant-attended nectar and honeydew sources from an Australian rainforest. *Austral Ecology* 29:418–29.

Blüthgen, N., F. Menzel, and N. Blüthgen. 2006. Measuring specialization in species interaction networks. *BMC Ecology* 6, in press.

Blüthgen, N., and K. Reifenrath. 2003. Extrafloral nectaries in an Australian rainforest—structure and distribution. *Australian Journal of Botany* 51:515–27.

Blüthgen, N., N. E. Stork, and K. Fiedler. 2004. Bottom-up control and co-occurrence in complex communities: Honeydew and nectar determine a rainforest ant mosaic. *Oikos* 106:344–58.

Blüthgen, N., M. Verhaagh, W. Goitía, K. Jaffé, W. Morawetz, and W. Barthloot. 2000. How plants shape the ant community in the Amazonian rainforest canopy: The key role of extrafloral nectaries and hemipteran honeydew. *Oecologia* 125:229–40.

Blüthgen, N., and J. Wesenberg. 2001. Ants induce domatia in a rain forest tree (*Vochysia vismiaefolia*). *Biotropica* 33:637–42.

Böhning-Gaese, K., B. H. Gaese, and S. B. Rabemanantsoa. 1999. Importance of primary and secondary seed dispersal in the Malagasy tree *Commiphora guillaumini*. *Ecology* 80:821–32.

Bond, W. J., and P. Slingsby. 1983. Seed dispersal by ants in shrublands of the Cape Province and its evolutionary implications. *South African Journal of Science* 79:231–33.

———. 1984. Collapse of an ant-plant mutualism: The argentine ant (*Iridomyrmex humilis*) and myrmecochorous Proteaceae. *Ecology* 65:1031–37.

Bond, W. J., and W. D. Stock. 1989. The costs of leaving home: Ants disperse myrmecochorous seeds to low nutrient sites. *Oecologia* 81:412–17.

Bond, W. J., R. Yeaton, and W. D. Stock. 1991. Myrmecochory in Cape fynbos. In Huxley and Cutler 1991, 448–62.

Bondini, A., and G. Giavelli. 1989. The qualitative approach in investigating the role of species interactions on stability of natural communities. *BioSystems* 22:289–99.

Bono, J. M., and E. R. Heithaus. 2002. Sex ratios and the distribution of elaiosomes in colonies of the ant, *Aphaenogaster rudis*. *Insectes Sociaux* 49:320–25.

Borger, G. A., and M. E. Hoenecke. 1984. Extra-floral nectaries of bracken fern (*Pteridium aquilinum*). *Bulletin of the Botanical Club of Wisconsin* 16:42–46.

Borror, D. J., D. M. De Long, and C. A. Triplehorn. 1981. *An introduction to the study of insects*. Philadelphia: Saunders College Publishing.

Bosch, J., J. Retana, and X. Cerdá. 1997. Flowering phenology, floral traits, and pollinator composition in a herbaceous Mediterranean plant community. *Oecologia* 109:583–91.

Boucher, D. H., S. James, and K. H. Keeler. 1982. The ecology of mutualism. *Annual Review of Ecology and Systematics* 13:315–47.

Boyd, R. S. 1996. Ant-mediated seed dispersal of the rare chaparral shrub *Fremontodendron decumbens* (Sterculiaceae). *Madroño* 43:299–315.

Brantjes, N. B. M. 1981. Ant, bee, and fly pollination in *Epipactis palustris* (L.) Crantz (Orchidaceae). *Acta Botanica Neerlandica* 30:59–68.

Breton, L. M., and J. F. Addicott. 1992. Density-dependent mutualism in an aphid-ant interaction. *Ecology* 73:2175–80.

Brew, C. R., D. J. O'Dowd, and Y. D. Rae. 1989. Seed dispersal by ants: Behaviour-relating compounds in elaiosomes. *Oecologia* 80:490–97.

Briese, D. T. 1982. Damage to saltbush by the coccid *Pulvinaria maskelli* Olliff, and the role played by an attendant ant. *Journal of the Australian Entomological Society* 21:293–94.

Bristow, C. M. 1983. Treehoppers transfer parental care to ants: A new benefit of mutualism. *Science* 220:532–33.

———. 1984. Differential benefits from ant attendance to two species of Homoptera on New York iron weed. *Journal of Animal Ecology* 53:715–26.

———. 1991. Are ant-aphid associations a tritrophic interaction? Oleander aphids and Argentine ants. *Oecologia* 87:514–21.

Bronstein, J. L. 1994a. Conditional outcomes in mutualistic interactions. *Trends in Ecology and Evolution* 9:214–17.

———. 1994b. Our current understanding of mutualism. *Quarterly Review of Biology* 69:31–51.

———. 1995. The plant-pollinator landscape. In *Mosaic landscapes and ecological processes,* ed. L. Hansson, L. Fahrig, and G. Merriam, 256–88. London: Chapman and Hall.

———. 1998. The contribution of ant-plant protection studies to our understanding of mutualism. *Biotropica* 30:150–61.

Bronstein, J. L., and P. Barbosa. 2002. Multitrophic/multispecies mutualistic interactions: The role of nonmutualists in shaping and mediating mutualisms. In *Multitrophic level interactions,* ed. T. Tscharntke and B. A. Hawkins, 44–66. Cambridge: Cambridge University Press.

Brothers, D. J., and J. M. Carpenter. 1993. Phylogeny of Aculeata: Chrysidoidea and Vespoidea (Hymenoptera). *Journal of Hymenopteran Research* 2:227–304.

Brouat, C., and D. McKey. 2001. Leaf-stem allometry, hollow stems, and the evolution of caulinary domatia in myrmecophytes. *New Phytologist* 151:391–406.

Brouat, C., D. McKey, and E. J. P. Douzery. 2004. Differentiation in a geographical mosaic of plants coevolving with ants: Phylogeny of the *Leonardoxa africana* complex (Fabaceae: Caesalpinioideae) using amplified fragment length polymorphism markers. *Molecular Ecology* 13:1157–71.

Brown, E. S. 1959. Immature nutfall of coconuts in the Solomon Islands. II. Changes in ant populations, and their relation to vegetation. *Bulletin of Entomological Research* 50:523–58.

Brown, J. H. 1975. Geographical ecology of desert rodents. In *Ecology and evolution of communities,* ed. M. L. Cody and J. M. Diamond, 315–41. Cambridge, MA: Harvard University Press.

Brown, J. H., D. W. Davidson, J. C. Munger, and R. S. Inouye. 1986. Experimental community ecology: The desert granivore system. In *Community ecology,* ed. J. Diamond and T. J. Case, 41–61. New York: Harper and Row.

Brown, J. H., O. J. Reichman, and D. W. Davidson. 1979. Granivory in desert ecosystems. *Annual Review of Ecology and Systematics.* 10:201–27.

Brown, M. J. F., and K. G. Human. 1997. Effects of harvester ants on plant species distribution and abundance in a serpentine grassland. *Oecologia* 112:237–43.

Brühl, C. A., G. Gunsalam, and K. E. Linsenmair. 1998. Stratification of ants (Hymenoptera, Formicidae) in a primary rain forest in Sabah, Borneo. *Journal of Tropical Ecology* 14:285–97.

Bruna, E. M., D. M. Lapola, and H. L. Vasconcelos. 2004. Interspecific variation in the defensive responses of obligate plant-ants: Experimental tests and consequences for herbivory. *Oecologia* 138:558–65.

Bruna, E. M., H. L. Vasconcelos, and S. Heredia. 2005. The effect of habitat fragmentation on communities of mutualists: Amazonian ants and their host plants. *Biological Conservation* 124:209–16.

Buckley, R. C. 1982a. Ant-plant interactions: A world review. In Buckley 1982b, 111–41.

———, ed. 1982b. *Ant-plant interactions in Australia*. The Hague: Dr. W. Junk. 1983. Interaction between ants and membracid bugs decreases growth and seed set of host plant bearing extrafloral nectaries. *Oecologia* 58:132–36.

———. 1987a. Ant-plant-homopteran interactions. *Advance Ecological Research* 16:53–85.

———. 1987b. Interactions involving plants, Homoptera, and ants. *Annual Review of Ecology and Systematics* 18:111–35.

———. 1990. Ants protect tropical Homoptera against nocturnal spider predation. *Biotropica* 22:207–9.

Buckley, R. C., and P. Gullan. 1991. More aggressive ant species (Hymenoptera: Formicidae) provide better protection for soft scales and mealybugs (Homoptera: Coccidae, Pseudococcidae). *Biotropica* 23:282–86.

Bullock, S. H., and A. Solís-Magallanes. 1990. Phenology of canopy trees of a tropical deciduous forest in México. *Biotropica* 22:22–35.

Burd, M. 1996a. Foraging performance by *Atta colombica,* a leaf-cutting ant. *American Naturalist* 148:597–612.

———. 1996b. Server system and queuing models of leaf harvesting by leaf-cutting ants. *American Naturalist* 148:613–29.

———. 2000. Body size effects on locomotion and load carriage in the highly polymorphic leaf-cutting ants *Atta colombica* and *Atta Cephalotes. Behavioral Ecology* 11:125–31.

Burdon, J. J., and P. H. Thrall. 1999. Spatial and temporal patterns in coevolving plant and pathogen associations. *American Naturalist* 153:S15–S33.

Burslem, D. F. R. P., M. A. Pinard, and S. E. Hartley, eds. 2005. *Biotic Interactions in the Tropics.* Cambridge: Cambridge University Press.

Byrne, M. M., and D. J. Levey. 1993. Removal of seeds from frugivore defecations by ants in a Costa Rican rain forest. *Vegetatio* 107–8:363–74.

Cabrera, M., and K. Jaffe. 1994. A trophic mutualism between the myrmecophytic Melastomataceae *Tococa guianensis* Aublet and an *Azteca* ant species. *Ecotropicos* 7:1–10.

Camargo, J. M. F., and S. R. M. Pedro. 2002. Mutualistic association between a tiny Amazonian stingless bee and a wax-producing scale insect. *Biotropica* 34:446–51.

Campbell, B. C. 1986. Host-plant oligosaccharins in the honeydew of *Schizaphis graminum* (Rondani) (Insecta, Aphididae). *Experientia* 42:451–52.

Cardillo, M. 2002. The life-history basis of latitudinal diversity gradients: How do species traits vary from the poles to the equator? *Journal of Animal Ecology* 71:79–87.

Carney, S. E., M. B. Byerley, and D. A. Holway. 2003. Invasive Argentine ants (*Linepithema humile*) do not replace native ants as seed dispersers of *Dendromecon rigida* (Papaveraceae) in California, USA. *Oecologia* 235:576–82.

Carpenter, F. M. 1930. The fossil ants of North America. *Bulletin of the Museum of Comparative Zoology* 70:1–66.

Carpenter, J. M., and A. P. Rasnitsyn. 1990. Mesozoic Vespidae. *Psyche* 97:1–20.

Carroll, C. R., and D. H. Janzen. 1973. Ecology of foraging by ants. *Annual Review of Ecology and Systematics* 4:231–257.

Carvalho, K. S., and H. L. Vasconcelos. 1999. Forest fragmentation in central Amazonia and its effects on litter-dwelling ants. *Biological Conservation* 91:151–57.

Castillo, S., and J. Carabias. 1982. Ecología de la vegetación de dunas costeras: Fenología. *Biotica* 7:551–68.

Catling, P. M. 1997. Influence of aerial *Azteca* nests on the epiphyte community of some Belizean orange orchards. *Biotropica* 29:237–42.

Chambers, J. C., and J. A. MacMahon. 1994. A day in the life of a seed: Movements and fates of seeds and their implications for natural and managed systems. *Annual Review of Ecology and Systematics* 25:263–92.

Chapela, Y. H., S. A. Rehner, T. R. Schultz, and U. G. Mueller. 1994. Evolutionary history of the symbiosis between fungus-growing ants and their fungus. *Science* 266:1691–94.

Chapin, F. S., III, E. S. Zavaleta, V. T. Eviner, R. L. Naylor, P. M. Vitousek, H. L. Reynolds, D. U. Hooper, S. Lavorel, O. E. Sala, S. E. Hobbie, M. C. Mack, and S. Diaz. 2000. Consequences of changing biodiversity. *Nature* 405:234–42.

Chemás, A., and V. Rico-Gray. 1991. Apiculture and management of associated vegetation by the Maya of Tixcacaltuyub, Yucatán, México. *Agroforestry Systems* 13:13–25.

Cherrett, J. M. 1968. The foraging behaviour of *Atta cephalotes* L. (Hymenoptera, Formicidae). 1. Foraging pattern and plant species attacked in tropical rain forest. *Journal of Animal Ecology* 37:387–403.

———. 1972. Some factors involved in the selection of vegetable substrate by *Atta cephalotes* (L.) (Hymenoptera, Formicidae) in tropical rain forest. *Journal of Animal Ecology* 41:647–60.

———. 1989. Leaf-cutting ants. In *Ecosystems of the world 14B,* ed. H. Lieth and M. J. A. Werger, 437–86. Amsterdam: Elsevier.

Chiotis, M., L. S. Jermiin, and R. H. Crozier. 2000. A molecular framework for the phylogeny of the ant subfamily Dolichoderinae. *Molecular Phylogenetics and Evolution* 17:108–16.

Christian, C. E. 2001. Consequences of a biological invasion reveal the importance of mutualism for plant communities. *Nature* 413:635–39.

Christian, C. E., and M. L. Stanton. 2004. Cryptic consequences of a dispersal mutualism: Seed burial, elaiosome removal, and seed-bank dynamics. *Ecology* 85:1101–10.

Christianini, A. V., and G. Machado. 2004. Induced responses to herbivory and associated cues in the Amazonian ant-plant *Maieta poeppigii*. *Entomologia Experimentalis et Applicata* 112:81–88.

Clausing, G. 1998. Observations on ant-plant interactions in *Pachycentria* and other genera of the Dissochaeteae (Melastomataceae) in Sabah and Sarawak. *Flora* 193:361–68.

Clifford, H. T., and G. B. Monteith. 1989. A three phase seed dispersal mechanism in Australian Quinine bush. *Biotropica* 21:284–86.

Cocroft, R. B. 1999. Parent-offspring communication in response to predators in a subsocial treehopper (Hemiptera: Membracidae: *Umbonia crassicornis*). *Ethology* 105:553–68.

Cogni, R., and A. V. L. Freitas. 2002. The ant assemblage visiting the extrafloral nectaries of *Hibiscus pernambucensis* (Malvaceae) in a mangrove forest in Southeast Brazil. *Sociobiology* 40:373–83.

Cogni, R., A. V. L. Freitas, and P. S. Oliveira. 2003. Interhabitat differences in ant activity on plant foliage: Ants at extrafloral nectaries of *Hibiscus pernambucensis* in sandy and mangrove forests. *Entomologia Experimentalis et Applicata* 107:125–31.

Cogni, R., R. L. G. Raimundo, and A. V. L. Freitas. 2000. Daily activity of ants associated with the extrafloral nectaries of *Turnera ulmifolia* L. (Turneraceae) in a suburban area in Southeast Brazil. *Entomologist's Monthly Magazine* 136:141–47.

Cole, F. R., A. C. Medeiros, L. L. Loope, and W. W. Zuehlke. 1992. Effects of the Argentine ant on arthropod fauna of a Hawaiian high-elevation shrubland. *Ecology* 73:1313–22.

Coley, P. D. 1986. Costs and benefits of defense by tannins in a neotropical tree. *Oecologia* 70:238–41.

Coley, P. D., and T. M. Aide. 1991. Comparison of herbivory and plant defenses in temperate and tropical broad-leaved forests. In Price et al. 1991, 25–49.

Compton, S. G., and H. G. Robertson. 1988. Complex interactions between mutualisms: Ants tending homopterans protecting fig seeds and pollinators. *Ecology* 69:1302–5.

———. 1991. Effects of ant-homopteran systems on fig–fig wasp interactions. In Huxley and Cutler 1991, 120–30.

Cook, O. F. 1904. *Report on habits of kelep, or Guatemalan cotton-boll-weevil ant.* Bureau of Entomology Bulletin, n.s., 49. Washington, DC: Government Printing Office.

———. 1905. *Social organization and breeding habits of cotton-protecting kelep of Guatemala.* Bureau of Entomology, Technical Series 10. Washington, DC: Government Printing Office.

Cooper-Driver, G. A. 1978. Insect fern associations. *Entomologia Experimentalis et Applicata* 24:310–16.

Cooper-Driver, G. A., S. Finch, and T. Swain. 1977. Seasonal variation in secondary plant compounds in relation to the palatability of *Pteridium aquilinum. Biochemical Systematics and Ecology* 5:177–83.

Corlett, R. T. 1996. Characteristics of vertebrate-dispersed fruits in Hong Kong. *Journal of Tropical Ecology* 12:819–33.

Costa, F. M. C. B., A. T. Oliveira-Filho, and P. S. Oliveira. 1992. The role of extrafloral nectaries in *Qualea grandiflora* (Vochysiaceae) in limiting herbivory: An experiment of ant protection in cerrado vegetation. *Ecolological Entomology* 17:363–65.

Cox, C. B., and P. D. Moore. 1993. *Biogeography.* Oxford: Blackwell Scientific Publications.

Crane, P. R. 1998. The phylogenetic position and fossil history of the Magnoliaceae. In *Magnolias and their allies,* ed. D. Hunt, 21–36. Milborne Port, UK: International Dendrology Society and Magnolia Society.

Crane, P. R., E. M. Friis, and K. R. Pedersen. 1995. The origin and early diversification of angiosperms. *Nature* 374:27–33.

Crepet, W. L. 1996. Timing in the evolution of derived floral characters: Upper Cretaceous (Turonian) taxa with tricolpate and tricolpate-derived pollen. *Review of Palaeobotany and Palynology* 90:339–59.

Crist, T. O., and J. A. MacMahon. 1992. Harvester ant foraging and shrub-steppe seeds: Interactions of seed resources and seed use. *Ecology* 73:1768–79.

Crist, T. O., and J. A. Wiens. 1994. Scale effects of vegetation on forager movement and seed harvesting by ants. *Oikos* 69:37–46.

Cronin, G. 1998. Between-species and temporal variation in *Acacia*-ant-herbivore interactions. *Biotropica* 30:135–39.

Crossley, D. A., Jr., G. J. House, R. M. Snider, R. J. Snider, and B. R. Stinner. 1984. Positive interactions in agroecosystems. In *Agricultural ecosystems,* ed. R. Lowrance, B. R. Stinner, and G. J. House, 73–82. New York: Wiley.

Cuautle, M., J. G. García-Franco, and V. Rico-Gray. 1998. Estructura poblacional y características foliares de *Agave kerchovei:* Relación con la presencia de homópteros y hongos en el valle de Zapotitlán, Puebla. *Cactáceas y Suculentas Mexicanas* 43:75–80.

Cuautle, M., and V. Rico-Gray. 2003. The effect of wasps and ants on the reproductive success of the extrafloral nectaried plant *Turnera ulmifolia* (Turneraceae). *Functional Ecology* 17:417–23.

Cuautle, M., V. Rico-Gray, and C. Díaz-Castelazo. 2005. Effects of ant behaviour and extrafloral nectaries presence on seed dispersal of the neotropical myrmecochore *Turnera ulmifolia* L. (Turneraceae), in a sand dune matorral. *Biological Journal of the Linnean Society* 86:67–77.

Cuautle, M., V. Rico-Gray, J. G. García-Franco, J. López-Portillo, and L. B. Thien. 1999. Description and seasonality of a plant-ant-Homoptera interaction in the semiarid Zapotitlán Valley, México. *Acta Zoológica Mexicana,* n.s., 78:73–83.

Culver, D. C., and A. J. Beattie. 1978. Myrmecochory in *Viola:* Dynamics of seed-ant interactions in some West Virginia species. *Journal of Ecology* 66:53–72.

———. 1980. The fate of *Viola* seeds dispersed by ants. *American Journal of Botany* 67:710–14.

Currie, C. R. 2001. A community of ants, fungi, and bacteria: A multilateral approach to studying symbiosis. *Annual Review of Microbiology* 55:357–80.

Currie, C. R., U. G. Mueller, and D. Malloch. 1999. Fungus-growing ants use antibiotic-producing bacteria to control garden parasites. *Nature* 398:701–4.

Cushman, J. H. 1991. Host-plant mediation of insect mutualisms: Variable outcomes in herbivore-ant interactions. *Oikos* 61:138–44.

Cushman, J. H., and J. F. Addicott. 1991. Conditional interactions in ant-plant-herbivore mutualisms. In Huxley and Cutler 1991, 92–103.

Cushman, J. H., and A. J. Beattie. 1991. Mutualisms: Assessing the benefits to hosts and visitors. *Trends in Ecology and Evolution* 6:193–95.

Cushman, J. H., S. G. Compton, C. Zachariades, A. B. Ware, R. J. C. Nefdt, and V. K. Rashbrook. 1998. Geographic and taxonomic distribution of a positive interaction: Ant-tended homopterans indirectly benefit figs across southern Africa. *Oecologia* 116:373–80.

Cushman, J. H., J. H. Lawton, and B. F. J. Manly. 1993. Latitudinal patterns in European ant assemblages: Variation in species richness and body size. *Oecologia* 95:30–37.

Cushman, J. H., V. K. Rashbrook, and A. J. Beattie. 1994. Assessing benefits to both participants in a lycaenid-ant association. *Ecology* 75:1031–41.

Cushman, J. H., and T. G. Whitham. 1989. Conditional mutualism in a membracid-ant association: Temporal, age-specific, and density-dependent effects. *Ecology* 70:1040–47.

———. 1991. Competition mediating the outcome of a mutualism: Protective services of ants as a limiting resource for membracids. *American Naturalist* 138:851–65.

Dalling, J. W., and R. Wirth. 1998. Dispersal of *Miconia argentea* seeds by the leaf-cutting ant *Atta colombica*. *Journal of Tropical Ecology* 14:705–10.

Dansa, C. V. A., and C. F. D. Rocha. 1992. An ant-membracid-plant interaction in a cerrado area of Brazil. *Journal of Tropical Ecology* 8:339–48.

Davidson, D. W. 1977a. Foraging ecology and community organization in desert seed-eating ants. *Ecology* 58:725–37.

———. 1977b. Species diversity and community organization in desert seed-eating ants. *Ecology* 58:711–24.

———. 1978. Size variability in the worker caste of a social insect (*Veromessor pergandei* Mayr) as a function of the competitive environment. *American Naturalist* 112:523–32.

———. 1988. Ecological studies of neotropical ant gardens. *Ecology* 69:1138–52.

———. 1993. The effects of herbivory and granivory on terrestrial plant succession. *Oikos* 68:23–35.

———. 1997. The role of resource imbalances in the evolutionary ecology of tropical arboreal ants. *Biological Journal of the Linnean Society* 61:153–81.

———. 1998. Resource discovery versus resource domination in ants: A functional mechanism for breaking the trade-off. *Ecological Entomology* 23: 484–90.

Davidson, D. W., S. C. Cook, and R. R. Snelling. 2004. Liquid-feeding performances of ants (Formicidae): Ecological and evolutionary implications. *Oecologia* 139:255–66.

Davidson, D. W., S. C. Cook, R. R. Snelling, and T. H. Chua. 2003. Explaining the abundance of ants in lowland tropical rainforest canopies. *Science* 300:969–72.

Davidson, D. W., and W. W. Epstein. 1989. Epiphytic associations with ants. In *Vascular plants as epiphytes,* ed. U. Lüttge, 200–233. Berlin: Springer-Verlag.

Davidson, D. W., and B. L. Fisher. 1991. Symbiosis of ants with *Cecropia* as a function of light regime. In Huxley and Cutler 1991, 289–309.

Davidson, D. W., R. B. Foster, R. R. Snelling, and P. W. Lozada. 1991. Variable composition of some tropical ant-plant symbioses. In Price et al. 1991, 145–62.

Davidson, D. W., J. T. Longino, and R. R. Snelling. 1988. Pruning of host plant neighbors by ants: An experimental approach. *Ecology* 69:801–8.

Davidson, D. W., and D. McKey. 1993. The evolutionary ecology of symbiotic ant-plant relationships. *Journal of Hymenopteran Research* 2:13–83.

Davidson, D. W., and S. R. Morton. 1981a. Competition for dispersal in ant-dispersed plants. *Science* 213:1259–61.

———. 1981b. Myrmecochory in some plants (*F. chenopodiaceae*) of the Australian arid zone. *Oecologia* 50:357–66.

———. 1984. Dispersal adaptations of some *Acacia* species in the Australian arid zone. *Ecology* 65:1038–51.

Davidson, D. W., and L. Patrell-Kim. 1996. Tropical arboreal ants: Why so abundant? In *Neotropical Biodiversity and Conservation,* ed. A. C. Gibson, 127–140. Los Angeles: Mildred E. Mathias Botanical Garden, University of California.

Davidson, D. W., R. R. Snelling, and J. T. Longino. 1989. Competition among ants for myrmecophytes and the significance of plant trichomes. *Biotropica* 21:64–73.

Davies, T. J., T. G. Barraclough, M. W. Chase, P. S. Soltis, D. E. Soltis, and V. Savolainen. 2004. Darwin's abominable mystery: Insights from a supertree of the angiosperms. *Proceedings of the National Academy of Sciences* 101:1904–9.

Davison, E. A. 1982. Seed utilization by harvester ants in western NSW. In Buckley 1982b, 1–6.

Dean, W. R. J., and R. I. Yeaton. 1993. The influence of harvester ant *Messor capensis* nest-mounds on the productivity and distribution of some plant species in the southern Karoo, South Africa. *Vegetatio* 106:21–35.

De Andrade, J. C., and J. P. P. Carauta. 1982. The *Cecropia-Azteca* association: A case of mutualism? *Biotropica* 14:15.

Dejean, A., A. Akoa, C. Djiéto-Lordon, and A. Lenoir. 1994. Mosaic ant territories in an African secondary forest (Hymenoptera: Formicidae). *Sociobiology* 23:275–92.

Dejean, A., and B. Corbara. 2003. Review of mosaics of dominant ants in rainforests and plantations. In *Arthropods of tropical forests: Spatio-temporal dynamics and resource use in the canopy,* ed. Y. Basset, V. Novotny, S. E. Miller, and R. L. Kitching, 341–47. Cambridge: Cambridge University Press.

Dejean, A., D. McKey, M. Gibernau, and M. Belin-Depoux. 2000. The arboreal ant mosaic in a Cameroonian rainforest. *Sociobiology* 35: 403–23.

Dejean, A., I. Olmsted, and R. R. Snelling. 1995. Tree-epiphyte-ant relationships in the low inundated forest of Sian Ka'an biosphere reserve, Quintana Roo, México. *Biotropica* 27:57–70.

Dejean, A., A. Quilichini, J. H. Delabie, J. Orivel, B. Corbara, and M. Gibernau. 2004. Influence of its associated ant species on the life history of the myrmecophyte *Cordia nodosa* in French Guiana. *Journal of Tropical Ecology* 20:701–4.

Dejean A., P. J. Solano, J. Ayroles, B. Corbara, and J. Orivel. 2005. Arboreal ants build a trap to ambush and capture prey. *Nature* 434:973.

Delabie, J. H. C. 1990. The ant problems of cocoa farms in Brazil. In Vander Meer, Jaffe, and Cedeno 1990, 555–69.

———. 2001. Trophobiosis between Formicidae and Hemiptera (Sternorrhyncha and Auchenorrhyncha): An overview. *Neotropical Entomology* 30:501–16.

De la Fuente, M. A. S., and R. J. Marquis. 1999. The role of ant-tended extrafloral nectaries in the protection and benefit of a neotropical rainforest tree. *Oecologia* 118:192–202.

Del-Claro, K. 2004. Multitrophic relationships, conditional mutualisms, and the study of interaction biodiversity in tropical savannas. *Neotropical Entomology* 33:665–72.

Del-Claro, K., V. Berto, and W. Réu. 1996. Effect of herbivore deterrence by ants on the fruit set of an extrafloral nectary plant, *Qualea multiflora* (Vochysiaceae). *Journal of Tropical Ecology* 12:887–92.

Del-Claro, K., and P. S. Oliveira. 1993. Ant-Homoptera interaction: Do alternative sugar sources distract tending ants? *Oikos* 68:202–6.

———. 1996. Honeydew flicking by treehoppers provides cues to potential tending ants. *Animal Behavior* 51:1071–75.

———. 1999. Ant-Homoptera interactions in a neotropical savanna: The honeydew-producing treehopper, *Guayaquila xiphias* (Membracidae), and its associated ant fauna on *Didymopanax vinosum* (Araliaceae). *Biotropica* 31:135–44.

———. 2000. Conditional outcomes in a neotropical treehopper-ant association: Temporal and species-specific variation in ant protection and homopteran fecundity. *Oecologia* 124:156–65.

Delpino, F. 1875. Rapporti tra insetti e tra nettari estranuzali in alcune piante. *Bolletino della Societa Entomologica (Firenze)* 7:69–90.

DeNiro, M. J., and S. Epstein. 1981. Influence of diet on the distribution of nitrogen isotopes in animals. *Geochimica et Cosmochimica Acta* 45:341–51.

Denno, R. F., M. S. McClure, and J. R. Ott. 1995. Interspecific interactions in phytophagous insects: Competition reexamined and resurrected. *Annual Review of Entomology* 40:297–331.

Denslow, J. S., and A. E. Gomez-Dias. 1990. Seed rain to treefall gaps in a neotropical rain forest. *Journal of Canadian Forest Research* 20:815–17.

Deslippe, R. J., and R. Savolainen. 1995. Mechanisms of competition in a guild of formicine ants. *Oikos* 72:67–73.

Deuth, D. 1977. The function of extra-floral nectaries in *Aphelandra deppeana* Schl. and Cham. (Acanthaceae). *Brenesia* 10–11:135–45.

DeVries, P. J. 1990. Enhancement of symbiosis between butterfly caterpillars and ants by vibrational communication. *Science* 248:1104–6.

———. 1991. Mutualism between *Thisbe irenea* butterflies and ants, and the role of ant ecology in the evolution of larval-ant associations. *Biological Journal of the Linnean Society* 43:179–95.

DeVries, P. J., and I. Baker. 1989. Butterfly exploitation of an ant-plant mutualism: Adding insult to herbivory. *Journal of the New York Entomological Society* 97:332–40.

Diamond, J. 1998. Ants, crops, and history. *Science* 281:1974–75.

Díaz-Castelazo, C., and V. Rico-Gray. 1998. Número y variación estacional de asociaciones hormiga-planta en un bosque montano bajo de Veracruz, México. *Acta Zoológica Mexicana,* n.s., 73:45–55.

Díaz-Castelazo, C., V. Rico-Gray, P. S. Oliveira, and M. Cuautle. 2004. Extrafloral nectary-mediated ant-plant interactions in the coastal vegetation of Veracruz, México: Richness, occurrence, seasonality, and ant foraging patterns. *Ecoscience* 11:472–81.

Díaz-Castelazo, C., V. Rico-Gray, F. Ortega, and G. Angeles. 2005. Morphological and secretory characterization of extrafloral nectaries in plants of coastal Veracruz, Mexico. *Annals of Botany* 96:1175–89.

Dieckmann, U., B. O'Hara, and W. Welsser. 1999. The evolutionary ecology of dispersal. *Trends in Ecology and Evolution* 14:88–90.

Di Giusto, B., M.-C. Anstett, E. Dounias, and D. B. McKey. 2001. Variation in the effectiveness of biotic defense: The case of an opportunistic ant-plant protection mutualism. *Oecologia* 129:367–75.

Doebeli, M., and N. Knowlton. 1998. The evolution of interspecific mutualism. *Proceedings of the National Academy of Sciences* 95:8687–98.

Domínguez, C. A., R. Dirzo, and S. H. Bullock. 1989. On the function of floral nectar in *Croton suberosus* (Euphorbiaceae). *Oikos* 56:109–14.

Douglas, A. E. 2003. The nutritional physiology of aphids. *Advances in Insect Physiology* 31:73–140.

Downes, W. L., Jr., and G. A. Dahlem. 1987. Keys to the evolution of Diptera: Role of Homoptera. *Forum: Environmental Entomology* 16:847–54.

Downhower, J. F. 1975. The distribution of ants on *Cecropia* leaves. *Biotropica* 7:59–62.

Drake, W. E., 1981. Ant-seed interaction in dry sclerophyll forest on North Stradbroke Island, Queensland. *Australian Journal of Botany* 29:292–309.

Dutra, H. P., A. V. L. Freitas, and P. S. Oliveira. 2006. Dual ant attraction in the neotropical shrub *Urera baccifera* (Urticaceae): The role of ant visitation to pearl bodies and fruits in herbivore deterrence and leaf longevity. *Functional Ecology* 20:252–60.

Dyer, L. A., and D. K. Letourneau. 1999a. Relative strengths of top-down and bottom-up forces in a tropical forest community. *Oecologia* 119:265–74.

———. 1999b. Trophic cascades in a complex terrestrial community. *Proceedings of the United States National Academy of Sciences* 96:5072–76.

Edward, E. P. 1982. Hummingbirds feeding on an excretion produced by scale insects. *Condor* 84:182.

Eisner, T. 1957. A comparative morphological study of the proventriculus of ants (Hymenoptera: Formicidae). *Bulletin of the Museum of Comparative Zoology* 116:429–90.

Eisner, T., and E. O. Wilson. 1952. The morphology of the proventriculus of a formicine ant. *Psyche* 59:47–60.

Elias, T. S. 1983. Extrafloral nectaries: Their structure and functions. In *The biology of nectaries,* ed. B. Bentley and T. S. Elias, 174–203. New York: Columbia University Press.

Elias, T. S., and H. Gelband. 1975. Nectar: Its production and function in trumpet creeper. *Science* 189:289–91.

———. 1976. Morphology and anatomy of floral and extrafloral nectaries in *Campsis* (Bignoniaceae). *American Journal of Botany* 63:1349–53.

Elias, T. S., and G. T. Prance. 1978. Nectaries on the fruit of *Crescentia* and other Bignoniaceae. *Brittonia* 30:175–81.

Elias, T. S., W. R. Rozich, and L. Newcombe. 1975. The foliar and floral nectaries of *Turnera ulmifolia* L. *American Journal of Botany* 62:570–76.

Espadaler, X., and C. Gómez. 1996. Seed production, predation, and dispersal in the Mediterranean myrmecochore *Euphorbia characias* (Euphorbiaceae). *Ecography* 19:7–15.

Estrada, A., and T. H. Fleming, eds. 1986. *Frugivores and seed dispersal.* Dordrecht: Dr. W. Junk.

Eubanks, M. D., K. A. Nesci, M. K. Petersen, Z. W. Liu, and H. B. Sanchez. 1997. The exploitation of an ant-defended host plant by a shelter-building herbivore. *Oecologia* 109:454–60.

Evans, H. E., and M. J. W. Eberhard. 1970. *The wasps.* Ann Arbor: University of Michigan Press.

Ewuim, S. C., M. A. Badejo, and O. O. Ajayi. 1997. Ants of forest and fallow plots in Nigeria. *Biotropica* 29:93–99.

Fagan, W. F., R. S. Cantrell, and C. Cosner. 1999. How habitat sedges change species interactions. *American Naturalist* 152:165–82.

Farji-Brener, A. G. 2001. Why are leaf-cutter ants more common in early secondary forests than in old-growth tropical forests? An evaluation of the palatable forage hypothesis. *Oikos* 92:169–77.

———. 2005. The role of abandoned leaf-cutting ant nests on plant assemblage composition in a tropical rainforest of Costa Rica. *Ecoscience* 12: 554–60.

Farji-Brener, A. G., and L. Ghermandi. 2000. Influence of nests of leaf-cutting ants on plant species diversity in road verges of northern Patagonia. *Journal of Vegetation Science* 11:453–60.

———. 2004. Seedling recruitment in a semi-arid Patagonian steppe: Facilitative effects of refuse dumps of leaf-cutting ants. *Journal of Vegetation Science* 15:823–30.

Farji-Brener, A. G., and A. E. Illes. 2000. Do leaf-cutting ant nests make "bottom-up" gaps in neotropical rain forests? A critical review of the evidence. *Ecology Letters* 3:219–27.

Farji-Brener, A. G., and C. A. Medina. 2000. The importance of where to dump the refuse: Seed banks and fine roots in nests of the leaf-cutting ants *Atta cephalotes* and *A. colombica. Biotropica* 32:120–26.

Farji-Brener, A. G., and A. Ruggiero. 1994. Leaf-cutting ants (*Atta* and *Acromyrmex*) inhabiting Argentina: Patterns in species richness and geographical range sizes. *Journal of Biogeography* 21:391–99.

Farji-Brener, A. G., and Y. Sasal. 2003. Is dump material an effective small-scale deterrent to herbivory by leaf-cutting ants? *Ecoscience* 10:151–54.

Farji-Brener, A. G., and J. F. Silva. 1995a. Leaf-cutting ant nests and soil fertility in a well-drained savanna in western Venezuela. *Biotropica* 27:250–53.

———. 1995b. Leaf-cutting ants and forest groves in a tropical parkland savanna of Venezuela: Facilitated succession? *Journal of Tropical Ecology* 11:651–69.

———. 1996. Leaf-cutter ants' (*Atta laevigata*) aid to the establishment success of *Tapirira velutinifolia* (Anacardiaceae) seedlings in a parkland savanna. *Journal of Tropical Ecology* 12:163–68.

Farrell, B. D. 1998. "Inordinate fondness" explained: Why are there so many beetles? *Science* 281:555–59.

Farrell, B. D., and C. Mitter. 1993. Phylogenetic determinants of insect/plant community diversity. In *Species diversity in ecological communities,* ed. R. E. Ricklefs and D. Schluter, 253–66. Chicago: University of Chicago Press.

Federle, W., U. Maschwitz, and B. Fiala. 1998. The two-partner ant-plant system of *Camponotus* (*Colobopsis*) sp.1 and *Macaranga puncticulata* (Euphorbiaceae): Natural history of the exceptional ant partner. *Insectes Sociaux* 45:1–16.

Federle, W., U. Maschwitz, B. Fiala, M. Riederer, and B. Hölldobler. 1997. Slippery ant-plants and skilful climbers: Selection and protection of specific ant partners by epicuticular wax blooms in *Macaranga* (Euphorbiaceae). *Oecologia* 112:217–24.

Fedriani, J. M., P. J. Rey, J. L. Garrido, J. Guitián, C. M. Herrera, M. Medrano, A. M. Sánchez-Lafuente, and X. Cerdá. 2004. Geographical variation in the potential of mice to constrain an ant-seed dispersal mutualism. *Oikos* 105:181–91.

Feener, D. H., Jr., and E. W. Schupp. 1998. Effect of treefall gaps on the patchiness and species richness of neotropical ant assemblages. *Oecologia* 116:191–201.

Feinsinger, P., and L. A. Swarm. 1978. How common are ant-repellent nectars? *Biotropica* 10:238–39.

Feldhaar, H., B. Fiala, J. Gadau, M. Mohamed, and U. Maschwitz. 2003. Molecular phylogeny of *Crematogaster* subgenus *Decacrema* ants (Hymenoptera: Formicidae) and the colonization of Macaranga (Euphorbiaceae) trees. *Molecular Phylogenetics and Evolution* 27:441–52.

Feldhaar, H., B. Fiala, R. Hashim, and U. Maschwitz. 2003. Patterns of the *Crematogaster-Macaranga* association: The ant partner makes the difference. *Insectes Sociaux* 50:9–19.

Fellers, J. H. 1987. Interference and exploitation in a guild of woodland ants. *Ecology* 68:1466–78.

Fernandes, W. D., P. S. Oliveira, S. L. Carvalho, and M. E. M. Habib. 1994. *Pheidole* ants as potential biological control agents of the boll weevil, *Anthonomus grandis* (Coleoptera: Curculionidae), in Southeast Brazil. *Journal of Applied Entomology* 118:437–41.

Ferreira, S. O. 1994. Nectários extraflorais de *Ouratea spectabilis* (Ochnaceae) e a comunidade de formigas associadas: Um estudo em vegetação de cerrado, no sudeste do Brasil. Master's thesis, Universidade Estadual de Campinas.

Fiala, B. 1990. Extrafloral nectaries vs ant-Homoptera mutualisms: A comment on Becerra and Venable. *Oikos* 59:281–82.

Fiala, B., H. Grunsky, U. Maschwitz, and K. E. Linsenmair. 1994. Diversity of ant-plant interactions: Protective efficacy in *Macaranga* species with different degrees of ant association. *Oecologia* 97:186–92.

Fiala, B., A. Jakob, U. Maschwitz, and K. E. Linsenmair. 1999. Diversity, evolutionary specialization, and geographic distribution of a mutualistic ant-plant complex: *Macaranga* and *Crematogaster* in south east Asia. *Biological Journal of the Linnean Society* 66:305–31.

Fiala, B., and K. E. Linsenmair. 1995. Distribution and abundance of plants with extrafloral nectaries in the woody flora of a lowland primary forest in Malaysia. *Biodiversity and Conservation* 4:165–82.

Fiala, B., and U. Maschwitz. 1991. Extrafloral nectaries in the genus *Macaranga* (Euphorbiaceae) in Malaysia: Comparative studies of their possible significance as predispositions for myrmecophytism. *Biological Journal of the Linnean Society* 44:287–305.

———. 1992. Food bodies and their significance for obligate ant-association in the tree genus *Macaranga* (Euphorbiaceae). *Botanical Journal of the Linnean Society* 110:61–75.

Fiala, B., U. Maschwitz, and K. E. Linsenmair. 1996. *Macaranga caladifolia,* a new type of ant-plant among southeast Asian myrmecophitic *Macaranga* species. *Biotropica* 28:408–12.

Fiala, B., U. Maschwitz, and T. Y. Pong. 1991. The association between *Macaranga* trees and ants in south-east Asia. In Huxley and Cutler 1991, 263–70.

Fiala, B., U. Maschwitz, T. Y. Pong, and A. J. Helbig. 1989. Studies of a south east Asian ant-plant association: Protection of *Macaranga* trees by *Crematogaster borneensis. Oecologia* 79:463–70.

Fillman, D. A., and W. L. Sterling. 1985. Inaction levels for the red imported fire ant, *Solenopsis invicta* (Hym.: Formicidae): A predator of the boll weevil, *Anthonomus grandis* (Col.: Curculionidae). *Agricultural Ecosystems and Environment* 13:93–102.

Finnegan, R. J. 1974. Ants as predators of forest pests. *Entomophoga Memoires Hors Series* 7:53–59.

Fischer, M. K., and A. W. Shingleton. 2001. Host plant and ants influence the honeydew sugar composition of aphids. *Functional Ecology* 15:544–50.

Fischer, R. C., A. Richter, W. Wanek, and V. Mayer. 2002. Plants feed ants: Food bodies of myrmecophitic *Piper* and their significance for the interaction with *Pheidole bicornis* ants. *Oecologia* 133:186–92.

Fischer, R. C., W. Wanek, A. Richter, and V. Mayer. 2003. Do ants feed plants? A ^{15}N labelling study of nitrogen fluxes from ants to plants in the mutualism of *Pheidole* and *Piper. Journal of Ecology* 91:126–34.

Fisher, B. L. 1992. Facultative ant association benefits a neotropical orchid. *Journal of Tropical Ecology* 8:109–14.

Fisher, B. L., L. da S. L. Sternberg, and D. Price. 1990. Variation in the use of orchid extrafloral nectar by ants. *Oecologia* 83:263–66.

Fisher, B. L., and J. K. Zimmerman. 1988. Ant/orchid associations in the Barro Colorado National Monument, Panama. *Lindleyana* 3:12–16.

Fleming, T. H., and A. Estrada, eds. 1993. *Frugivory and seed dispersal: Ecological and evolutionary aspects.* Dordrecht: Kluwer.

Floren, A., and K. E. Linsenmair. 1997. Diversity and recolonization dynamics of selected arthropod groups on different tree species in a lowland rainforest in Shaba, Malaysia with special reference to Formicidae. In *Canopy arthropods,* ed. N. E. Stork, J. Adis, and R. K. Didham, 344–81. London: Chapman and Hall.

———. 2000. Do ant mosaics exist in pristine lowland rain forests? *Oecologia* 123:129–37.

Folgarait, P. J., and D. W. Davidson. 1994. Antiherbivore defenses of myrmecophytic *Cecropia* under different light regimes. *Oikos* 71:305–20.

———. 1995. Myrmecophytic *Cecropia:* Antiherbivore defenses under different nutrient treatments. *Oecologia* 104:189–206.

Folgarait, P. J., L. A. Dyer, R. J. Marquis, and H. E. Braker. 1996. Leaf-cutting ant preferences for five native tropical plantation tree species and different light conditions. *Entomologia Experimentalis et Applicata* 80:521–30.

Folgarait, P. J., H. L. Johnson, and D. W. Davidson. 1994. Responses of *Cecropia* to experimental removal of Müllerian bodies. *Functional Ecology* 8:22–28.

Fonseca, C. R. 1993. Nesting space limits colony size of the plant-ant *Pseudomyrmex concolor. Oikos* 67:473–82.

———. 1994. Herbivory and the long-lived leaves of an Amazonian ant-tree. *Journal of Ecology* 82:833–42.

———. 1999. Amazonian ant-plant interactions and the nesting space limitation hypothesis. *Journal of Tropical Ecology* 15:807–25.

Fonseca, C. R., and G. Ganade. 1996. Asymmetries, compartments, and null interactions in an Amazonian ant-plant community. *Journal of Animal Ecology* 65:339–47.

Forget, P.-M., and D. S. Hammond. 2005. Rainforest vertebrates and food plant diversity in the Guiana shield. In *Tropical forests of the Guiana shield,* ed. D. S. Hammond. Wallingford, UK: CABI International.

Fourcassié, V., and P. S. Oliveira. 2002. Foraging ecology of the giant Amazonian ant *Dinoponera gigantea* (Hymenoptera, Formicidae, Ponerinae): Activity schedule, diet, and spatial foraging patterns. *Journal of Natural History* 36:2211–27.

Fowler, H. G. 1977. Some factors influencing colony spacing and survival in the grass-cutting ant *Acromyrmex landolti fracticornis* (Forel) (Formicidae: Attini) in Paraguay. *Revista de Biología Tropical* 25:88–99.

———. 1993. Herbivory and assemblage structure of myrmecophytous understory plants and their associated ants in the central Amazon. *Insectes Sociaux* 40:137–45.

Fowler, H. G., and S. Claver. 1991. Leaf-cutter ant assemblies: Effects of latitude, vegetation, and behaviour. In Huxley and Cutler 1991, 51–59.

Frankie, G. W., H. G. Baker, and P. A. Opler. 1974. Comparative phenological studies of trees in tropical wet and dry forests in the lowlands of Costa Rica. *Journal of Ecology* 62:881–919.

Freitas, A. V. L., and P. S. Oliveira. 1992. Biology and behavior of *Eunica bechina* (Lepidoptera: Nymphalidae) with special reference to larval defense against ant predation. *Journal of Research on the Lepidoptera* 31:1–11.

———. 1996. Ants as selective agents on herbivore biology: Effects on the behaviour of a non-myrmecophilous butterfly. *Journal of Animal Ecology* 65: 205–10.

Freitas, L., L. Galetto, G. Bernardello, and A. A. S. Paoli. 2000. Ant exclusion and reproduction of *Croton sarcopetalus* (Euphorbiaceae). *Flora* 195:398–402.

Friis, E. M., W. G. Chaloner, and P. R. Crane. 1987. Introduction to angiosperms. In *The origins of angiosperms and their biological consequences,* ed. E. M. Friis, W. G. Chaloner, and P. R. Crane, 1–15. Cambridge: Cambridge University Press.

Fritz, R. S. 1983. Ant protection of a host plant's defoliator: Consequence of an ant-membracid mutualism. *Ecology* 64:789–97.

Fritz, R. S., and D. H. Morse. 1981. Nectar parasitism of *Asclepias syriaca* by ants: Effect on nectar levels, pollinia insertion, pollinaria removal, and pod production. *Oecologia* 50:316–19.

Fuller, B. W., and T. E. Reagan. 1988. Comparative predation of the sugarcane borer (Lepidoptera: Pyralidae) on sweet sorghum and sugarcane. *Journal of Economic Entomology* 81:713–17.

Futuyma, D. J. 1986. *Evolutionary biology.* Sunderland, MA: Sinauer.

Futuyma, D. J., and C. Mitter. 1997. Insect-plant interactions: The evolution of component communities. In *Plant life histories,* ed. J. Silvertown, M. Franco, and J. L. Harper, 253–64. Cambridge: Cambridge University Press.

Futuyma, D. J., and M. Slatkin. 1983. Introduction to *Coevolution,* ed. D. J. Futuyma, and M. Slatkin, 1–13. Sunderland, MA: Sinauer.

Galen, C. 1983. The effects of nectar thieving ants on seed set in floral scent morphs of *Polemonius viscosum. Oikos* 41:245–49.

Galetti, M., R. Laps, and M. A. Pizo. 2000. Frugivory by toucans (Ramphastidae) at two altitudes in the Atlantic forest of Brazil. *Biotropica* 32:842–50.

García, M. B., R. J. Antor, and X. Espadaler. 1995. Ant pollination of the palaeoendemic dioecious *Borderea pyrenaica* (Dioscoreaceae). *Plant Systematics and Evolution* 198:17–27.

García-Franco, J. G., and V. Rico-Gray. 1997. Reproductive biology of the holoparasite *Bdallophyton bambusarum* (Rafflesiaceae). *Botanical Journal of the Linnean Society* 123:237–47.

Garrido, J. L., P. J. Rey, X. Cerdá, and C. M. Herrera. 2002. Geographical variation in diaspore traits of an ant-dispersed plant (*Helleborus foetidus*): Are ant community composition and diaspore traits correlated? *Journal of Ecology* 90:446–55.

Gaston, K. J. 2000. Global patterns in biodiversity. *Nature* 405:220–27.

Gates, B. N. 1943. Carunculate seed dissemination by ants. *Rhodora* 45:438–45.

Gaume, L., D. McKey, and M. C. Anstett. 1997. Benefits conferred by "timid" ants: Active anti-herbivore protection of the rainforest tree *Leonardoxa africana* by the minute ant *Petalomyrmex phylax. Oecologia* 112:209–16.

Gaume, L., D. McKey, and S. Terrin. 1998. Ant-plant-homopteran mutualism: How the third partner affects the interaction between a plant-specialist ant and its myrmecophyte host. *Proceedings of the Royal Society of London, Series B* 265:569–75.

Gay, H. 1993. Animal-fed plants: An investigation into uptake of ant-derived nutrients by the far-eastern epiphytic fern *Lecantopteris* Reinw. (Polypodiaceae). *Biological Journal of the Linnean Society* 50:221–33.

Gay, H., and R. Hensen. 1992. Ant specificity and behaviour in mutualisms with epiphytes: The case of *Lecantopteris* (Polypodiaceae). *Biological Journal of the Linnean Society* 47:261–84.

Gaze, P. D., and M. N. Clout. 1983. Honeydew and its importance to birds in beech forests of South Island, New Zealand. *New Zealand Journal of Ecology* 6:33–37.

Gerling, D. 1990. *Whiteflies: Their bionomics, pest status, and management.* Andover, UK: Intercept.

Gerson, U. 1979. The associations between pteridophytes and arthropods. *Fern Gazette* 12:29–45.

Ghazoul, J. 2001. Can floral repellents pre-empt potential ant-plant conflicts? *Ecology Letters* 4:295–99.

Giavelli, G., and A. Bodini. 1990. Plant-ant-fungus communities investigated through qualitative modeling. *Oikos* 57:357–65.

Giladi, I. 2006. Choosing benefits or partners: A review of the evidence for the evolution of myrmecochory. *Oikos* 112:481–92.

Gilbert, L. E. 1980. Food web organization and the conservation of neotropical diversity. In *Conservation biology,* ed. M. E. Soulé and B. A. Wilcox, 11–33. Sunderland, MA: Sinauer.

Godfray, H. C. J. 1994. *Parasitoids.* Princeton, NJ: Princeton University Press.

Goebel, R., E. Fernandez, J. M. Begue, and C. Alauzet. 1999. Predation by *Pheidole megacephala* (Fabricius) (Hym: Formicidae) on eggs of the sugarcane Stern borer *Chilo sacchariphagus* (Bojer) (Lep.: Pyralidae) in Reunion Island. *Annales de la Société Entomologique de France* 35: 440–42.

Gómez, J. M. 2000. Effectiveness of ants as pollinators of *Lobularia maritima:* Effects on main sequential fitness components of the host plant. *Oecologia* 122:90–97.

Gómez, J. M., and R. Zamora. 1992. Pollination by ants: Consequences of the quantitative effects on a mutualistic system. *Oecologia* 91:410–18.

Gómez, J. M., R. Zamora, J. A. Hódar, and D. García. 1996. Experimental study of pollination by ants in Mediterranean high mountain and arid habitats. *Oecologia* 105:236–42.

Gorb, E. V., and S. N. Gorb. 2003. *Seed dispersal by ants in a deciduous forest ecosystem: Mechanisms, strategies, adaptations.* The Hague: Kluwer.

Gorb, S. N., and E. V. Gorb. 1995. Removal rates of seeds of five myrmecochorous plants by the ant *Formica polyctena* (Hymenoptera: Formicidae). *Oikos* 73:367–74.

Gorb, S. N., E. V. Gorb, and P. Punttila. 2000. Effects of redispersion of seeds by ants on the vegetation pattern in a deciduous forest: A case study. *Acta Oecologica* 21:293–301.

Gosswald, K. 1990. *Die Waldameise im Oekosystem Wald, ihr Nutzen und ihre Hege*. Vol. 2 of *Die Waldameise*. Wiesbaden: Aula-Verlag.

Gotwald, W. H., Jr. 1986. The beneficial economic role of ants. In *Economic impact and control of social insects,* ed. S. B. Vinson, 290–313. New York: Praeger Scientific.

Gove, A. D., J. D. Majer, and V. Rico-Gray. 2005. Methods for conservation outside of formal reserve systems: The case of ants in the seasonally dry tropics of Veracruz, Mexico. *Biological Conservation* 126:328–38.

Gove, A. D., and V. Rico-Gray. 2006. What determines conditionality in ant-Hemiptera interactions? Hemiptera habitat preference and the role of local ant activity. *Ecological Entomology* 31:568–79.

Graham, S. W., and R. G. Olmstead. 2000. Utility of 17 chloroplast genes for inferring the phylogeny of the basal angiosperms. *American Journal of Botany* 87:1712–30.

Green, G. W., and C. R. Sullivan. 1950. Ants attacking larvae of the forest tent caterpillar, *Malacosoma disstria* Hbn. (Lepidoptera: Lasiocampidae). *Canadian Entomologist* 82:194–95.

Greenberg, R., C. Macias-Caballero, and P. Bichier. 1993. Defense of homopteran honeydew by birds in the Mexican highlands and other warm temperate forests. *Oikos* 68:519–24.

Grimaldi, D. 1999. The co-radiation of pollinating insects and angiosperms in the Cretaceous. *Annals of the Missouri Botanical Garden* 86:373–406.

Grimaldi, D., and D. Agosti. 2000. A formicine in New Jersey Cretaceous amber (Hymenoptera: Formicidae) and early evolution of the ants. *Proceedings of the National Academy of Sciences* 97:13678–83.

Grimaldi, D., D. Agosti, and J. M. Carpenter. 1997. New and rediscovered primitive ants (Hymenoptera: Formicidae) in Cretaceous amber from New Jersey, and their phylogenetic relationships. *American Museum Novitates* 3208:1–43.

Grimaldi, D., and M. S. Engel. 2005. *Evolution of the insects.* Cambridge: Cambridge University Press.

Guerrant, E. O., and P. L. Fiedler. 1981. Flower defenses against nectar-pilferage by ants. *Biotropica* (suppl.) 13:25–33.

Guimarães, P. R., Jr., M. A. M. de Aguiar, J. Bascompte, P. Jordano, and S. F. dos Reis. 2005. Random initial condition in small Barabasi-Albert networks and deviations from the scale-free behavior. *Physical Review E* 71:037101.

Guimarães, P. R., Jr., and R. Cogni. 2002. Seed cleaning of *Cupania vernalis* (Sapindaceae) by ants: Edge effect in a highland forest in south-east Brazil. *Journal of Tropical Ecology* 18:303–7.

Guimarães, P. R., Jr., V. Rico-Gray, S. F. Reis, and J. N. Thompson. 2006. Asymmetries in specialization in ant-plant mutualistic networks. *Proceedings of the Royal Society of London, Series B*, 273:2041–47.

Gullan, P. J., R. C. Buckely, and P. S. Ward. 1993. Ant-tended scale insects (Hemiptera: Coccidae: *Myzolecanium*) within lowland rain forest trees in Papua New Guinea. *Journal of Tropical Ecology* 9:81–91.

Gunther, R. W., and J. Lanza. 1989. Variation in attractiveness of *Trillium* diaspores to a seed-dispersing ant. *American Midland Naturalist* 122:321–28.

Haber, W. A., G. W. Frankie, H. G. Baker, I. Baker, and S. Koptur. 1981. Ants like flower nectar. *Biotropica* 13:211–14.
Haines, B. 1975. Impact of leaf-cutting ants on vegetation development at Barro Colorado Island. In *Tropical ecological systems,* ed. F. B. Golley and E. Medina, 99–111. Berlin: Springer-Verlag.
———. 1988. Element and energy flows through colonies of the leaf-cutting ant *Atta colombica* in Panama. *Biotropica* 10:270–77.
Hallwacks, W. 1986. Agoutis (*Dasyprocta punctata*), the inheritors of guapinol (*Hymenaea courbaril:* Leguminosae). In Estrada and Fleming 1986, 285–305.
Handel, S. N. 1976. Dispersal ecology of *Carex pedunculata* (Cyperaceae), a new North American myrmecochore. *American Journal of Botany* 63:1071–79.
———. 1978. The competitive relationship of three woodland sedges and its bearing on the evolution of ant-dispersal of *Carex pedunculata. Evolution* 32:151–63.
Handel, S. N., and A. J. Beattie. 1990. Seed dispersal by ants. *Scientific American* 263:76–83.
Handel, S. N., S. B. Fisch, and G. E. Schatz. 1981. Ants disperse a majority of herbs in a mesic forest community in New York state. *Bulletin of the Torrey Botanical Club* 108:430–37.
Harper, J. L. 1977. *Population biology of plants.* London: Academic Press.
Hassell, M. P., H. N. Comins, and R. M. May. 1991. Spatial structure and chaos in insect population dynamics. *Nature* 353:255–58.
Hassell, M. P., and R. M. May. 1974. Aggregation in predators and insect parasites and its effect on stability. *Journal of Animal Ecology* 43:567–94.
Haynes, I. H., and J. B. Haynes. 1978a. Colony structure, seasonality, and food requirements of the crazy ant, *Anoplolepis longipes* (Jerd.), in the Seychelles. *Ecological Entomology* 3:109–18.
———. 1978b. Pest status of the crazy ant, *Anoplolepis longipes* (Jerdon), (Hymenoptera: Formicidae) in the Seychelles. *Bulletin of Entomological Research* 68:627–38.
Heads, P. A., and J. H. Lawton. 1984. Bracken, ants, and extrafloral nectaries. II. The effect of ants on the insect herbivores of bracken. *Journal of Animal Ecology* 53:1015–32.
———. 1985. Bracken, ants, and extrafloral nectaries. III. How insect herbivores avoid predation. *Ecological Entomology* 10:29–42.
Heckroth, H.-P., B. Fiala, P. J. Gullan, A. H. J. Idris, and U. Maschwitz. 1998. The soft scale (Coccidae) associates of Malaysian ant-plants. *Journal of Tropical Ecology* 14:427–43.
Heil, M. 1999. Systemic acquired resistance: Available information and open ecological questions. *Journal of Ecology* 87:341–46.
———. 2004. Induction of two indirect defences benefits Lima bean (*Phaseolus lunatus,* Fabaceae) in nature. *Journal of Ecology* 92:527–36.
Heil, M., B. Baumann, C. Andary, K. E. Linsenmair, and D. McKey. 2002. Extraction and quantification of "condensed tannins" as a measure of plant anti-herbivore defence? Revisiting an old problem. *Naturwissenschaften* 89: 519–24.

Heil, M., B. Baumann, R. Kruger, and K. E. Linsenmair. 2004. Main nutrient compounds in food bodies of Mexican *Acacia* ant-plants. *Chemoecology* 14:45–82.

Heil, M., B. Fiala, W. Kaiser, and K. E. Linsenmair. 1998. Chemical contents of *Macaranga* food bodies: Adaptations to their role in ant attraction and nutrition. *Functional Ecology* 12:117–22.

Heil, M., B. Fiala, K. E. Linsenmair, G. Zotz, P. Menke, and U. Maschwitz. 1997. Food body production in *Macaranga triloba* (Euphorbiaceae): A plant investment in anti-herbivore defence via symbiotic ant partners. *Journal of Ecology* 85:847–61.

Heil, M., B. Fiala, U. Maschwitz, and K. E. Linsenmair.2001. On benefits of indirect defence: Short- and long-term studies of antiherbivore protection via mutualistic ants. *Oecologia* 126:395–403.

Heil, M., S. Greiner, H. Meimberg, R. Kruger, J.-L. Noyer, G. Heubl, K. E. Linsenmair, and W. Boland. 2004. Evolutionary change from induced to constitutive expression of an indirect plant resistance. *Nature* 430:205–8.

Heil, M., A. Hilpert, B. Fiala, R. Bin Hashim, E. Strohm, G. Zotz, and K. E. Linsenmair. 2002. Nutrient allocation of *Macaranga triloba* ant plants to growth, photosynthesis, and indirect defence. *Functional Ecology* 16:475–83.

Heil, M., A. Hilpert, B. Fiala, and K. E. Linsenmair. 2001. Nutrient availability and indirect (biotic) defence in a Malaysian ant-plant. *Oecologia* 126:404–8.

Heil, M., T. Koch, A. Hilpert, B. Fiala, W. Boland, and K. E. Linsenmair. 2001. Extrafloral nectar production of the ant-associated plant, *Macaranga tanarius,* is an induced, indirect, defensive response elicited by jasmonic acid. *Proceedings of the National Academy of Sciences* 98:1083–88.

Heil, M., and D. McKey. 2003. Protective ant-plant interactions as model systems in ecological and evolutionary research. *Annual Review of Ecology and Systematics* 34:425–53.

Heil, M., J. Rattke, and W. Boland. 2005. Postsecretory hydrolysis of nectar sucrose and specialization in ant/plant mutualism. *Science* 308:560–63.

Heithaus, E. R. 1981. Seed predation by rodents on three ant-dispersed plants. *Ecology* 62:136–45.

Hendrix, D. L., Y. Wei, and J. E. Leggerr. 1992. Homopteran honeydew sugar composition is determined by both the insect and plant species. *Comparative Biochemistry and Physiology* 101B:23–27.

Hendrix, S. D. 1977. The resistance of *Pteridium aquilinum* (L.) Kuhn to insect attack by *Trichoplusia ni* (Hubn.). *Oecologia* 26:347–61.

———. 1980. An evolutionary and ecological perspective of the insect fauna of ferns. *American Naturalist* 115:171–96.

Hendrix, S. D., and R. J. Marquis. 1983. Herbivore damage to three tropical ferns. *Biotropica* 15:108–11.

Herbers, J. M. 1989. Community structure in north temperate ants: Temporal and spatial variation. *Oecologia* 81:201–11.

Herrera, C. M. 1982. Season variation in the quality of fruits and diffuse coevolution between plants and avian dispersers. *Ecology* 63:773–85.

———. 1984. A study of avian frugivores, bird-dispersed plants, and their interaction in Mediterranean scrublands. *Ecological Monographs* 54:1–23.

———. 2000. Measuring the effects of pollinators and herbivores: Evidence for non-additivity in a perennial herb. *Ecology* 81:2170–76.

———. 2005. Plant generalization on pollinators: Species property or local phenomenon? *American Journal of Botany* 92:13–20.

Herrera, C. M., J. Herrera, and X. Espadaler. 1984. Nectar thievery by ants from southern Spanish insect-pollinated flowers. *Insectes Sociaux* 31:142–54.

Herrera, C. M., P. Jordano, L. López-Soria, and J. A. Amat. 1994. Recruitment of a mast-fruiting, bird-dispersed tree: Bridging frugivore activity and seedling establishment. *Ecological Monographs* 64:315–44.

Herrera, C. M., and O. Pellmyr, eds. *Plant animal interactions: An evolutionary approach*. Oxford: Blackwell Science.

Hespenheide, H. A. 1985. Insect visitors to extrafloral nectaries of *Byttneria aculeata* (Sterculiaceae): Relative importance and roles. *Ecological Entomology* 10:191–204.

Hickman, J. C. 1974. Pollination by ants: A low-energy system. *Science* 184: 1290–92.

Higashi, S., and F. Ito. 1991. Ground beetles and seed dispersal of the myrmecochorous plant *Trillium tschonoskii* (Trillaceae). In Huxley and Cutler 1991, 486–92.

Hinkle, G., J. K. Wetterer, T. R. Schultz, and M. L. Sogin. 1994. Phylogeny of the Attine ant fungi based on analysis of small subunit ribosomal RNA gene sequences. *Science* 266:1695–97.

Holland, J. N., J. H. Ness, A. Boyle, and J. L. Bronstein. 2005. Mutualisms as consumer-resource interactions. In *Ecology of predator-prey interactions,* ed. P. Barbosa and I. Castellanos, 17–35. Oxford: Oxford University Press.

Hölldobler, B. 1983. Territorial behavior of the green tree ant (*Oecophylla longinoda*). *Biotropica* 15:241–50.

———. 1987. Communication and competition in ant communities. In *Evolution and coadaptation in ant communities,* ed. S. Kawano, J. H. Connell, and T. Hidaka, 95–124. Tokyo: Tokyo University Press.

Hölldobler, B., and C. J. Lumsden. 1980. Territorial strategies in ants. *Science* 210:732–39.

Hölldobler, B., and E. O. Wilson. 1990. *The ants.* Cambridge, MA: Harvard University Press.

Horvitz, C. C. 1981. Analysis of how ant behaviors affect germination in a tropical myrmecochore *Calathea microcephala* (P. and E.) Koernicke (Marantaceae): Microsite selection and aril removal by neotropical ants, *Odontomachus, Pachycondyla,* and *Solenopsis* (Formicidae). *Oecologia* 51:47–52.

———. 1991. Light environments, stage structure, and dispersal syndromes of Costa Rican Marantaceae. In Huxley and Cutler 1991, 463–85.

Horvitz, C. C., and A. J. Beattie. 1980. Ant dispersal of *Calathea* (Marantaceae) seeds by carnivorous ponerines (Formicidae) in a tropical rain forest. *American Journal of Botany* 67:321–26.

Horvitz, C. C., and J. LeCorff. 1993. Spatial scale and dispersion pattern of ant- and bird-dispersed herbs in two tropical lowland rain forests. *Vegetatio* 107–8:351–62.

Horvitz, C. C., M. A. Pizo, B. Bello y Bello, J. LeCorff, and R. Dirzo. 2002. Are plant species that need gaps for recruitment more attractive to seed-dispersing birds and ants than other species? In *Seed dispersal and frugivory: Ecology, evolution, and conservation,* ed. D. J. Levey, W. R. Silva, and M. Galetti, 145–59. Wallingford, UK: CABI Publishing.

Horvitz, C. C., and D. W. Schemske. 1984. Effects of ants and an ant-tended herbivore on seed production of a neotropical herb. *Ecology* 65:1369–78.

———. 1986a. Ant-nest soil and seedling growth in a neotropical ant-dispersed herb. *Oecologia* 70:318–20.

———. 1986b. Seed dispersal of a neotropical myrmecochore: Variation in removal rates and dispersal distance. *Biotropica* 18:319–23.

———. 1990. Spatiotemporal variation in insect mutualists of a neotropical herb. *Ecology* 71:1085–97.

Hossaert-McKey, M., J. Orivel, E. Labeyrie, L. Pascal, J. H. C. Delabie, and A. Dejean. 2001. Differential associations with ants of three co-occurring extrafloral nectary-bearing plants. *Ecoscience* 8:325–35.

Howard, J. J. 1991. Resource quality and cost in the foraging of leaf-cutter ants. In Huxley and Cutler 1991, 42–50.

Howe, H. F. 1980. Monkey dispersal and waste of a neotropical fruit. *Ecology* 61:944–59.

———. 1989. Scatter- and clump-dispersal and seedling demography: Hypothesis and implications. *Oecologia* 79:417–26.

Howe, H. F., and E. W. Schupp. 1985. Early consequences of seed dispersal for a neotropical tree (*Virola surinamensis*). Ecology 66:781–91.

Howe, H. F., and J. Smallwood. 1982. Ecology of seed dispersal. *Annual Review of Ecology and Systematics* 13:201–28.

Howe, H. F., and L. C. Westley. 1988. *Ecological relationships of plants and animals.* New York: Oxford University Press.

Huang, H. T., and P. Yang. 1987. The ancient cultured citrus ant. *BioScience* 37:665–71.

Hubbell, S. P., J. J. Howard, and D. F. Wiemer. 1984. Chemical leaf repellancy to an attine ant: Seasonal distribution among potential host plant species. *Ecology* 65:1067–76.

Hubbell, S. P., D. F. Wiemer, and A. Adejare. 1983. An antifungal terpenoid defends a neotropical tree (*Hymenaea*) against attack by fungus-growing ants (*Atta*). *Oecologia* 60:321–27.

Hughes, L., and M. Westoby. 1990. Removal rates of seeds adapted for dispersal by ants. *Ecology* 71:138–48.

———. 1992a. Capitula on stick insect eggs and elaiosomes on seeds: Convergent adaptations for burial by ants. *Functional Ecology* 6:642–48.

———. 1992b. Effect of diaspore characteristics on removal of seeds adapted for dispersal by ants. *Ecology* 73:1300–1312.

———. 1992c. Fate of seeds adapted for dispersal by ants in Australian sclerophyll vegetation. *Ecology* 73:1285–99.

Hughes, L., M. Westoby, and A. D. Johnson. 1993. Nutrient costs of vertebrate- and ant-dispersed fruits. *Functional Ecology* 7:54–62.

Hughes, L., M. Westoby, and E. Jurado. 1994. Convergence of elaiosomes and insect prey: Evidence from ant foraging behavior and fatty acid composition. *Functional Ecology* 8:358–65.

Hughes, S. J. 1976. Sooty moulds. *Mycologia* 68:693–820.

Hull, D. A., and A. J. Beattie. 1988. Adverse effects on pollen exposed to *Atta texana* and other North American ants: Implications for ant pollination. *Oecologia* 75:153–55.

Huston, M. A. 1994. *Biological diversity.* Cambridge: Cambridge University Press.

Huxley, C. R. 1978. The ant-plants *Myrmecodia* and *Hydnophytum* (Rubiaceae), and the relationships between their morphology, ant occupants, physiology, and ecology. *New Phytologist* 80:231–68.

———. 1980. Symbiosis between ants and epiphytes. *Biological Review* 55: 321–40.

———. 1986. Evolution of benevolent ant-plant relationships. In Juniper and Southwood 1986, 257–82.

Huxley, C. R., and D. F. Cutler, eds. 1991. *Ant-plant interactions.* Oxford: Oxford University Press.

Ibarra-Manríquez, G., and R. Dirzo. 1990. Plantas mirmecófilas arbóreas de la estacion de biología "Los Tuxtlas," Veracruz, México. *Revista de Biología Tropical* 38:79–82.

Inouye, B. D., and A. A. Agrawal. 2004. Ant mutualists alter the composition and attack rate of the parasitoid community for the gall wasp *Disholcaspis eldoradensis* (Cynipidae). *Ecological Entomology* 29:692–96.

Inouye, D. W., and O. R. Taylor. 1979. A temperate region plant-ant-seed predator system: Consequences of extrafloral nectar secretion by *Helianthella quinquenervis. Ecology* 60:1–7.

Itino, T., and T. Itioka. 2001. Interspecific variation and ontogenetic change in antiherbivore defense in myrmecophytic *Macaranga* species. *Ecological Research* 16:765–74.

Itino, T., T. Itioka, A. Hatada, and A. A. Hamid. 2001. Effects of food rewards offered by ant-plant *Macaranga* on the colony size of ants. *Ecological Research* 16:775–86.

Itioka, T., and T. Inoue. 1996. Density-dependent ant attendance and its effects on the parasitism of a honeydew-producing scale insect, *Ceroplastes rubens. Oecologia* 105:448–54.

Itioka, T., M. Nomura, Y. Iuni, T. Itino, and T. Inoue. 2000. Difference in intensity of ant defense among three species of *Macaranga* myrmecophytes in a southeast Asian dipterocarp forest. *Biotropica* 32:318–26.

Izzo, T. J., and H. L. Vasconcelos. 2002. Cheating the cheater: Domatia loss minimizes the effects of ant castration in an Amazonian ant-plant. *Oecologia* 133:200–205.

Jackson, D. A. 1984. Ant distribution patterns in a Cameroonian cocoa plantation: Investigation of the ant mosaic hypothesis. *Oecologia* 62:318–24.

Jaffe, K., C. Pavis, G. Vansuyt, and A. Kermarrec. 1989. Ants visit extrafloral nectaries of the orchid *Spathoglotis plicata* Blume. *Biotropica* 21:278–79.

Jaffe, K., and E. Vilela. 1989. On nest densities of the leaf-cutting ant *Atta cephalotes* in tropical primary forest. *Biotropica* 21:234–36.

Jahn, G. C., and J. W. Beardsley. 1996. Effects of *Pheidole megacephala* (Hymenoptera: Formicidae) on survival and dispersal of *Dysmicoccus neobrevipis* (Homoptera: Pseudococcidae). *Journal of Economic Entomology* 89: 1124–29.

Janzen, D. H. 1966. Coevolution of mutualism between ants and acacias in Central America. *Evolution* 20:249–75.

———. 1967a. Fire, vegetation structure, and the ant x *Acacia* interaction in Central America. *Ecology* 48:26–35.

———. 1967b. Interaction of the bull's-horn acacia (*Acacia cornigera* L.) with an ant inhabitant (*Pseudomyrmex ferruginea* F. Smith) in eastern Mexico. *University of Kansas Scientific Bulletin* 47:315–558.

———. 1969a. Allelopathy by myrmecophytes: The ant *Azteca* as an allelopathic agent of *Cecropia. Ecology* 50:147–53.

———. 1969b. Birds and the ant x *Acacia* interaction in Central America, with notes on birds and other myrmecophytes. *Condor* 71:240–56.

———. 1970. Herbivores and the number of tree species in tropical forests. *American Naturalist* 104:501–29.

———. 1971. Seed predation by animals. *Annual Review of Ecology and Systematics* 2:465–92.

———. 1972. Protection of *Barteria* (Passifloraceae) by *Pachysima* ants (Pseudomyrmecinae) in a Nigerian rain forest. *Ecology* 53:885–92.

———. 1973a. Dissolution of a mutualism between *Cecropia* and its *Azteca* ants. *Biotropica* 5:15–28.

———. 1973b. Evolution of polygynous obligate *Acacia*-ants in western Mexico. *Journal of Animal Ecology* 42:727–50.

———. 1973c. Sweep samples of tropical foliage insects: Effects of seasons, vegetation types, elevation, time of day, and insularity. *Ecology* 54:687–701.

———. 1974. Epiphytic myrmecophytes in Sarawak: Mutualism through the feeding of plants by ants. *Biotropica* 6:237–59.

———. 1975. *Pseudomyrmex nigropilosa:* A parasite of a mutualism. *Science* 188:936–37.

———. 1977. Why don't ants visit flowers? *Biotropica* 9:252.

———. 1980. When is it coevolution? *Evolution* 34:611–12.

———. 1983. Seed and pollen dispersal by animals: Convergence in the ecology of contamination and sloppy harvest. *Biological Journal of the Linnean Society* 20:103–13.

Janzen, D. H., M. Ataroff, M. Farinas, S. Reyes, N. Rincon, A. Soler, P. Soriano, and M. Vera. 1976. Changes in the arthropod community along an elevational transect in the Venezuelan Andes. *Biotropica* 8:193–203.

Jeanne, R. L. 1979. A latitudinal gradient in rates of ant predation. *Ecology* 60:1211–24.

Jeffrey, D. C., J. Arditti, and H. Koopowitz. 1970. Sugar content in floral and extrafloral exudates of orchids: Pollination, myrmecology, and chemotaxonomy implication. *New Phytologist* 69:187–95.

Jervis, M. A., and N. A. C. Kidd, eds. 1996. *Insect natural enemies: Practical approaches to their study and evaluation.* London: Chapman and Hall,.

Jirón, L. F., and S. Salas. 1975. Simbiosis entre "cochinillas de cola" (Coccoidea: Margarodidae) y otros insectos. 1. Los componentes del sistema simbiótico en la tierra alta de Costa Rica. *Brenesia* 5:67–71.

Jolivet, P. 1986. *Les fourmis et les plantes.* Paris: Boubée.

Jones, C. G., and R. D. Firm. 1979. Resistance of *Pteridium aquilinum* to attack by non-adapted phytophagous insects. *Biochemical Systematics and Ecology* 7:95–101.

Jones, D., and W. L. Sterling. 1979. Manipulation of red imported fire ants in a trap crop for boll weevil suppression. *Environmental Entomology* 8:1073–77.

Jones, D. L. 1975. Pollination of *Microtis parviflora* R.Br. *Annals of Botany* 39:585–89.

Jordano, D., J. Rodríguez, C. D. Thomas, and J. Fernández-Haeger. 1992. The distribution and density of a lycaenid butterfly in relation to *Lasius* ants. *Oecologia* 91:439–46.

Jordano, D., and C. D. Thomas. 1992. Specificity of an ant-lycaenid interaction. *Oecologia* 91:431–38.

Jordano, P. 1982. Migrant birds are the main dispersers of blackberries in southern Spain. *Oikos* 38:183–93.

———. 1987. Patterns of mutualistic interactions in pollination and seed dispersal: Connectance, dependence asymmetries, and coevolution. *American Naturalist* 129:657–77.

———. 1989. Pre-dispersal biology of *Psitacia lentiscus* (Anacardiaceae): Cumulative effects on seed removal by birds. *Oikos* 55:375–86.

———. 1993. Fruits and frugivory. In *Seeds: The ecology of regeneration in plant communities,* ed. M. Fenner, 105–56. Wallingford, UK: CAB International.

Jordano, P., J. Bascompte, and J. L. Olesen. 2003. Invariant properties in coevolutionary networks of plant-animal interactions. *Ecology Letters* 6:69–81.

Juenger, T., and J. Bergelson. 1998. Pairwise versus diffuse natural selection and the multiple herbivores of Scarlet gilia, *Ipomopsis aggregata. Evolution* 52:1583–92.

Jules, E. S. 1996. Yellow jackets (*Vespula vulgaris*) as a second seed disperser for the myrmecochorous plant, *Trillium ovatum. American Midland Naturalist* 135:367–69.

Juniper, B., and R. Southwood, eds. 1986. *Insects and the plant surface.* London: Edward Arnold.

Jutsum, A. R., J. M. Cherrett, and M. Fisher. 1981. Interactions between the fauna of citrus trees in Trinidad and the ants *Atta cephalotes* and *Azteca* sp. *Journal of Applied Ecology* 18:187–95.

Kacelnik, A. 1993. Leaf-cutting ants tease optimal foraging theorists. *Trends in Ecology and Evolution* 8:346–48.

Kaiser, J. 1998. Of mice and moths—and Lyme disease? *Science* 279:984–85.

Kalisz, S., F. M. Hanzawa, S. J. Tonsor, D. A. Thiede, and S. Voigt. 1999. Ant-mediated seed dispersal alters pattern of relatedness in a population of *Trillium grandiflorum. Ecology* 80:2620–34.

Karban, R., and Y. T. Baldwin. 1997. *Induced responses to herbivory.* Chicago: University of Chicago Press.

Karhu, K. J., and S. Neuvonen. 1998. Wood ants and a geometrid defoliator of birch: Predation outweighs beneficial effects through the host plant. *Oecologia* 113:509–16.

Kaspari, M. 1993. Removal of seeds from neotropical frugivore droppings: Ant responses to seed number. *Oecologia* 95:81–88.

———. 1996. Testing resource-based models of patchiness in four neotropical litter ant assemblages. *Oikos* 76:443–54.

Kaspari, M., M. Yuan, and L. Alonso. 2003. Spatial grain and causes of regional diversity gradients in ants. *American Naturalist* 161:459–77.

Katayama, N., and N. Suzuki. 2003. Changes in the use of extrafloral nectaries of *Vicia faba* (Leguminosae) and honeydew of aphids by ants with increasing aphid density. *Annals of the Entomological Society of America* 96: 579–84.

———. 2004. Role of extrafloral nectaries of *Vicia faba* in attraction of ants and herbivore exclusion of ants. *Entomological Science* 7:119–24.

Kaufmann, S., D. McKey, M. Hossaert-McKey, and C. C. Horvitz. 1991. Adaptations for a two-phase seed dispersal system involving vertebrates and ants in a hemiepiphytic fig (*Ficus microcarpa:* Moraceae). *American Journal of Botany* 78:971–77.

Kawano, S., H. Azuma, M. Ito, and K. Suzuki. 1999. Extrafloral nectaries and chemical signals of *Fallopia japonica* and *Fallopia sachalinensis* (Polygonaceae), and their roles as defense systems against insect herbivory. *Plant Species Biology* 14:167–78.

Keeler, K. H. 1977. The extrafloral nectaries of *Ipomoea carnea* (Convolvulaceae). *American Journal of Botany* 64:1182–88.

———. 1979a. Distribution of plants with extrafloral nectaries and ants at two elevations in Jamaica. *Biotropica* 11:152–54.

———. 1979b. Distribution of plants with extrafloral nectaries in a temperate flora (Nebraska). *Prairie Naturalist* 11:33–37.

———. 1980a. Distribution of plants with extrafloral nectaries in temperate communities. *American Midland Naturalist* 104:274–80.

———. 1980b. The extrafloral nectaries of *Ipomoea leptophylla* (Convolvulaceae). *American Journal of Botany* 67:216–22.

———. 1981a. Cover of plants with extrafloral nectaries at four northern California sites. *Madroño* 28: 26–29.

———. 1981b. Function of *Mentzelia nuda* (Loasaceae) postfloral nectaries in seed defense. *American Journal of Botany* 68:295–99.

———. 1981c. Infidelity by *Acacia*-ants. *Biotropica* 13:79–80.

———. 1981d. A model of selection for facultative nonsymbiotic mutualism. *American Naturalist* 118:488–98.

———. 1985. Extrafloral nectaries on plants in communities without ants: Hawaii. *Oikos* 44:407–14.

———. 1989. Ant-plant interactions. In *Plant-animal interactions,* ed. W. G. Abrahamson, 207–42. New York: McGraw-Hill.

Kelly, C. A. 1986. Extrafloral nectaries: Ants, herbivores, and fecundity in *Cassia fasciculata. Oecologia* 69:600–605.

Kelly, C. A., and R. J. Dyer. 2002. Demographic consequences of inflorescence-feeding insects for *Liatris cylindracea,* an iteroparous perennial. *Oecologia* 132:350–60.

Kersh, M. F., and C. R. Fonseca. 2005. Abiotic factors and the conditional outcome of an ant-plant mutualism. *Ecology* 86:2117–26.

Kiester, A. R., R. Lande, and D. W. Schemske. 1984. Models of coevolution and speciation in plants and their pollinators. *American Naturalist* 124:220–43.

King, T. J. 1976. The viable seed contents of ant-hill and pasture soil. *New Phytologist* 77:143–47.

———. 1977a. The plant ecology of ant-hills in calcareous grasslands. I. Patterns of species in relation to ant-hills in southern England. *Journal of Ecology* 65:235–56.

———. 1977b. The plant ecology of ant-hills in calcareous grasslands. II. Succession on the mounds. *Journal of Ecology* 65:257–78.

———. 1977c. The plant ecology of ant-hills in calcareous grasslands. III. Factors affecting the population sizes of selected species. *Journal of Ecology* 65:279–315.

Kleinfeldt, S. E. 1978. Ant-gardens: The interaction of *Codonanthe crassifolia* (Gesneriaceae) and *Crematogaster longispina* (Formicidae). *Ecology* 59: 449–56.

———. 1986. Ant-gardens: Mutual exploitation. In Juniper and Southwood 1986, 283–94.

Klink, C. A., and A. G. Moreira. 2002. Past and current human occupation, and land use. In *The Cerrados of Brazil: Ecology and Natural History of a Neotropical Savanna,* ed. P. S. Oliveira and R. J. Marquis, 69–88. New York: Columbia University Press.

Kluge, R. L. 1991. Biological control of triffid weed, *Chromolaena odorata* (Asteraceae), in South Africa. *Agriculture, Ecosystems, and Environment* 37:193–97.

Knoch, T. R., S. H. Faeth, and D. L. Arnott. 1993. Endophytic fungi alter foraging and dispersal by desert seed-harvesting ants. *Oecologia* 95:470–73.

Knox, R. B., R. Marginson, J. Kenrick, and A. J. Beatie. 1986. The role of extrafloral nectaries in *Acacia.* In Juniper and Southwood 1986, 295–307.

Koptur, S. 1979. Facultative mutualism between weedy vetches bearing extrafloral nectaries and weedy ants in California. *American Journal of Botany* 66:1016–20.

———. 1984. Experimental evidence for defense of *Inga* saplings (Mimosoideae) by ants. *Ecology* 65:1787–93.

———. 1985. Alternative defenses against herbivory in *Inga* (Fabaceae: Mimosoideae) over an elevational gradient. *Ecology* 66:1639–50.

———. 1989. Is extrafloral nectar production an inducible defense? In *The evolutionary ecology of plants,* ed. J. H. Block and Y. B. Linhart, 323–39. Boulder: Westview Press.

———. 1991. Extrafloral nectaries of herbs and trees: Modelling the interaction with ants and parasitoids. In Huxley and Cutler 1991, 213–30.

———. 1992a. Extrafloral nectary-mediated interactions between insects and plants. In *Insect-plant interactions,* ed. E. Bernays, 4:81–129. Boca Raton, FL: CRC Press.

———. 1992b. Plants with extrafloral nectaries and ants in Everglades habitats. *Florida Entomologist* 75:39–50.

———. 1996. Extrafloral nectaries in California plants. *Fremontia* 24:23–26.

Koptur, S., and J. H. Lawton. 1988. Interactions among vetches bearing extrafloral nectaries, their biotic protective agents, and herbivores. *Ecology* 69:278–83.

Koptur, S., V. Rico-Gray, and M. Palacios-Rios. 1998. Ant protection of the nectaried fern *Polypodium plebeium* in central Mexico. *American Journal of Botany* 85:736–39.

Koptur, S., A. R. Smith, and I. Baker. 1982. Nectaries in some neotropical species of *Polypodium* (Polypodiaceae): Preliminary observations and analyses. *Biotropica* 14:108–13.

Koptur, S., and N. Truong. 1998. Facultative ant-plant interactions: Nectar sugar preferences of introduced pest ant species in south Florida. *Biotropica* 30:179–89.

Kost, C., and M. Heil. 2005. Increased availability of extrafloral nectar reduces herbivory in lima bean plants (*Phaseolus lunatus,* Fabaceae). *Basic and Applied Ecology* 6:237–48.

———. 2006. Herbivore-induced plant volatiles induce an indirect defence in neighbouring plants. *Journal of Ecology* 94:619–28.

Kraaijeveld, A. R., and H. C. J. Godfray. 1999. Geographic patterns in the evolution of resistance and virulence in *Drosophila* and its parasitoids. *American Naturalist* 153:S61–S74.

Krombein, K. V. 1951. Wasp visitors of tuliptree honeydew at Dunn Loring, Virginia. *Annals of the Entomological Society of America* 44:141–43.

Kusnezov, N. 1957. Number of species of ants in faunae of different latitudes. *Evolution* 11:298–99.

Labandeira, C. C. 2002. The history of associations between plants and animals. In Herrera and Pellmyr 2002, 26–74.

Labandeira, C. C., and J. J. Sepkoski Jr. 1993. Insect diversity in the fossil record. *Science* 261:310–15.

Labeyrie, E., L. Pascal, J. H. C. Delabie, J. Orivel, A. Dejean, and M. Hossaert-Mckey. 2001. Protection of *Passiflora glandulosa* (Passifloraceae) against herbivory: Impact of ants exploiting extrafloral nectaries. *Sociobiology* 38: 317–21.

Lachaud, J. P. 1990. Foraging activity and diet in some neotropical ponerine ants. I. *Ectatomma ruidum* Roger (Hymenoptera: Formicidae). *Acta Entomológica Mexicana* 78:241–56.

Laman, T. G. 1996a. *Ficus* seed shadow in a Bornean rainforest. *Oecologia* 107:347–55.

———. 1996b. The impact of seed harvesting ants (*Pheidole* sp. nov.) on *Ficus* establishment in the canopy. *Biotropica* 28:777–81.

Lanza, J. 1988. Ant preferences for *Passiflora* nectar mimics that contain amino acids. *Biotropica* 20:341–44.

Lanza, J., M. A. Schmitt, and A. B. Awad. 1992. Comparative chemistry of elaiosomes of three species of *Trillium*. *Journal of Chemical Ecology* 18: 209–21.

Lapola, D. M., E. M. Bruna, C. Granara de Willink, and H. L. Vasconcelos. 2005. Ant-tended Hemiptera in Amazonian myrmecophytes: Patterns of abundance and implications for mutualism function. *Sociobiology* 46:1–10.

Lapola, D. M., E. M. Bruna, and H. L. Vasconcelos. 2003. Contrasting responses to induction cues by ants inhabiting *Maieta guianensis* (Melastomataceae) *Biotropica* 35:295–300.

Lawton, J. H. 1976. The structure of the arthropod comunity on bracken (*Pteridium aquilinum* (L.) Kuhn). *Botanical Journal of the Linnean Society* 73:187–216.

———. 1996. Patterns in ecology. *Oikos* 75:145–47.

Lawton, J. H., and P. A. Heads. 1984. Bracken, ants, and extrafloral nectaries. I. The components of the system. *Journal of Animal Ecology* 53:995–1014.

Lawton, J. H., M. MacGarvin, and P. A. Heads. 1987. Effects of altitude on the abundance and species richness of insect herbivores on bracken. *Journal of Animal Ecology* 56:147–60.

Lawton, J. H., and D. R. Strong Jr. 1981. Community patterns and competition in folivorous insects. *American Naturalist* 118:317–38.

Leal, I. R. 2003. Dispersão de sementes por formigas na caatinga. In *Ecologia e conservação da caatinga,* ed. I. R. Leal, M. Tabarelli, and J. M. C. Silva, 593–624. Recife, Brazil: Editora da Universidade Federal de Pernambuco.

Leal, I. R., and P. S. Oliveira. 1998. Interactions between fungus-growing ants (Attini), fruits, and seeds in cerrado vegetation in southeast Brazil. *Biotropica* 30:170–78.

———. 2000. Foraging ecology of attine ants in a neotropical savanna: Seasonal use of fungal substrate in the cerrado vegetation of Brazil. *Insectes Sociaux* 47:376–82.

LeCorff, J., and C. C. Horvitz. 1995. Dispersal of seeds from chasmogamous and cleistogamous flowers in an ant-dispersed neotropical herb. *Oikos* 73: 59–64.

Leston, D. 1970. Entomology of the cocoa farm. *Annual Review of Entomology* 15:273–94.

———. 1973a. The ant mosaic, tropical tree crops, and the limiting of pests and diseases. *Pest Articles and News Summaries* 19:311–41.

———. 1973b. Ants and tropical tree crops. *Proceedings of the Royal Entomological Society of London,* ser. C, 38:1.

———. 1978. A neotropical ant mosaic. *Annals of the Entomological Society of America* 71:649–53.

Letourneau, D. K. 1983. Passive aggression: An alternative hypothesis for the *Piper-Pheidole* association. *Oecologia* 60:122–26.

———. 1990. Code of ant-plant mutualism broken by a parasite. *Science* 248: 215–17.

———. 1998. Ants, stem-borers, and fungal pathogens: Experimental tests of fitness advantage in *Piper* plants. *Ecology* 79:593–603.

Letourneau, D. K., and L. A. Dyer. 1998a. Density patterns of *Piper* ant-plants and associated arthropods: Top predator trophic cascades in a terrestrial system? *Biotropica* 30:162–69.

———. 1998b. Experimental test in lowland tropical forest shows top-down effects through four trophic levels. *Ecology* 79:1678–87.

———. 2005. Multi-trophic interactions and biodiversity: Beetles, ants, caterpillars, and plants. In Burslem, Pinard, and Hartley 2005, 366–85.

Levey, D. J., and M. M. Byrne. 1993. Complex ant-plant interactions: Rain forest ants as secondary dispersers and post-dispersal seed predators. *Ecology* 74:1802–12.

Levins, R. 1969. Some demographic and genetical consequences of environmental heterogeneity for biological control. *Bulletin of the Entomological Society of America* 15:237–40.

———. 1970. Extinction. In *Some problems in biology,* ed. M. Gerstenhaber, 77–107. Providence, RI: American Mathematical Society.

Littledyke, M., and J. M. Cherrett. 1976. Direct ingestion of plant sap from cut leaves by the leaf-cutting ant *Atta cephalotes* (L.) and *Acromyrmex octospinosus* (Reich) (Formicidae, Attini). *Bulletin of Entomological Research* 66:205–17.

Lively, C. M. 1999. Migration, virulence, and the geographic mosaic of adaptation by parasites. *American Naturalist* 153:S34–S47.

Longino, J. T. 1989. Geographic variation and community structure in an ant-plant mutualism: *Azteca* and *Cecropia* in Costa Rica. *Biotropica* 21: 126–32.

———. 1991. *Azteca* ants in *Cecropia* trees: Taxonomy, colony structure, and behaviour. In Huxley and Cutler 1991, 271–88.

Longino, J. T., J. Coddington, and R. K. Colwell. 2002. The ant fauna of a tropical rain forest: Estimating species richness three different ways. *Ecology* 83:689–702.

Longino, J. T., and R. K. Colwell. 1997. Biodiversity assessment using structured inventory: Capturing the ant fauna of a tropical rain forest. *Ecological Applications* 7:1263–77.

López, F., J. M. Serrano, and F. J. Acosta. 1994. Parallels between the foraging strategies of ants and plants. *Trends in Ecology and Evolution* 9:150–53.

Louda, S. M., and M. A. Potvin. 1995. Effect of inflorescence-feeding insects on the demography and lifetime of a native plant. *Ecology* 76:229–45.

Lu, K. L., and M. R. Mesler. 1981. Ant dispersal of a neotropical forest floor gesneriad. *Biotropica* 13:159–60.

Lubin, Y. D. 1984. Changes in the native fauna of the Galapagos Islands following invasion by the little red fire ant, *Wasmannia auropuctata. Biological Journal of the Linnean Society* 21:229–42.

MacGarvin, M., J. H. Lawton, and P. A. Heads. 1986. The herbivorous insect communities of open and woodland bracken—observations, experiments, and habitat manipulations. *Oikos* 47:135–48.

Machado, G., and A. V. L. Freitas. 2001. Larval defence against ant predation in the butterfly *Smyrna blomfildia. Ecological Entomology* 26:436–39.

Mackay, D. A., and M. A. Whalen. 1991. Some associations between ants and euporbs in tropical Australia. In Huxley and Cutler 1991, 238–49.

———. 1998. Associations between ants (Hymenoptera: Formicidae) and *Adriana* Gaudich (Euphorbiaceae) in East Gippsland. *Australian Journal of Entomology* 37:335–39.

MacMahon, J. A., J. F. Mull, and T. O. Crist. 2000. Harvester ants (*Pogonomyrmex* spp.): Their community and ecosystem influences. *Annual Review of Ecology and Systematics* 31:265–91.

Madden, D., and T. P. Young. 1992. Symbiotic ants as an alternative defense against giraffe herbivory in spinescent *Acacia drepanolobium*. *Oecologia* 91:235–38.

Madison, M. 1979. Additional observations on ant-gardens in Amazonas. *Selbyana* 5:107–15.

Mahdi, T., and J. B. Whittaker. 1993. Do birch trees (*Betula pendula*) grow better if foraged by wood ants? *Journal of Animal Ecology* 62:101–16.

Majer, J. D. 1972. The ant-mosaic in Ghana cocoa farms. *Bulletin of Entomological Research* 62:151–60.

———. 1976a. The ant mosaic in Ghana cocoa farms: Further structural considerations. *Journal of Applied Ecology* 13:145–56.

———. 1976b. The maintenance of the ant-mosaic in Ghana cocoa farms. *Journal of Applied Ecology* 13:123–44.

———. 1982a. Ant manipulation in agro- and forest-ecosystems. In *The biology of social insects,* ed. M. D. Breed, C. D. Michener, and H. E. Evans, 90–97. Boulder, CO: Westview Press.

———. 1982b. Ant-plant interactions in the Darling botanical district of Western Australia. In Buckley 1982b, 45–61.

———. 1986. Utilizing economically beneficial ants. In *Economic impact and control of social insects,* ed. S. B. Vinson, 314–31. New York: Praeger Scientific.

———. 1990. The abundance and diversity of arboreal ants in northern Australia. *Biotropica* 22:191–99.

———. 1993. Comparison of the arboreal ant mosaic in Ghana, Brasil, Papua New Guinea, and Australia: Its structure and influence on ant diversity. In *Hymenoptera and biodiversity,* ed. J. LaSalle and I. D. Gauld, 115–41. Wallingford, UK: CAB International.

Majer, J. D., and P. Camer-Pesci. 1991. Ant species in tropical Australian tree crops and native ecosystems—is there a mosaic? *Biotropica* 23:173–81.

Majer, J. D., J. H. C. Delabie, and M. R. B. Smith. 1994. Arboreal ant community patterns in Brazilian cocoa farms. *Biotropica* 26:73–83.

Majer, J. D., S. O. Shattuck, A. N. Andersen, and A. J. Beattie. 2004. Australian ant research: Fabulous fauna, functional groups, pharmaceuticals, and the fatherhood. *Australian Journal of Entomology* 43:235–47.

Manly, B. F. J. 1998. Testing for latitudinal and other body-size radients. *Ecology Letters* 1:104–11.

Mansfield, S., N. V. Elias, and J. A. Lytton-Hitchins. 2003. Ants as egg predators of *Helicoverpa armigera* (Hubner) (Lepidoptera: Noctuidae) in Australian cotton crops. *Australian Journal of Entomology* 42:349–51.

Margules, C. R., and R. L. Pressey. 2000. Systematic conservation planning. *Nature* 405:243–53.
Mark, S., and J. M. Olesen. 1996. Importance of elaiosome size to removal of ant-dispersed seeds. *Oecologia* 107:95–101.
Marquis, R. J. 1991. Herbivore fauna of *Piper* (Piperaceae) in a Costa Rican wet forest: Diversity, specificity, and impact. In Price et al. 1991, 179–208.
Marshall, D. L., A. J. Beattie, and W. E. Bollenbacher. 1979. Evidence for diglycerides as attractants in an ant-seed interaction. *Journal of Chemical Ecology* 5:335–44.
Martínez-Mota, R., J. C. Serio-Silva, and V. Rico-Gray. 2004. The role of canopy ants in removing *Ficus perforata* seeds from howler monkey (*Alouatta palliata mexicana*) feces at Los Tuxtlas, México. *Biotropica* 36:429–32.
Maschwitz, U., and B. Fiala. 1995. Investigations on ant-plant associations in the south-east-Asian genus *Neonauclea* Merr. (Rubiaceae). *Acta Oecologica* 16:3–18.
McCann, K. S. 2000. The diversity-stability debate. *Nature* 405:228–33.
McCoy, E. D. 1990. The distribution of insects along elevational gradients. *Oikos* 58:313–22.
McDade, L. A., and S. Kinsman. 1980. The impact of floral parasitism in two neotropical hummingbird-pollinated plant species. *Evolution* 34:944–58.
McDaniel, S. G., and W. L. Sterling. 1982. Predation of *Heliothis virescens* (F.) eggs on cotton in east Texas. *Environmental Entomology* 11:60–66.
McEvoy, P. B. 1979. Advantages and disadvantages to group living in treehoppers (Homoptera: Membracidae). *Miscellaneous Publications of the Entomological Society of America* 11:1–13.
McGinley, M. A., S. S. Dhillion, and J. C. Neumann. 1994. Environmental heterogeneity and seedling establishment: Ant-plant-microbe interactions. *Functional Ecology* 8:607–15.
McKey, D. 1974. Ant-plants: Selective eating of an unoccupied *Barteria* by *Colobus* monkeys. *Biotropica* 6:269–70.
———. 1979. The distribution of secondary compounds within plants. In *Coevolution of animals and plants,* ed. L. E. Gilbert and P. H. Raven, 55–133. Austin: University of Texas Press.
———. 1984. Interaction of the ant-plant *Leonardoxa africana* (Caesalpiniaceae) with its obligate inhabitants in a rainforest in Cameroon. *Biotropica* 16:81–99.
———. 1989. Interactions between ants and leguminous plants. In *Advances in legume biology,* ed. C. H. Stirton and J. L. Zarucchi, 673–718. Monographs in Systematic Botany 29. St. Louis: Missouri Botanical Garden.
———. 1991. Phylogenetic analysis of the evolution of a mutualism: *Leonardoxa* (Caesalpiniaceae) and its associated ants. In Huxley and Cutler 1991, 310–34.
McKey, D., and D. W. Davidson. 1993. Ant-plant symbioses in Africa and the neotropics: History, biogeography, and diversity. In *Biological relationships between Africa and South America,* ed. P. Goldblatt, 568–606. New Haven, CT: Yale University Press.

McKey, D., L. Gaume, C. Brouat, B. di Gíusto, L. Pascal, G. Debout, A. Dalecky, and M. Heil. 2005. The trophic structure of tropical ant-plant-herbivore interactions: Community consequences and coevolutionary dynamics. In Burslem, Pinard, and Hartley 2005, 386–413.

McLain, D. K. 1983. Ants, extrafloral nectaries, and herbivory on the passion vine, *Passiflora incarnata. American Midland Naturalist* 110:433–39.

McLaughlin, J. F., and J. Roughgarden. 1993. Species interactions in space. In *Species diversity in ecological communities,* ed. R. E. Ricklefs and D. Schluter, 89–98. Chicago: University of Chicago Press.

McNeil, J. N., J. Deslisle, and R. J. Finnegan. 1977. Inventory of aphids on seven conifer species in association with the introduced red wood ant Formica lugubris (Hymenoptera: Formicidae). *Canadian Entomologist* 109:1199–1202.

Messina, F. J. 1981. Plant protection as a consequence of an ant-membracid mutualism: Interactions on goldenrod (*Solidago* sp.). *Ecology* 62:1433–40.

Milesi, F. A., and J. Lopez de Casenave. 2004. Unexpected relationships and valuable mistakes: Non-myrmecochorous *Prosopis* dispersed by messy leaf-cutting ants in harvesting their seeds. *Austral Ecology* 29:558–67.

Milewski, A. V., and W. J. Bond. 1982. Convergence of myrmecochory in Mediterranean Australia and South Africa. In Buckley 1982b, 89–98.

Milewski, A. V., T. P. Young, and D. Madden. 1991. Thorns as induced defenses: Experimental evidence. *Oecologia* 86:70–75.

Mody, K., and K. E. Linsenmair. 2004. Plant-attracted ants affect arthropod community structure but not necessarily herbivory. *Ecological Entomology* 29:217–25.

Moller, H., and J. A. V. Tilley. 1989. Beech honeydew: Seasonal variation and use by wasps, honeybees, and other insects. *New Zealand Journal of Ecology* 16:289–302.

Mondor, E. B., and J. F. Addicott. 2003. Conspicuous extra-floral nectaries are inducible in *Vicia faba. Ecology Letters* 6:495–97.

Morais, H. C. 1994. Coordinated group ambush—a new predatory behavior in *Azteca* ants (Dolichoderinae). *Insectes Sociaux* 41:339–42.

Morales, M. A., and E. R. Heithaus. 1998. Food from seed dispersal mutualism shifts sex ratios in colonies of the ant *Aphaenogaster rudis. Ecology* 79:734–39.

Moran, N. A., C. Dale, H. Dunbar, W. A. Smith, and H. Ochman. 2003. Intracellular symbionts of sharpshooters (Insecta: Hemiptera: Cicadellinae) form a distinct clade with a small genome. *Environmental Microbiology* 5:116–26.

Moreau, C. S., C. D. Bell, R. Vila, S. B. Archibald, and N. E. Pierce. 2006. Phylogeny of the ants: Diversification in the age of angiosperms. *Science* 312: 101–4.

Moreira, V. S. S., and K. Del-Claro. 2005. The outcomes of an ant-treehopper association on *Solanum lycocarpum* St. Hill: Increased membracid fecundity and reduced damage by chewing herbivores. *Neotropical Entomology* 34:881–87.

Morellato, L. P. C. 1992. Sazonalidade e dinâmica de ecossistemas florestais na Serra do Japi. In *Historia natural da Serra do Japi: Ecologia e preservação de uma area florestal no sudeste do Brasil,* ed. L. P. C. Morellato, 98–110. Campinas, Brazil: Editora da UNICAMP.

Morellato, L. P. C., and P. S. Oliveira. 1991. Distribution of extrafloral nectaries in different vegetation types of Amazonian Brazil. *Flora* 185:33–38.

———. 1994. Extrafloral nectaries in the tropical tree *Guarea macrophylla* (Meliaceae). *Canadian Journal of Botany* 72:157–60.

Morrison, L. W. 1996. Community organization in a recently assembled fauna: The case of Polynesian ants. *Oecologia* 107:243–56.

———. 1998. The spatiotemporal dynamics of insular ant metapopulations. *Ecology* 79:1135–46.

———. 2006. Mechanisms of coexistence and competition between ants and land hermit crabs in a Bahamian archipelago. *Acta Oecologica* 29:1–8.

Moutinho, P. 1991. Note of the foraging activity and diet of two *Pheidole* Westwood species (Hymenoptera, Formicidae) in an area of "shrub canga" vegetation in Amazonian Brazil. *Revista Brasileira de Biologia* 51:403–6.

Moutinho, P., D. C. Nepstad, and E. A. Davidson. 2003. Influence of leaf-cutting ant nests on secondary forest growth and soil properties in Amazonia. *Ecology* 84:1265–76.

Mueller, U. G., S. A. Rehner, and T. R. Schultz. 1998. The evolution of agriculture in ants. *Science* 281:2034–38.

Mulcahy, D. L. 1979. The rise of angiosperms: A genecological factor. *Science* 206:20–23.

Mull, J. F., and J. A. MacMahon. 1997. Spatial variation rates of seed removal by harvester ants (*Pogonomyrmex occidentalis*) in a shrub-steppe ecosystem. *American Midland Naturalist* 138:1–13.

Müller, C. B., and H. C. J. Godfray. 1999. Predators and mutualists influence the exclusion of aphid species from natural communities. *Oecologia* 119:120–25.

Nakanishi, H. 1994. Myrmecochorous adaptations of *Corydalis* species (Papaveraceae) in southern Japan. *Ecological Research* 9:1–8.

Nascimento, M. T., and J. Proctor. 1996. Seed attack by beetles and leaf-cutter ants on *Peltogyne gracilipes* Ducke (Caesalpiniaceae) on Maracá Island, Brazilian Amazonia. *Journal of Tropical Ecology* 12:723–27.

Nathan, R., and H. C. Muller-Landau. 2000. Spatial patterns of seed dispersal, their determinants, and consequences for recruitment. *Trends in Ecology and Evolution* 15:278–85.

Ness, J. H. 2003a. *Catalpa bignonioides* alters extrafloral nectar production after herbivory and attracts ant bodyguards. *Oecologia* 134:210–18.

———. 2003b. Contrasting exotic *Solenopsis invicta* and native *Forelius pruinosus* ants as mutualists with *Catalpa bignonioides,* a native plant. *Ecological Entomology* 28:247–51.

———. 2004. Forest edges and fire ants alter the seed shadow of an ant-dispersed plant. *Oecologia* 138:448–54.

Ness, J. H., J. L. Bronstein, A. N. Andersen, and J. N. Holland. 2004. Ant body size predicts dispersal distance of ant-adapted seeds: Implications of small-ant invasions. *Ecology* 85:1244–50.

Nichols-Orians, C. M. 1991a. Condensed tannins, attine ants, and the performance of symbiotic fungus. *Journal of Chemical Ecology* 17:1177–95.

———. 1991b. The effects of light on foliar chemistry, growth, and susceptibility of seedlings of a canopy tree to an attine ant. *Oecologia* 86:552–60.

Nichols-Orians, C. M., and J. C. Schultz. 1989. Leaf toughness affects leaf harvesting by leaf cutter ant, *Atta cephalotes* (L.) (Hymenoptera: Formicidae). *Biotropica* 21:80–83.

———. 1990. Interactions among leaf toughness, chemistry, and harvesting by attine ants. *Ecological Entomology* 15:311–20.

Niklas, K. J. 1997. *The evolutionary biology of plants.* Chicago: University of Chicago Press.

Nilsson, L. A. 1978. Pollination ecology of *Epipactis palustris* (Orchidaceae). *Botanischer Notiser* 131:355–68.

Nomura, M., T. Itioka, and T. Itino. 2000. Variations in abiotic defense within myrmecophytic and non-myrmecophytic species of *Macaranga* in Bornean dipterocarp forest. *Ecological Research* 15:1–11.

Norment, C. J. 1988. The effect of nectar-thieving ants on the reproductive success of *Frasera speciosa* (Gentianaceae). *American Midland Naturalist* 120:331–36.

North, R. D., C. W. Jackson, and P. E. Howse. 1997. Evolutionary aspects of ant-fungus interactions in leaf-cutting ants. *Trends in Ecology and Evolution* 12:386–89.

Nuismer, S. L., J. N. Thompson, and R. Gomulkiewicz. 1999. Gene flow and geographically structured coevolution. *Proceedings of the Royal Society of London B* 266:605–9.

O'Dowd, D. J. 1979. Foliar nectar production and ant activity on a neotropical tree, *Ochroma pyramidale. Oecologia* 43:223–48.

———. 1980. Pearl bodies of a neotropical tree, *Ochroma pyramidale:* Ecological implications. *American Journal of Botany* 67:543–49.

———. 1982. Pearl bodies as ant food: An ecological role for some leaf emergences of tropical plants. *Biotropica* 14:40–49.

O'Dowd, D. J., and E. A. Catchpole. 1983. Ants and extrafloral nectaries: No evidence for plant protection in *Helichrysum* spp.–ant interactions. *Oecologia* 59:191–200.

O'Dowd, D. J., and M. E. Hay. 1980. Mutualism between harvester ants and a desert ephemeral: Seed escape from rodents. *Ecology* 6:531–40.

Offenberg, J., S. Havanon, S. Aksornkoae, D. J. MacIntosh, and M. G. Nielsen. 2004. Observations on the ecology of weaver ants (*Oecophylla smaragdina* Fabricius) in a Thai mangrove ecosystem and their effect on herbivory of *Rhizophora mucronata* Lam. *Biotropica* 36:344–51.

Ohara, M., and S. Higashi. 1987. Interference by ground beetles with the dispersal by ants of seeds of *Trillium* species (Liliaceae). *Journal of Ecology* 75:1091–98.

Ohkawara, K. 1995. Seed dispersal of some myrmecochorous plant species, with effects of elaiosomes and ground beetles. Ph.D. diss., Hokkaido University.

Ohkawara, K., and S. Higashi. 1994. Relative importance of ballistic and ant dispersal in two diplochorous *Viola* species (Violaceae). *Oecologia* 100:135–40.

Ohkawara, K., S. Higashi, and M. Ohara. 1996. Effects of ants, ground beetles, and the seed-fall patterns on myrmecochory of *Erythronium japonicum* Decne. (Liliaceae). *Oecologia* 106:500–506.

Olesen, J. M., and P. Jordano. 2002. Geographic patterns in the plant-pollinator mutualistic networks. *Ecology* 83:2416–24.

Oliveira, P. S. 1997. The ecological function of extrafloral nectaries: Herbivore deterrence by visiting ants and reproductive output in *Caryocar brasiliense* (Caryocarpaceae). *Functional Ecology* 11:323–30.

Oliveira, P. S., and C. R. S. Brandão. 1991. The ant community associated with extrafloral nectaries in the Brazilian cerrados. In Huxley and Cutler 1991, 198–212.

Oliveira, P. S., A. F. da Silva, and A. B. Martins. 1987. Ant foraging on extrafloral nectaries of *Qualea grandiflora* (Vochysiaceae) in cerrado vegetation: Ants as potential antiherbivore agents. *Oecologia* 74:228–30.

Oliveira, P. S., and K. Del-Claro. 2005. Multitrophic interactions in a neotropical savanna: Ant-hemipteran systems, associated insect herbivores, and a host plant. In Burslem, Pinard, and Hartley 2005, 414–38.

Oliveira, P. S., and A. V. L. Freitas. 2004. Ant-plant-herbivore interactions in the neotropical cerrado savanna. *Naturwissenschaften* 91:557–70.

Oliveira, P. S., A. V. L. Freitas, and K. Del-Claro. 2002. Ant foraging on plant foliage: Contrasting effects on the behavioral ecology of insect herbivores. In *The Cerrados of Brazil: Ecology and Natural History of a Neotropical Savanna,* ed. P. S. Oliveira and R. J. Marquis, 287–305. New York: Columbia University Press.

Oliveira, P. S., M. Galetti, F. Pedroni, and L. P. C. Morellato. 1995. Seed cleaning by *Mycocepurus goeldii* ants (Attini) facilitates germination in *Hymenaea courbaril* (Caesalpiniaceae). *Biotropica* 27:518–22.

Oliveira, P. S., C. Klitzke, and E. Vieira. 1995. The ant fauna associated with the extrafloral nectaries of *Ouratea hexasperma* (Ochnaceae) in an area of cerrado vegetation in central Brazil. *Entomologist's Monthly Magazine* 131:77–82.

Oliveira, P. S., and H. F. Leitão-Filho. 1987. Extrafloral nectaries: Their taxonomic distribution and abundance in the woody flora of Cerrado vegetation in southeast Brazil. *Biotropica* 19:140–48.

Oliveira, P. S., and A. T. Oliveira-Filho. 1991. Distribution of extrafloral nectaries in the woody flora of tropical communities in western Brazil. In Price et al. 1991, 163–75.

Oliveira, P. S., A. T. Oliveira-Filho, and R. Cintra. 1987. Ant foraging on ant-inhabited *Triplaris* (Polygonaceae) in western Brazil: A field experiment using live termite-baits. *Journal of Tropical Ecology* 3:193–200.

Oliveira, P. S., and M. R. Pie. 1998. Interaction between ants and plants bearing extrafloral nectaries in cerrado vegetation. *Anais da Sociedade Entomológica do Brasil* 27:161–76.

Oliveira, P. S., V. Rico-Gray, C. Díaz-Castelazo, and C. Castillo-Guevara. 1999. Interaction between ants, extrafloral nectaries, and insect herbivores in neotropical coastal sand dunes: Herbivore deterrence by visiting ants increases fruit set in *Opuntia stricta* (Cactaceae). *Functional Ecology* 13:623–31.

Olmstead, K. L., and T. K. Wood. 1990. Altitudinal patterns in species richness of neotropical treehoppers (Homoptera: Membracidae): The role of ants. *Proceedings of the Entomological Society of Washington* 92:552–60.

Olofsson, J., J. Moen, and L. Oksanen. 1999. On the balance between positive and negative plant interactions in harsh environments. *Oikos* 86:539–49.

Orivel, J., and A. Dejean. 1998. Selection of epiphyte seeds by ant-garden ants. *Ecoscience* 6:51–55.

———. 1999. L'adaptation à la vie arboricole chez les fourmis. *L'Année Biologique* 38:131–48.

Orivel, J., A. Dejean, and C. Errard. 1998. Active role of two ponerine ants in the elaboration of ant gardens. *Biotropica* 30:487–91.

Orivel, J., M. C. Malherbe, and A. Dejean. 2001. Relationships between pretarsus morphology and arboreal life in ponerine ants of the genus *Pachycondyla* (Formicidae : Ponerinae). *Annals of the Entomological Society of America* 94:449–56.

Overal, W. L., and P. A. Posey. 1984. Uso de formigas do genero *Azteca* por controle de saúvas entre os indios kaiapos do Brasil. *Attini* 16:2.

Palmer, T. M., T. P. Young, and M. L. Straton. 2002. Burning bridges: Priority effects and the persistence of a competitive subordinate acacia-ant in Laikipia, Kenya. *Oecologia* 133:372–79.

Parr, Z. J. E., C. L. Parr, and S. L. Chown. 2003. The size-grain hypothesis: A phylogenetic and field test. *Ecological Entomology* 28:475–81.

Parra-Tabla, V., V. Rico-Gray, and M. Carbajal. 2004. Effect of herbivory on leaf growth, sexual expression, and reproductive success of *Cnidoscolus aconitifolius* (Euphorbiaceae). *Plant Ecology* 173:153–60.

Passos, L., and S. O. Ferreira. 1996. Ant dispersal of *Croton priscus* (Euphorbiaceae) seeds in a tropical semideciduous forest in southeastern Brazil. *Biotropica* 28:697–700.

Passos, L., and P. S. Oliveira. 2002. Ants affect the distribution and performance of *Clusia criuva* seedlings, a primarily bird-dispersed rainforest tree. *Journal of Ecology* 90:517–28.

———. 2003. Interactions between ants, fruits, and seeds in a restinga forest in south-eastern Brazil. *Journal of Tropical Ecology* 19:261–70.

———. 2004. Interaction between ants and fruits of *Guapira opposita* (Nyctaginaceae) in a Brazilian sandy plain rainforest: Ant effects on seeds and seedlings. *Oecologia* 139:376–82.

Paterson, S. 1982. Observations on ant associations with rain forest ferns in Borneo. *Fern Gazette* 12:243–45.

Paton, D. C. 1980. The importance of manna, honeydew, and lerp in the diet of honeyeaters. *Emu* 80:213–26.

Paulson, G. S., and R. D. Akre. 1991. Behavioral interactions among formicid species in the ant mosaic of an organic pear orchard. *Pan-Pacific Entomologist* 67:288–97.

Peakall, R. 1989. The unique pollination of *Leporella fimbriata* (Orchidaceae): Pollination by pseudocopulating male ants (*Myrmecia urens,* Formicidae). *Plant Systematics and Evolution* 167:137–48.

Peakall, R., C. J. Angus, and A. J. Beattie. 1990. The significance of ant and plant traits for ant pollination in *Leporella fimbriata*. *Oecologia* 84:457–60.

Peakall, R., and A. J. Beattie. 1989. Pollination of the orchid *Microtis parviflora* R. Br. by flightless worker ants. *Functional Ecology* 3:515–22.

———. 1991. The genetic consequences of worker ant pollination in a self-compatible, clonal plant. *Evolution* 45:1837–48.

———. 1995. Does ant dispersal of seeds in *Sclerolaena diacantha* (Chenopodiaceae) generate local spatial genetic structure? *Heredity* 75:351–61.

Peakall, R., A. J. Beattie, and S. H. James. 1987. Pseudocopulation of an orchid by male ants: A test of two hypotheses accounting for the rarity of ant pollination. *Oecologia* 73:522–24.

Peakall, R., S. N. Handel, and A. J. Beattie. 1991. The evidence for, and importance of, ant pollination. In Huxley and Cutler 1991, 421–29.

Peakall, R., and S. H. James. 1989. Outcrossing in an ant pollinated clonal orchid. *Heredity* 62:161–67.

Pellmyr, O., and L. B. Thien. 1986. Insect reproduction and floral fragrances: Keys to the evolution of the angiosperms? *Taxon* 35:76–85.

Pellmyr, O., L. B. Thien, G. Bergstrom, and I. Groth. 1990. Pollination of New Caledonian Winteraceae: Opportunistic shifts or parallel radiation with their pollinators? *Plant Systematics and Evolution* 173:143–57.

Pemberton, R. W. 1988. The abundance of plants bearing extrafloral nectaries in Colorado and Mojave desert communities of southern California. *Madroño* 35:238–46.

———. 1990. The occurrence of extrafloral nectaries in Korean plants. *Korean Journal of Ecology* 13:251–66.

———. 1992. Fossil extrafloral nectaries, evidence for the ant-guard antiherbivore defense in an oligocene *Populus*. *American Journal of Botany* 79: 1242–46.

———. 1998. The occurrence and abundance of plants with extrafloral nectaries, the basis for antiherbivore defensive mutualisms along a latitudinal gradient in east Asia. *Journal of Biogeography* 25:661–68.

Pemberton, R. W., and J. H. Lee. 1996. The influence of extrafloral nectaries on parasitism of an insect herbivore. *American Journal of Botany* 83:1187–94.

Peñaloza, C., and A. G. Farji-Brener. 2003. The importance of treefall gaps as foraging sites for leaf-cutting ants depends on forest age. *Journal of Tropical Ecology* 19:603–5.

Pereira, R. M. 2003. Areawide suppression of fire ant populations in pastures: Project update. *Journal of Agricultural and Urban Entomology* 20:123–30.

Perfecto, I. 1990. Indirect and direct effects in a tropical agroecosystem: The maize-pest-ant system in Nicaragua. *Ecology* 71:2125–34.

———. 1991. Ants (Hymenoptera: Formicidae) as natural control agents of pests in irrigated maize in Nicaragua. *Journal of Economic Entomology* 84:65–70.

———. 1994. Foraging behavior as a determinant of asymmetric competitive interaction between two ant species in a tropical agroecosystem. *Oecologia* 98:184–92.

Perfecto, I., and I. Armbrecht. 2003. The coffee agroecosystem in the neotropics: Combining ecological and economic goals. In *Tropical agroecosystems*, ed. J. Vandermeer, 157–92. Boca Raton, FL: CRC Press.

Perfecto, I., and A. Sediles. 1992. Vegetational diversity, ants (Hymenoptera: Formicidae), and herbivorous pests in a neotropical agroecosystem. *Environmental Entomology* 21:61–67.

Perfecto, I., and J. H. Vandermeer. 1994. Understanding biodiversity loss in agroecosystems: Reduction of ant diversity resulting from transformation of the coffee ecosystem in Costa Rica. *Entomological Trends in Agricultural Science* 2:7–13.

———. 1996. Microclimatic changes and the indirect loss of ant diversity in a tropical agroecosystem. *Oecologia* 108:577–82.

———. 2002. Quality of agroecological matrix in a tropical montane landscape: Ants in coffee plantations in southern Mexico. *Conservation Biology* 16:174–82.

Petersen, B. 1977a. Pollination by ants in the alpine tundra of Colorado, U.S.A. *Transactions Illinois State Academy of Science* 70:349–55.

———. 1977b. Pollination of *Thlaspi alpina* by selfing and by insects in the alpine zone of Colorado. *Arctic and Alpine Research* 9:211–15.

Philipott, S. M., and I. Armbrecht. 2006. Biodiversity in tropical agroforests and the ecological role of ants and ant diversity in predatory function. *Ecological Entomology* 31:369–77.

Philipott, S. M., R. Greenberg, P. Bichier, and I. Perfecto. 2004. Impacts of major predators on tropical agroforest arthropods: Comparisons within and across taxa. *Oecologia* 140:140–49.

Philipott, S. M., J. Maldonado, J. Vandermeer, and I. Perfecto. 2004. Taking trophic cascades up a level: Behaviorally-modified effects of phorid flies on ants and ant prey in coffee agroecosystems. *Oikos* 105:141–47.

Pickett, C. H., and W. D. Clark. 1979. The function of extrafloral nectaries in *Opuntia acanthocarpa* (Cactaceae). *American Journal of Botany* 66: 618–25.

Pielou, E. C. 1979. *Biogeography*. New York: Wiley.

Pierce, N. E. 1985. Lycaenid butterflies and ants: Selection for nitrogen fixing and other protein rich food plants. *American Naturalist* 125:888–95.

Pierce, N. E., M. F. Braby, A. Heath, D. J. Lohman, J. Mathew, D. B. Rand, and M. A. Travassos. 2002. The ecology and evolution of ant association in the Lycaenidae (Lepidoptera). *Annual Review of Entomology* 47:733–71.

Pierce, N. E., and M. A. Elgar. 1985. The influence of ants on host plant selection by *Jalmenus evagora*, a myrmecophilous lycaenid butterfly. *Behavioral Ecology and Sociobiology* 16:209–22.

Pizo, M. A. 1997. Seed dispersal and predation in two populations of *Cabralea canjerana* (Meliaceae) in the Atlantic forest of southeastern Brazil. *Journal of Tropical Ecology* 13:559–78.

Pizo, M. A., P. R. Guimarães Jr., and P. S. Oliveira. 2005. Seed removal by ants from vertebrate faeces produced by different vertebrate species. *Ecoscience* 12:136–40.

Pizo, M. A., and P. S. Oliveira. 1998. Interaction between ants and seeds of a nonmyrmecochorous neotropical tree, *Cabralea canjerana* (Meliaceae), in the Atlantic forest of southeast Brazil. *American Journal of Botany* 85: 669–74.

———. 1999. Removal of seeds from vertebrate faeces by ants: Effects of seed species and deposition site. *Canadian Journal of Zoology* 77:1595–1602.

———. 2000. The use of fruits and seeds by ants in the Atlantic forest of southeast Brazil. *Biotropica* 32:851–61.

———. 2001. Size and lipid content of nonmyrmecochorous diaspores: Effects on the interaction with litter-foraging ants in the Atlantic rain forest of Brazil. *Plant Ecology* 157:37–52.

Pizo, M. A., L. Passos, and P. S. Oliveira. 2005. Ants as seed dispersers of fleshy diaspores in Brazilian Atlantic forests. In *Seed fate: Predation and secondary dispersal,* ed. P.-M. Forget, J. E. Lambert, P. E. Hulme, and S. B. Vander Wall, 315–29. Wallingford, UK: CABI.

Pollard, S. D., M. W. Beck, and G. N. Dodson. 1995. Why do male crab spiders drink nectar? *Animal Behaviour* 49:1443–48.

Porter, E. E., and B. A. Hawkins. 2001. Latitudinal gradients in colony size for social insects: Termites and ants show different patterns. *American Naturalist* 157:97–106.

Post, D. M. 2002. Using stable isotopes to estimate trophic position: Models, methods, and assumptions. *Ecology* 83:703–18.

Powell, R. J., and D. J. Stradling. 1991. The selection and detoxification of plant material by fungus-growing ants. In Huxley and Cutler 1991, 19–41.

Price, P. W. 1991. Patterns in communities along latitudinal gradients. In Price et al. 1991, 51–69.

Price, P. W., C. E. Bouton, P. Gross, B. A. McPheron, J. N. Thompson, and A. E. Weis. 1980. Interaction among three trophic levels: Influence of plants on interactions between insect herbivores and natural enemies. *Annual Review of Ecology and Systematics* 11:41–65.

Price, P. W., T. M. Lewinsohn, G. W. Fernandes, and W. W. Benson, eds. 1991. *Plant-animal interactions: Evolutionary ecology in tropical and temperate regions.* New York: Wiley.

Price, P. W., M. Westoby, B. Rice, P. R. Atsatt, R. S. Fritz, J. N. Thompson, and K. Mobley. 1986. Parasite mediation in ecological interactions. *Annual Review of Ecology and Systematics* 17:487–505.

Pudlo, R. J., A. J. Beattie, and D. C. Culver. 1980. Population consequences of changes in an ant-seed mutualism in *Sanguinaria canadensis. Oecologia* 146:32–37.

Purvis, A., and A. Hector. 2000. Getting the measure of biodiversity. *Nature* 405:212–19.

Puterbaugh, M. N. 1998. The roles of ants as flower visitors: Experimental analysis in three alpine plant species. *Oikos* 83:36–46.

Putz, F. E., and N. M. Holbrook. 1988. Further observations on the dissolution of mutualism between *Cecropia* and its ants: The Malaysian case. *Oikos* 53:121–25.

Qiu, Y.-L., J. Lee, F. Bernasconi-Quadroni, D. E. Soltis, P. S. Soltis, M. Zanis, E. A. Zimmer, Z. Chen, V. Savolainen, and M. W. Chase. 1999. The earliest angiosperms: Evidence from mitochondrial, plastid, and nuclear genomes. *Nature* 404–7.

Queiroz, J. M., and P. S. Oliveira. 2001. Tending-ants protect honeydew-producing whiteflies (Homoptera: Aleyrodidae). *Environmental Entomology* 30:295–97.

Quental, T. B., J. R. Trigo, and P. S. Oliveira. 2005. Host-plant flowering status and sugar concentration of phloem sap: Effects on an ant-treehopper interaction. *European Journal of Entomology* 102:201–8.

Raimundo, R. L. G., P. R. Guimarães Jr., M. Almeida-Neto, and M. A. Pizo. 2004. The influence of fruit morphology and habitat structure on ant-seed interactions: A study with artificial fruits. *Sociobiology* 44:261–70.

Raine, N. E., N. Gammans, I. J. MacFadyen, G. K. Scrivner, and G. N. Stone. 2004. Guards and thieves: Antagonistic interactions between two ant species coexisting on the same ant-plant. *Ecological Entomology* 29:345–52.

Raine, N. E., P. Willmer, and G. N. Stone. 2002. Spatial structuring and floral avoidance behavior prevent ant-pollinator conflict in a Mexican ant-acacia. *Ecology* 83:3086–96.

Ramsey, M. 1995. Ant pollination of the perennial herb *Blandfordia grandiflora* (Liliaceae). *Oikos* 74:265–72.

Rashbrook, V. K., S. G. Compton, and J. H. Lawton. 1991. Bracken and ants: Why is there no mutualism? In Huxley and Cutler 1991, 231–37.

———. 1992. Ant-herbivore interactions: Reasons for the absence of benefits to a fern with foliar nectaries. *Ecology* 73:2167–74.

Raven, P. H., R. F. Evert, and S. E. Eichhorn. 1986. *Biology of plants.* New York: Worth.

Reagan, T. E. 1986. Beneficial aspects of the imported fire ant: A field ecology approach. In *Fire ants and leaf-cutting ants,* ed. C. S. Lofgren and R. K. Vander Meer, 58–71. Boulder, CO: Westview Press.

Reid, W. V. 1998. Biodiversity hotspots. *Trends in Ecology and Evolution* 13:275–80.

Renner, S. S., and R. E. Ricklefs. 1998. Herbicidal activity of domatia-inhabiting ants in patches of *Tococa guianensis* and *Clidemia heterophylla. Biotropica* 30:324–27.

Retana, J., F. X. Picó, and A. Rodrigo. 2004. Dual role of harvesting ants as seed predators and dispersers of a non-myrmechorous Mediterranean perennial herb. *Oikos* 105:377–85.

Rhoades, D. F., and J. C. Bergdahl. 1981. Adaptative significance of toxic nectar. *American Naturalist* 117:798–803.

Ribas, C. R., and J. H. Schoereder. 2002. Are all ant mosaics caused by competition? *Oecologia* 131:606–11.

———. 2004. Determining factors of arboreal ant mosaics in cerrado vegetation (Hymenoptera: Formicidae). *Sociobiology* 44:49–68.

Rice, B., and M. Westoby. 1986. Evidence against the hypothesis that ant-dispersed seeds reach nutrient-enriched microsites. *Ecology* 67:1270–74.

Ricklefs, R. E. 1984. *Ecology*. New York: Chiron Press.

Rickson, F. R. 1969. Developmental aspects of the shoot apex, leaf, and Beltian bodies of *Acacia cornigera*. *American Journal of Botany* 56:196–200.

———. 1971. Glycogen plastids in Müllerian body cells of *Cecropia peltata*—a higher green plant. *Science* 173:344–47.

———. 1973. Review of glycogen plastid differentiation in Müllerian body cells of *Cecropia peltata*. *Annals of the New York Academy of Sciences* 210: 104–14.

———. 1975. The ultrastructure of *Acacia cornigera* L. Beltian body tissue. *American Journal of Botany* 62:913–22.

———. 1976. Anatomical development of the leaf trichilium and Müllerian bodies of *Cecropia peltata* L. *American Journal of Botany* 63:1266–71.

———. 1977. Progressive loss of ant-related traits of *Cecropia peltata* on selected Caribbean islands. *American Journal of Botany* 64:585–92.

———. 1979. Absorption of animal tissue breakdown products into a plant stem—the feeding of a plant by ants. *American Journal of Botany* 66: 87–90.

———. 1980. Developmental anatomy and ultrastructure of the ant food bodies (Beccariian bodies) of *Macaranga triloba* and *M. hypoleuca* (Euphorbiaceae). *American Journal of Botany* 67:285–92.

Rickson, F. R., and M. M. Rickson. 1998. The cashew nut, *Anacardium occidentale* (Anacardiaceae), and its perennial association with ants: Extrafloral nectary location and the potential for ant defense. *American Journal of Botany* 85:835–49.

Rickson, F. R., and S. J. Risch. 1984. Anatomical and ultrastructural aspects of the ant-food cell of *Piper cenocladum* C.DC. (Piperaceae). *American Journal of Botany* 71:1268–74.

Rico-Gray, V. 1980. Ants and tropical flowers. *Biotropica* 12:223–24.

———. 1989. The importance of floral and circum-floral nectar to ants inhabiting dry tropical lowlands. *Biological Journal of the Linnean Society* 38:173–81.

———. 1993. Use of plant-derived food resources by ants in the dry tropical lowland of coastal Veracruz, Mexico. *Biotropica* 25:301–15.

———. 2001. Interspecific interaction. In *Encyplopedia of life sciences*, 1–6. London: Macmillan. Also available at www.els.net.

Rico-Gray, V., J. T. Barber, L. B. Thien, E. G. Ellgaard, and J. J. Toney. 1989. An unusual animal-plant interaction: Feeding of *Schomburgkia tibicinis* by ants. *American Journal of Botany* 76:603–8.

Rico-Gray, V., and G. Castro. 1996. Effect of an ant-aphid-plant interaction on the reproductive fitness of *Paullinia fuscecens* (Sapindaceae). *Southwestern Naturalist* 41:434–40.

Rico-Gray, V., A. Chemás, and S. Mandujano. 1991. Uses of tropical deciduous forest species by the Yucatecan Maya. *Agroforestry Sy*stems 14:149–61.

Rico-Gray, V., and J. G. García-Franco. 1991. The Maya and the vegetation of the Yucatan Peninsula. *Journal of Ethnobiology* 11:135–42.

———. 1992. Vegetation and soil seed bank of successional stages in tropical lowland deciduous forest. *Journal of Vegetation Science* 3:617–24.

Rico-Gray, V., J. G. García-Franco, and A. Chemás. 1988. Yucatecan Mayas knowledge of pollination and breeding systems. *Journal of Ethnobiology* 8:203–4.

Rico-Gray, V., J. G. García-Franco, A. Chemás, A. Puch, and P. Simá. 1990. Species composition, similarity, and structure, of Mayan homegardens in Tixpeual and Tixcacaltuyub, Yucatan, Mexico. *Economic Botany* 44: 470–87.

Rico-Gray, V., J. G. García-Franco, M. Palacios-Rios, C. Díaz-Castelazo, V. Parra-Tabla, and J. A. Navarro. 1998. Geographical and seasonal variation in the richness of ant-plant interactions in Mexico. *Biotropica* 30:190–200.

Rico-Gray, V., J. G. García-Franco, A. Puch, and P. Simá. 1988. Composition and structure of a tropical dry forest in Yucatan, Mexico. *International Journal of Ecology and Environmental Scienes* 14:21–29.

Rico-Gray, V., and H. C. Morais. 2006. Efecto de una fuente de alimento experimental sobre una asociación hormiga-hemíptero. *Acta Zoológica Mexicana* 22:23–28.

Rico-Gray, V., P. S. Oliveira, V. Parra-Tabla, M. Cuautle, and C. Díaz-Castelazo. 2004. Ant-plant interactions: Their seasonal variation and effects on plant fitness. In *Coastal dunes: Ecology and conservation,* ed. M. L. Martínez and N. P. Psuty, 221–39. Ecological Studies, 171. Berlin: Springer-Verlag.

Rico-Gray, V., M. Palacios-Rios, J. G. García-Franco, and W. P. Mackay. 1998. Richness and seasonal variation of ant-plant associations mediated by plant-derived food resources in the semiarid Zapotitlán valley, Mexico. *American Midland Naturalist* 140:21–26.

Rico-Gray, V., and L. da S. L. Sternberg. 1991. Carbon isotopic evidence for seasonal change in feeding habits of *Camponotus planatus* Roger (Formicidae) in Yucatan, Mexico. *Biotropica* 23:93–95.

Rico-Gray, V., and L. B. Thien. 1989a. Ant-mealybug interaction decreases reproductive fitness of *Schomburgkia tibicinis* Bateman (Orchidaceae) in Mexico. *Journal of Tropical Ecology* 5:109–12.

———. 1989b. Effect of different ant species on the reproductive fitness of *Schomburgkia tibicinis* (Orchidaceae). *Oecologia* 81:487–89.

Rios-Casanova, L., A. Valiente-Banuet, and V. Rico-Gray. 2004. Las hormigas del Valle de Tehuacán: Comparación con otras zonas áridas de México. *Acta Zoológica Mexicana* 20:37–54.

———. 2006. Ant diversity and its relationship with vegetation and soil factors in an alluvial fan of the Tehuacán Valley, Mexico. *Acta Oecologica* 29:316–23.

Risch, S. J. 1981. Ants as important predators of rootworm eggs in the Neotropics. *Journal of Economic Entomology* 74:88–90.

Risch, S. J., and C. R. Carroll. 1982a. The ecological role of ants in two Mexican agroecosystems. *Oecologia* 55:114–19.

———. 1982b. Effect of a keystone predacious ant, *Solenopsis geminata,* on arthropods in a tropical agroecosystem. *Ecology* 63:1979–83.

Risch, S., M. McClure, J. Vandermeer, and S. Waltz. 1977. Mutualism between three species of tropical *Piper* (Piperaceae) and their ant inhabitants. *American Midland Naturalist* 98:433–44.

Risch, S. J., and F. R. Rickson. 1981. Mutualism in which ants must be present before plants produce food bodies. *Nature* 291:149–50.

Rissing, S. W. 1986. Indirect effects of granivory by harvester ants: Plant species composition and reproductive increase near ant nests. *Oecologia* 68: 231–34.

Roberts, J. T., and E. R. Heithaus. 1986. Ants rearrange the vertebrate-generated seed shadow of a neotropical fig tree. *Ecology* 67:1046–51.

Roces, F., and B. Hölldobler. 1995. Vibrational communication between hitchhikers and foragers in leaf-cutting ants (*Atta cephalotes*). *Behavioral Ecology and Sociobiology* 37:297–302.

———. 1996. Use of stridulation in foraging leaf-cutting ants: Mechanical support during cutting or short-range recruitment signal? *Behavioral Ecology and Sociobiology* 39:293–99.

Roces, F., and J. R. B. Lighton. 1995. Larger bites of leaf-cutting ants. *Nature* 373:392–93.

Rocha, C. F., and H. G. Bergallo. 1992. Bigger ant colonies reduce herbivory and herbivore residence time on leaves of an ant-plant: *Azteca muelleri* vs. *Coelomera ruficornis* on *Cecropia pachystachya*. *Oecologia* 91:249–52.

Rockwood, L. L. 1975. The effects of seasonality on foraging in two species of leaf-cutting ants (*Atta*) in Guanacaste Province, Costa Rica. *Biotropica* 7:176–93.

Rodgerson, L. 1998. Mechanical defense in seeds adapted for ant dispersal. *Ecology* 79:1669–77.

Romero, G. Q., and T. J. Izzo. 2004. Leaf damage induces ant recruitment in the Amazonian ant-plant *Hirtella myrmecophylla*. *Journal of Tropical Ecology* 20:675–82.

Room, P. M. 1971. The relative distribution of ant species in Ghana's cocoa farms. *Journal of Animal Ecology* 40:735–51.

———. 1972. The fauna of the mistletoe *Tapinanthus bangwensis* growing on cocoa in Ghana: Relationships between fauna and mistletoe. *Journal of Animal Ecology* 41:611–21.

———. 1975. Relative distributions of ant species in cocoa plantations in Papua New Guinea. *Journal of Applied Ecology* 12:47–61.

Root, R. B. 1973. Organization of plant-arthropod association in simple and diverse habitats: The fauna of collards (*Brassica oleracea*). *Ecological Monographs* 43:95–124.

Röschard, J., and F. Roces. 2002. The effect of load length, width, and mass in the grass-cutting ant *Atta vollenweideri*. *Oecologia* 131:319–24.

Rosengren, R., and L. Sundström. 1991. The interaction between red wood ants, *Cinara* aphids, and pines: A ghost of mutualism past? In Huxley and Cutler 1991, 80–91.

Roth, D. S., I. Perfecto, and B. Rathcke. 1994. The effects of management systems on ground-foraging ant diversity in Costa Rica. *Ecological Applications* 4:423–36.

Roughgarden, J. 1975. Evolution of marine symbiosis: A simple cost-benefit model. *Ecology* 56:1201–8.

———. 1979. *Theory of population genetics and evolutionary ecology: An introduction*. New York: Macmillan.

Rudgers, J. A. 2004. Enemies of herbivores can shape plant traits: Selection in a facultative ant-plant mutualism. *Ecology* 85:195–205.

Rudgers, J. A., and M. Gardener. 2004. Extrafloral nectar as a resource mediating multispecies interactions. *Ecology* 85:1495–1502.

Rudgers, J. A., and S. Y. Strauss. 2004. A selection mosaic in the facultative mutualism between ants and wild cotton. *Proceedings of the Royal Society of London Series B, Biological Sciences* 271:2481–88.

Ruffner, G. A., and W. D. Clark. 1986. Extrafloral nectar of *Ferocactus acanthodes* (Cactaceae): Composition and its importance to ants. *American Journal of Botany* 73:185–89.

Ruhren, S., and M. R. Dudash. 1996. Consequences of the timing of seed release of *Erythronium americanum* (Liliaceae), a deciduous forest myrmecochore. *American Journal of Botany* 83:633–40.

Ruhren, S., and S. N. Handel. 1999. Jumping spiders (Salticidae) enhance the seed production of a plant with extrafloral nectaries. *Oecologia* 119: 227–30.

Ryti, R. T., and T. J. Case. 1992. The role of neighborhood competition in the spacing and diversity of ant communities. *American Naturalist* 139:355–74.

Sagers, C. L., S. M. Ginger, and R. D. Evans. 2000. Carbon and nitrogen isotopes trace nutrient exchange in an ant-plant mutualism. *Oecologia* 123:582–86.

Salazar, B. A., and D. W. Whitman. 2001. Defensive tactics of caterpillars against predators and parasitoids. In *Insects and plant defence dynamics,* ed. T. N. Ananthakrishnan, 161–207. Enfield, NH: Science Publishers.

Samson, D. S., E. A. Rickart, and P. C. Gonzales. 1997. Ant diversity and abundance along an elevational gradient in the Philippines. *Biotropica* 29: 349–63.

Sanders, C. J., and A. Pang. 1992. Carpenter ants as predators of spruce budworm in the boreal forest of northwestern Ontario. *Canadian Entomologist* 124:1093–1100.

Sanderson, M. J., and M. J. Donoghue. 1994. Shifts in diversification rate with the origin of angiosperms. *Science* 264:1590–93.

Santos, J. C., and K. Del-Claro. 2001. Interação entre formigas, herbívoros e nectários extraflorais em *Tocoyena formosa* (Cham. & Schlechdt.) K. Schum. (Rubiaceae) na vegetação do cerrado. *Revista Brasileira de Zoociências* 3: 77–92.

Sauer, C., E. Stackebrandt, J. Gadau, B. Hölldobler, and R. Gross. 2000. Systematic relationships and cospeciation of bacterial endosymbionts and their carpenter ant host species: Proposal of the new taxon *Candidatus Blochmannia* gen. nov. *International Journal of Systematic and Evolutionary Microbiology* 50:1877–86.

Savolainen, R., and K. Vepsäläinen. 1989. Niche differentiation of ant species within territories of the wood ant *Formica polyctena. Oikos* 56:3–16.

Scheiner, S. M., and J. M. Rey-Benayas. 1994. Global patterns of plant diversity. *Evolutionay Ecology* 8:331–47.

Schemske, D. W. 1980. The evolutionary significance of extrafloral nectar production by *Costus woodsonii* (Zingiberaceae): An experimental analysis of ant protection. *Journal of Ecology* 68:959–67.

———. 1982. Ecological correlates of a neotropical mutualism: Ant assemblages at *Costus* extrafloral nectaries. *Ecology* 63:932–41.

———. 1983. Limits to specialization and coevolution in plant-animal mutualisms. In *Coevolution,* ed. M. H. Nitecki, 67–109. Chicago: University of Chicago Press.

Schmitz, O. J., P. A. Hamback, and A. P. Beckerman. 2000. Trophic cascades in terrestrial systems: A review of the effects of carnivore removals on plants. *American Naturalist* 155:141–53.

Schneider, H., E. Schuettpeiz, K. M. Pryer, R. Cranfill, S. Magallón, and R. Lupia. 2004. Ferns diversified in the shadow of angiosperms. *Nature* 428:553–57.

Schoener, T. 1974. Resource partitioning in ecological communities. *Science* 185:27–39.

———. 1983. Field experiments on interspecific interactions. *American Naturalist* 122:240–85.

Schowalter, T. D. 2000. *Insect ecology: An ecosystem approach.* San Diego: Academic Press.

Schubart, H. O. R., and A. B. Anderson. 1978. Why don't ants visit flowers? A reply to D.H. Janzen. *Biotropica* 10:310–11.

Schupp, E. W. 1986. *Azteca* protection of *Cecropia:* Ant occupation benefits juvenile trees. *Oecologia* 70:379–85.

———. 1995. Seed-seedling conflicts, habitat choices, and patterns of plant recruitment. *American Journal of Botany* 82:399–409.

Schupp, E. W., and D. H. Feener. 1991. Phylogeny, lifeform, and habitat dependence of ant-defended plants in a Panamanian forest. In Huxley and Cutler 1991, 175–97.

Schürch, S., M. Pfunder, and B. A. Roy. 2000. Effects of ants on the reproductive success of *Euphorbia cyparissias* and associated pathogenic rust fungus. *Oikos* 88:6–12.

Segraves, K. A., and J. N. Thompson. 1999. Plant polyploidy and pollination: Floral traits and insect visits to diploid and tetraploid *Heuchera grossulariifolia. Evolution* 53:1114–27.

Segraves, K. A., J. N. Thompson, P. S. Soltis, and D. E. Soltis. 1999. Multiple origins of polyploidy and the geographic mosaic structure of *Heuchera grossulariifolia. Molecular Ecology* 8:253–62.

Shattuck, S. O. 1992. Higher classification of the ant subfamilies Aneuretinae, Dolichoderinae, and Formicinae (Hymenoptera, Formicidae). *Systematic Entomology* 17:199–206.

Shepherd, J. D. 1985. Adjusting foraging effort to resources in adjacent colonies on the leaf-cutter ant, *Atta colombica. Biotropica* 17:245–52.

Sherbrooke, W. C., and J. C. Scheerens. 1979. Ant-visited extrafloral (calyx and foliar) nectaries and nectar sugars of *Erythrina flabelliformis* Kearney in Arizona. *Annals of the Missouri Botanical Garden* 66:427–81.

Shuter, E., and A. Westoby. 1992. Herbivorous arthropods on bracken *Pteridium aquilinum* (L.) Kuhn in Australia compared with elsewhere. *Australian Journal of Ecology* 17:329–39.

Skidmore, B. A., and E. R. Heithaus. 1988. Lipid cues for seed-carrying by ants in *Hepatica americana. Journal of Chemical Ecology* 14:2185–96.

Skogsmyr, I., and T. Fagerström. 1992. The cost of anti-herbivory defence: An evaluation of some ecological and physiological factors. *Oikos* 64:451–57.

Slingsby, P., and W. J. Bond. 1981. Ants—friends of the fynbos. *Veld and Flora* 67:39–45.

Smiley, J. 1986. Ant constancy at *Passiflora* extrafloral nectaries: Effects on caterpillar survival. *Ecology* 67:516–21.

Smith, B. H., C. E. deRivera, C. L. Bridgman, and J. J. Woida. 1989. Frequency-dependent seed dispersal by ants of two deciduous forest herbs. *Ecology* 70:1645–48.

Smith, B. H., P. D. Forman, and A. E. Boyd. 1989. Spatial patterns of seed dispersal and predation of two myrmecochorous forest herbs. *Ecology* 70: 1649–56.

Smith, J. H., and D. O. Atherton. 1944. Seed-harvesting and other ants in the tobacco-growing districts of North Queensland. *Queensland Journal of Agricultural Science* 1:33–61.

Smith, L. L., J. Lanza, and G. C. Smith. 1990. Amino acid concentrations in extrafloral nectar of *Impatiens sultani* increase after simulated herbivory. *Ecology* 71:107–15.

Smith, W. 1903. *Macaranga triloba:* A new myrmecophilous plant. *New Phytologist* 2:79–82.

Smythe, N. 1982. The seasonal abundance of night-flying insects in a neotropical forest. In *The ecology of a tropical forest,* ed. E. G. Leigh Jr., A. S. Rand, and D. M. Windor, 309–18. Washington, DC: Smithsonian Institution Press.

Snow, A. A., and M. L. Stanton. 1988. Aphids limit fecundity of a weedy annual (*Raphanus sativus*). *American Journal of Botany* 75:589–93.

Sobrinho, T. G., J. H. Schoereder, L. L. Rodrigues, and R. G. Collevatti. 2002. Ant visitation (Hymenoptera: Formicidae) to extrafloral nectaries increases seed set and seed viability in the tropical weed *Triumfetta semitriloba. Sociobiology* 39:353–68.

Solano, P., and A. Dejean. 2004. Ant-fed plants: Comparison between three geophytic myrmecophytes. *Biological Journal of the Linnean Society* 83:433–39.

Soltis, D. E., P. S. Soltis, M. W. Chase, M. E. Mort, D. C. Albach, M. Zanis, V. Savolainen, et al. 2000. Angiosperm phylogeny inferred from 18S rDNA, *rbcL,* and *atpB* sequences. *Botanical Journal of the Linnean Society* 133: 381–461.

Soltis, P. S., D. E. Soltis, and M. W. Chase. 1999. Angiosperm phylogeny inferred from multiple genes as a tool for comparative biology. *Nature* 402:402–4.

Speight, M. R., M. D. Hunter, and A. D. Watt. 1999. *Ecology of insects: Concepts and applications.* Oxford: Blackwell Science.

Stadler, B., and A. F. G. Dixon. 1998. Costs of ant attendance for aphids. *Journal of Animal Ecology* 67:454–59.

Stadler, B., K. Fiedler, T. J. Kawecki, and W. W. Weisser. 2001. Costs and benefits for phytophagous myrmecophiles: When ants are not always available. *Oikos* 92:467–78.

Stanton, M. L., T. M. Palmar, A. Evans, and M. L. Turner. 1999. Sterilization and canopy modification of a swollen thorn acacia tree by a plant-ant. *Nature* 401:378–81.

Stapley, L. 1998. The interaction of thorns and symbiotic ants as an effective defence mechanism of swollen-thorn acacias. *Oecologia* 115:401–5.

Stein, B. A. 1992. Sicklebill hummingbirds, ants, and flowers. *BioScience* 42: 27–33.

Stephenson, A. G. 1981. Toxic nectar deters nectar thieves of *Catalpa speciosa. American Midland Naturalist* 105:381–83.

———. 1982. The role of the extrafloral nectaries of *Catalpa speciosa* in limiting herbivory and increasing fruit production. *Ecology* 63:663–69.

Sterling, W. L. 1978. Fortuituos biological suppression of the boll weevil by the red imported fire ant. *Environmental Entomology* 7:564–68.

Steward, J. L., and K. H. Keeler. 1988. Are there trade-offs among antiherbivore defenses in *Ipomoea* (Convolvulaceae)? *Oikos* 53:79–86.

Steyn, J. J. 1954. *The pugnacious ant (*Anoplolepis custodiens *Smith) and its relation to the control of citrus scales at Letaba.* Memoirs of the Entomological Society of Southern Africa, no. 3. Hatfield, South Africa: Entomological Society of Southern Africa.

Stiefel, V. L., and D. C. Margolies. 1998. Is host plant choice by a clytrine leaf beetle mediated through interactions with the ant *Crematogaster lineolata? Oecologia* 115:434–38.

Stinner, B. R., and D. H. Stinner. 1989. Plant-animal interactions in agricultural systems. In *Plant-animal interactions,* ed. W. G. Abrahamson, 355–93. New York: McGraw-Hill.

Stone, L., T. Dayan, and D. Simberloff. 1996. Community-wide assembly patterns unmasked: The importance of species' differing geographical ranges. *American Naturalist* 148:997–1015.

Stork, N. E. 1991. The composition of the arthropod fauna of Bornean lowland rain forest trees. *Journal of Tropical Ecology* 7:161–80.

Stout, J. 1979. An association of an ant, a mealy bug, and an understory tree from a Costa Rican rain forest. *Biotropica* 11:309–11.

Stradling, D. J. 1991. An introduction to fungus-growing ants, Attini. In Huxley and Cutler 1991, 15–18.

Strauss, S. Y., H. Sahli, and J. K. Conner. 2005. Toward a more trait-centered approach to diffuse (co)evolution. *New Phytologist* 165:81–90.

Strong, D. R., J. H. Lawton, and R. Southwood. 1984. *Insects on plants.* Cambridge, MA: Harvard University Press.

Suarez, A. V., C. de Moraes, and A. Ippolito. 1998. Defense of *Acacia collinsii* by an obligate and nonobligate ant species: The significance of encroaching vegetation. *Biotropica* 30:480–82.

Sudd, J. H., and N. R. Franks. 1987. *The behavioral ecology of ants.* Cambridge: Cambridge University Press.

Suzuki, N., K. Ogura, and N. Katayama. 2004. Efficiency of herbivore exclusion by ants attracted to aphids on the vetch *Vicia angustifolia* L. (Leguminosae). *Ecological Research* 19:275–82.

Svensson, K. 1985. An estimate of pollen carryover by ants in a natural population of *Scleranthus perennis* L. (Caryophyllaceae). *Oecologia* 56:373–77.

Svoma, E., and W. Morawetz. 1992. Drüsenhaare, Emergenze, und Blattdomatien bei der Ameisenpflanze *Tococa occidentalis* (Melastomataceae). *Botanisches Jahrbucher fur Systematik* 114:185–200.

Swift, S., J. Bryant, and J. Lanza. 1994. Simulated herbivory on *Passiflora incarnata* causes increased ant attendance. *Bulletin of the Ecological Society of America* 75:225.

Swift, S., and J. Lanza. 1993. How do *Passiflora* vines produce more extrafloral nectar after simulated herbivory? *Bulletin of the Ecological Society of America* 74:451.

Taber, S. W. 1998. *The world of the harvester ants.* College Station: Texas A&M University Press.

Taylor, B. 1977. The ant mosaic on cocoa and other tree crops in western Nigeria. *Ecological Entomology* 2:245–55.

Tempel, A. S. 1981. Field studies of the relationship between herbivore damage and tannin concentration in bracken (*Pteridium aquilinum* (L.) Kuhn). *Oecologia* 51:97–106.

———. 1983. Bracken fern (*Pteridium aquilinum*) and nectar-feeding ants: A nonmutualistic interaction. *Ecology* 64:1411–22.

Thien, L. B. 1980. Patterns of pollination in the primitive angiosperms. *Biotropica* 12:1–14.

Thien, L. B., P. Bernhardt, G. W. Gibbs, O. Pellmyr, G. Bergstrom, I. Groth, and G. McPherson. 1985. The pollination of *Zygogynum* (Winteraceae) by a moth, *Sabatinca* (Micropterigidae): An ancient association? *Science* 227: 540–43.

Thien, L. B., S. Kawano, H. Azuma, S. Latimer, M. S. Devall, S. Rosso, S. Elakovich, V. Rico-Gray, and D. Jobes. 1998. The floral biology of the Magnoliaceae. In *Magnolias and their allies,* ed. D. Hunt, 37–58. Milborne Port, UK: International Dendrology Society and Magnolia Society.

Thien, L. B., A. E. Smalley, A. S. Bradburn, and V. Rico-Gray. 1987. Complexity of tropical rain forests. *Tulane Studies in Zoology and Botany* 26:5–18.

Thomas, D. W. 1988. The influence of aggressive ants on fruit removal in the tropical tree, *Ficus capensis* (Moraceae). *Biotropica* 20:49–53.

Thompson, J. N. 1981a. Elaiosomes and fleshy fruits: Phenology and selection pressures for ant-dispersed seeds. *American Naturalist* 117:104–8.

———. 1981b. Reversed animal-plant interactions: The evolution of insectivorous and ant-fed plants. *Biological Journal of the Linnean Society* 16:147–55.

———. 1982. *Interaction and coevolution.* New York: Wiley.

———. 1988. Variation in interspecific interactions. *Annual Review of Ecology and Systematics* 19:65–87.

———. 1994. *The coevolutionary process.* Chicago: University of Chicago Press.

———. 1996. Evolutionary ecology and the conservation of biodiversity. *Trends in Ecology and Evolution* 11:300–303.

———. 1997. Evaluating the dynamics of coevolution among geographically structured populations. *Ecology* 78:1619–23.

———. 1998. The population biology of coevolution. *Researches on Population Ecology* 40:159–66.

———. 1999a. Coevolution and escalation: Are ongoing coevolutionary meanderings important? *American Naturalist* 153:S92–S93.

———. 1999b. The evolution of species interactions. *Science* 284:2116–18.

———. 1999c. The raw material for coevolution. *Oikos* 84:5–16.

———. 1999d. Specific hypotheses on the geographic mosaic of coevolution. *American Naturalist* 153:S1–S14.

———. 2002. Plant-animal interactions: Future directions. In Herrera and Pellmyr 2002, 236–47.

———. 2005. *The geographic mosaic of coevolution.* Chicago: University of Chicago Press.

Tibbets, T. M., and S. H. Faeth. 1999. *Neotyphodium* endophytes in grasses: Deterrents or promoters of herbivory by leaf-cutting ants? *Oecologia* 118:297–305.

Tillberg, C. V. 2004. Friend or foe? A behavioral and stable isotopic investigation of an ant-plant symbiosis. *Oecologia* 140:506–15.

Tilman, D. 1978. Cherries, ants, and tent caterpillars: Timing of nectar production in relation to susceptibility of caterpillars to ant predation. *Ecology* 59: 686–92.

———. 2000. Causes, consequences, and ethics of biodiversity. *Nature* 405: 208–11.

Tobin, J. E. 1991. A neotropical rainforest canopy, ant community: Some ecological considerations. In Huxley and Cutler 1991, 536–38.

———. 1994. Ants as primary consumers: Diet and abundance in the Formicidae. In *Nourishment and evolution in insect societies,* ed. J. H. Hunt and C. A. Napela, 279–307. Boulder, CO: Westview Press.

———. 1995. Ecology and diversity of tropical forest canopy ants. In *Forest canopies,* ed. M. D. Lowman and N. M. Nadkarni, 129–47. London: Academic Press.

Torres, J. A. 1984a. Diversity and distribution of ant communities in Puerto Rico. *Biotropica* 16:296–303.

———. 1984b. Niches and coexistence of ant communities in Puerto Rico: Repeated patterns. *Biotropica* 16:284–95.

Torres-Hernández, L., V. Rico-Gray, C. Castillo-Guevara, and A. Vergara. 2000. Effect of nectar-foraging ants and wasps on the reproductive fitness of *Turnera ulmifolia* (Turneraceae) in a coastal sand dune, Mexico. *Acta Zoológica Mexicana,* n.s., 81:13–21.

Travassos, M. A., and N. E. Pierce. 2000. Acoustics, context, and function of vibrational signalling in a lycaenid butterfly-ant mutualism. *Animal Behaviour* 60:13–36.

Travis, J. 1996. The significance of geographical variation in species interactions. *American Naturalist* 148:S1–S8.

Trelease, W. 1881. The foliar nectar of *Populus*. *Botanical Gazette* 6:384–90.

Treseder, K. K., D. W. Davidson, and J. R. Ehleringer. 1995. Absorption of ant-provided carbon dioxide and nitrogen by a tropical epiphyte. *Nature* 375:137–39.

Trimble, S. T., and C. L. Sagers. 2004. Differential host use in two highly specialized ant-plant associations: Evidence from stable isotopes. *Oecologia* 138:74–82.

Tryon, A. F. 1985. Spores of myrmecophytic ferns. *Proceedings Royal Society of Edinburgh* 86B:105–10.

Ule, E. 1902. Ameisengarten in Amazonasgebeit. *Botanisches Jahrbucher* 30: 45–52.

———. 1905. Blumengarten der Ameisen am Amazonenstrome. *Vegetationsbilder,* ser. 3, pt. 1:1–14.

———. 1906. Ameisenflanzen. *Botanisches Jahrbucher* 37:335–52.

Ulloa-Chalon, P., and D. Cherix 1990. The little fire ant, *Wasmannia auropuctata* (R.) (Hymenoptera: Formicidae). In Vander Meer, Jaffe, and Cedeno 1990, 281–89.

Vandermeer, J., I. Perfecto, G. Ibarra Nuñez, S. Philpott, and A. García Ballinas. 2002. Ants (*Azteca* sp.) as potential biological control agents in shade coffee production in Chiapas, Mexico. *Agroforestry Systems* 56:271–76.

Vander Meer, R. K., K. Jaffe, and A. Cedeno, eds. 1990. *Applied myrmecology: A world perspective.* Boulder, CO: Westview Press.

Van der Pijl, L. 1982. *Principles of dispersal in higher plants.* Berlin: Springer.

Vander Wall, S. B., and W. S. Longland. 2004. Diplochory: Are two seed dispersers better than one? *Trends in Ecology and Evolution* 19:155–61.

Vanstone, V. A., and D. C. Paton. 1988. Extrafloral nectaries and pollination of *Acacia pycnantha* Benth. by birds. *Australian Journal of Botany* 36:519–31.

Vasconcelos, H. L. 1991. Mutualism between *Maieta guianensis* Aubl., a myrmecophytic melastome, and one of its ant inhabitants: Ant protection against insect herbivores. *Oecologia* 87:295–98.

———. 1993. Ant colonization of *Maieta guianensis* seedlings, an Amazon antplant. *Oecologia* 95:439–43.

———. 1997. Foraging activity of an Amazonian leaf-cutting ant: Responses to changes in the availability of woody plants and to previous plant damage. *Oecologia* 112:370–78.

———. 1999. Effects of forest disturbance on the structure of ground-foraging ant communities in central Amazonia. *Biodiversity and Conservation* 8:409–20.

Vasconcelos, H. L., and A. B. Casimiro. 1997. Influence of *Azteca alfari* ants on the exploitation of *Cecropia* trees by a leaf-cutting ant. *Biotropica* 29:84–92.

Vasconcelos, H. L., and J. M. Cherrett. 1997. Leaf-cutting ants and early forest regeneration in central Amazonia: Effects of herbivory on tree seedling establishment. *Journal of Tropical Ecology* 13:357–70.

Vasconcelos, H. L., and D. W. Davidson. 2000. Relationship between plant size and ant associations in two Amazonian plants. *Biotropica* 32:100–111.

Vasconcelos, H. L., and J. M. S. Vilhena. 2006. Species turnover and vertical partitioning of ant assemblages in the Brazilian Amazon: A comparison of forests and savannas. *Biotropica* 38:100–106.

Vázquez, D. P., and M. A. Aizen. 2004. Asymmetric specialization: A pervasive feature of plant-pollinator interactions. *Ecology* 85:1251–57.

Vázquez, D. P., W. F. Morris, and P. Jordano. 2005. Interaction frequency as a surrogate for the total effect of animal mutualists on plants. *Ecology Letters* 8:1088–94.

Venable, D. L., and J. S. Brown. 1993. The population-dynamic functions of seed dispersal. *Vegetatio* 107–8:31–55.

Vepsäläinen, K. 1982. Assembly of island ant communities. *Annales Zoologici Fennici* 19:327–35.

Vesprini, J. L., L. Galetto, and G. Bernardello. 2003. The beneficial effect of ants on the reproductive success of *Dyckia floribunda* (Bromeliaceae), an extrafloral nectary plant. *Canadian Journal of Botany* 81:24–27.

Vinson, S. B. 1976. Host selection by insect parasitoids. *Annual Review of Entomology* 21:109–23.

Vinson, S. B., S. T. O'Keefe, and G. W. Frankie. 2004. The conservation values of bees and ants in the Costa Rican dry forest. In *Biodiversity conservation in Costa Rica: Learning the lessons in a seasonal dry forest,* ed. G. W. Frankie, A. Mata, and S. B. Vinson, 67–79. Berkeley: University of California Press.

Völkl, W. 1994. The effect of ant-attendance on the foraging behaviour of the aphid parasitoid *Lysiphlebus cardui. Oikos* 70:149–55.

Völkl, W., J. Woodring, M. Fischer, M. W. Lorenz, and K. H. Hoffmann. 1999. Ant-aphid mutualisms: The impact of honeydew production and honeydew sugar composition on ant preferences. *Oecologia* 118:483–91.

Wagner, D. 1997. The influence of ant nests on *Acacia* seed production, herbivory, and soil nutrients. *Journal of Ecology* 85:83–93.

———. 2000. Pollen viability reduction as a potential cost of ant association for *Acacia constricta* (Fabaceae). *American Journal of Botany* 87:711–15.

Wagner, D., M. J. F. Brown, and D. M. Gordon. 1997. Harvester ant nests, soil biota, and soil chemistry. *Oecologia* 112:112–232.

Wagner, D., and A. Kay. 2002. Do extrafloral nectaries distract ants from visiting flowers? An experimental test of an overlooked hypothesis. *Evolutionary Ecology Research* 4:293–305.

Wang, B. C., and T. B. Smith 2002. Closing the seed dispersal loop. *Trends in Ecology and Evolution* 17:379–85.

Ward, P. S. 1991. Phylogenetic analysis of pseudomyrmecine ants associated with domatia-bearing plants. In Huxley and Cutler 1991, 335–52.

———. 1993. Systematic studies on *Pseudomyrmex* acacia-ants. *Journal of Hymenoptera Research* 2:117–68.

Ward, P. S., and S. G. Brady. 2003. Phylogeny and biogeography of the ant subfamily Myrmeciinae (Hymenoptera: Formicidae). *Invertebrate Systematics* 17:361–86.

Watt, A. D., N. E. Stork, and B. Bolton. 2002. The diversity and abundance of ants in relation to forest disturbance and plantation establishment in southern Cameroon. *Journal of Applied Ecology* 39:18–30.

Way, M. J. 1953. The relationship between certain ant species with particular reference to biological control of the coreid, *Theraptus* sp. *Bulletin of Entomological Research* 44:669–91.

———. 1954. Studies on the life history and ecology of the ant *Oecophylla longinoda* Latreille. *Bulletin of Entomological Research* 45:93–112.

———. 1963. Mutualism between ants and honeydew-producing Homoptera. *Annual Review of Entomology* 8:307–44.

Way, M. J., M. E. Cammell, and M. R. Paiva. 1992. Studies on egg predation by ants (Hymenoptera: Formicidae) especially on the eucalyptus borer *Phoracantha semipunctata* (Coleoptera: Cerambycidae) in Portugal. *Bulletin of Entomological Research* 82:425–32.

Way, M. J., and K. C. Khoo. 1991. Colony dispersion and nesting habits of *Dolichoderus thoracicus* and *Oecophylla smaragdina* (Hymenoptera: Formicidae), in relation to their success as biological control agents on cocoa. *Bulletin of Entomological Research* 81:341–50.

———. 1992. Role of ants in pest management. *Annual Review of Entomology* 37:479–503.

Weber, N. A. 1946. Two common ponerine ants of possible economic significance, *Ectatomma tuberculatum* (Oliver) and *E. ruidum* Roger. *Proceedings of the Entomological Society of Washington* 48:1–16.

———. 1966. Fungus growing ants. *Science* 153:587–604.

———. 1972. *Gardening ants: The attines.* Philadelphia: American Philosophical Society.

Wenny, D. G. 2001. Advantages of seed dispersal: A re-evaluation of directed dispersal. *Evolutionary Ecology Research* 3:51–74.

Westoby, M., L. Hughes, and B. L. Rice. 1991. Seed dispersal by ants: Comparing infertile with fertile soils. In Huxley and Cutler 1991, 435–47.

Westoby, M., and B. L. Rice. 1981. A note on combining two methods of dispersal-for-distance. *Australian Journal of Ecology* 6:23–27.

Westoby, M., B. Rice, J. M. Shelley, D. Haig, and J. L. Kohen. 1982. Plants' use of ants for dispersal at West Head. In Buckley 1982b, 75–87.

Wetterer, J. K. 1997. Ants on *Cecropia* in Hawaii. *Biotropica* 29:128–32.

Wetterer, J. K., X. Espadaler, A. L. Wetterer, D. Aguin-Pombo, and A. M. Franquinho-Aguiar. 2006. Long-term impact of exotic ants on the native ants of Madeira. *Ecological Entomology* 31:358–68.

Whalen, M. A., and D. A. Mackay. 1988. Patterns of ant and herbivore activity on five understory euphorbiaceous saplings in submontane Papua New Guinea. *Biotropica* 20:294–300.

Wheeler, W. M. 1910. *Ants: Their structure, development, and behavior.* New York: Columbia University Press.

———. 1921. The *Tachigali* ants. *Zoologica* 3:137–68.

———. 1942. Studies of neotropical ant-plants and their ants. *Bulletin of the Museum of Comparative Zoology* 90:3–262.

Wheeler, W. M., and J. C. Bequaert. 1929. Amazonian myrmecophytes and their ants. *Zoologischen Anzeiger* 82:10–39.

Whiffin, T. 1972. Observations on some upper Amazonian formicarial Melastomataceae. *Sida* 5:33–41.

Whitford, W. G. 1976. Foraging behavior of the Chihuahuan Desert harvester ant. *American Midland Naturalist* 95:455–58.

Whitham, T. G., A. G. Williams, and A. M. Robinson. 1984. The variation principle: Individual plants as temporal and spatial mosaics of resistance to rapidly evolving pests. In *A new ecology: Novel approaches to interactive systems,* ed. P. W. Price, C. N. Slobodchikoff, and W. S. Gaud, 15–51. New York: Wiley.

Whitman, D. 1994. Plant bodygards: Mutualistic interactions between plants and the third trophic level. In *Functional dynamics of phytophagous insects,* ed. T. N. Ananthakrishnan, 207–48. New Delhi: Oxford and IBH Publishing.

Wight, J. R., and J. T. Nichols. 1966. Effects of harvester ants on production of saltbush community. *Journal of Range Management* 19:68–71.

Wilcock, C. C., and C. J. de Almeida. 1988. The flora of the Algarve. *Plants Today* 5:151–57.

Wilf, P., C. C. Labandeira, W. J. Kress, C. L. Staines, D. M. Windsor, A. L. Allen, and K. R. Johnson. 2000. Timing the radiations of leaf beetles: Hispines on gingers from latest Cretaceous to recent. *Science* 289:291–94.

Wilkinson, D. M. 1999. Ants, agriculture, and antibiotics. *Trends in Ecology and Evolution* 14:459–60.

Williams-Linera, G., and J. Tolome. 1996. Litterfall, temperate and tropical dominant trees, and climate in a Mexican lower montane forest. *Biotropica* 28:649–56.

Willing, M. R., D. M. Kaufman, and R. D. Stevens. 2003. Latitudinal gradients of biodiversity: Pattern, process, scale, and synthesis. *Annual Review of Ecology and Systematics* 34:273–309.

Willmer, P. G., and G. N. Stone. 1997. How aggressive ant-guards assist seed-set in *Acacia* flowers. *Nature* 388:165–67.

Wilson, E. O. 1959. Some ecological characteristics of ants in New Guinea rain forests. *Ecology* 40:437–47.

———. 1971. *The insect societies.* Cambridge, MA: Harvard University Press.

———. 1987a. The arboreal ant fauna of Peruvian Amazon forests. *Biotropica* 19:245–51.

———. 1987b. Causes of ecological success: The case of ants. *Journal of Animal Ecology* 56:1–9.

———. 2003. Pheidole *in the New World: A dominant, hyperdiverse ant genus.* Cambridge, MA: Harvard University Press.

Wilson, E. O., and B. Hölldobler. 2005. The rise of the ants: A phylogenetical and ecological explanation. *Proceedings of the National Academy of Sciences* 102:7411–14.

Wilson, J. B. 1999. Guilds, functional types, and ecological groups. *Oikos* 86: 507–22.

Wirth, R., W. Beyschlag, R. J. Ryel, and B. Hölldobler. 1997. Annual foraging of the leaf-cutting ant *Atta colombica* in a semideciduous rain forest in Panama. *Journal of Tropical Ecology* 13:741–57.

Wirth, R., H. Herz, R. J. Ryel, W. Beyschlag, and B. Hölldobler. 2003. *Herbivory of leaf-cutting ants: A case study of Atta colombica in the tropical rainforest of Panama.* Ecological Studies, 164. Berlin: Springer-Verlag.

Wirth R., and I. R. Leal. 2001. Does rainfall affect temporal variability of ant protection in *Passiflora coccinea*? *Ecoscience* 8:450–53.

Wolff, A., and M. Debussche. 1999. Ants as seed dispersers in a Mediterranean old-field succession. *Oikos* 84:443–52.

Wood, T. K. 1982. Ant-attended nymphal aggregations in the *Enchenopa binotata* complex (Homoptera: Membracidae). *Annals of the Entomological Society of America* 75:649–53.

———. 1984. Life history patterns of tropical membracids (Homoptera: Membracidae). *Sociobiology* 8:299–344.

Woodell, S. R. J., and T. J. King. 1991. The influence of mound-building ants on British lowland vegetation. In Huxley and Cutler 1991, 521–35.

Woodring, J., R. Wiedemann, M. K. Fischer, K. H. Hoffmann, and W. Völkl. 2004. Honeydew amino acids in relation to sugars and their role in the establishment of ant-attendance hierarchy in eight species of aphids feeding on tansy (*Tanacetum vulgare*). *Physiological Entomology* 29:311–19.

Wright, S. 1943. Isolation by distance. *Genetics* 28:114–38.

Wyatt, R. 1981. Ant-pollination of the granite outcrop endemic *Diamorpha smallii* (Crassulaceae). *American Journal of Botany* 68:1212–17.

Wyatt, R., and A. Stoneburner. 1981. Patterns of ant-mediated pollen dispersal in *Diamorpha smallii* (Crassulaceae). *Systematic Botany* 6:1–7.

Yan, A. S., C. H. Martin, and H. F. Nijhout. 2004. Geographic variation of caste structure among ant populations. *Current Biology* 14:514–19.

Yano, S. 1994. Flower nectar of an autogamous perennial *Rorippa indica* as an indirect defense mechanism against herbivorous insects. *Researches in Population Ecology* 36:63–71.

Yanoviak, S. P., R. Dudley, and M. Kaspari. 2005. Directed aerial descent in canopy ants. *Nature* 433:624–26.

Young, T. P., C. H. Stubblefield, and L. A. Isbell. 1997. Ants on swollen-thorn acacias: Species coexistence in a simple system. *Oecologia* 109:98–107.

Yu, D. W. 1994. The structural role of epiphytes in ant gardens. *Biotropica* 26:222–26.

———. 2001. Parasites of mutualisms. *Biological Journal of the Linnean Society* 72:529–46.

Yu, D. W., and D. W. Davidson. 1997. Experimental studies of species-specificity in *Cecropia*-ant relationships. *Ecological Monographs* 67:273–94.

Yu, D. W., and N. E. Pierce. 1998. A castration parasite of an ant-plant mutualism. *Proceedings of the Royal Society of London, series B* 265:375–82.

Yu, D. W., H. B. Wilson, and N. E. Pierce. 2001. An empirical model of species coexistence in a spatially structured environment. *Ecology* 82:1761–71.

Zachariades, C. 1994. Complex interactions involving the Cape fig, Ficus sur Forssk;aral, and its associated insects. Ph.D. diss., Rhodes University.

Zoebelein, G. 1956a. Der Honigtau als Nahrung der Insekten. 1. Zeitschrift für Angewandte. *Entomologie* 38:369–416.

———. 1956b. Der Honigtau als Nahrung der Insekten. 2. Zeitschrift für Angewandte. *Entomologie* 39:129–67.

Index